T0213265

Cannabis Laboratory Fundamentals

Shaun R. Opie

Editor

Cannabis Laboratory
Fundamentals

 Springer

Editor
Shaun R. Opie
E4 Bioscience
Charlevoix, MI, USA

ISBN 978-3-030-62718-8   ISBN 978-3-030-62716-4   (eBook)
https://doi.org/10.1007/978-3-030-62716-4

This Springer imprint is published by the registered company Springer Nature Switzerland AG
The registered company address is: Gewerbestrasse 11, 6330 Cham, Switzerland

Preface

Thank you for purchasing this book.

Welcome to the first edition of *Cannabis Laboratory Fundamentals*. This book aims to help cannabis laboratory owners and operators, investors and investment firms, business professionals, financial experts, regulators, lobbyists, government officials, cannabis enthusiasts, and others who want to evaluate the opportunities and risks of cannabis safety testing. In addition, we believe that industries likely to become involved with cannabis in the future including pharmaceutical companies, food safety testing companies, beverage companies, and universities and colleges preparing to educate cannabis laboratory technicians and/or offer cannabis entrepreneurship courses will find the information in this book helpful. Cannabis safety testing is an area within the emerging cannabis market that is legally separated from cultivation, extraction, processing, distribution, and retail allowing individuals and organizations who may have been reluctant to enter previously a new entry route to the cannabis space. The current interest in cannabis laboratories is good for the industry by helping to bring both investor and consumer awareness to the importance of safety testing.

This book brings together a "who's who" of experts in the cannabis industry to share their working knowledge. Initially, we hoped to recruit practitioners employed by an established cannabis laboratory or a supporting ancillary field such as an instrument or consumable vendor, consulting agency, legal firm, investment bank, or venture capital group. As work progressed, we realized that it was necessary since there are a very limited number of potential contributors with a university affiliation. Further, academic groups we identified were primarily focused on a narrow research topic rather than comprehensive laboratory operations and were unable to dedicate the time necessary to participate. Thus, the contributing authors are dedicated subject matter experts and key opinion leaders who are at the forefront of commercial cannabis safety testing. Every effort was made to reduce commercial bias from vendors and manufacturers, but we acknowledge that these groups are the most knowledgeable about the products and services that they provide. We believe that this will provide a business-centric perspective on the industry and offer tangible, "real world" considerations and information that can be applied in a cannabis safety testing laboratory.

The cannabis industry is new and will cycle though many rounds of growing pains at both the state and federal level. With those changes in mind, it is imperative to appreciate that what is published today will become dated, and possibly obsolete, over the next few years as technology, standards, and regulations become more refined. We believe that this book contains the most current information available at the time of printing but acknowledge that portions of the book will need to be updated and revised.

While numerous books about cannabis cultivation, extraction, ethnobotany, phytochemicals, effect on consciousness are available, to our knowledge, this is the first reference book that addresses the spectrum of issues associated with setting up and maintaining a cannabis safety testing laboratory. This book aims to share with you some of the information, strategies, and tactics that directly apply to cannabis testing laboratories and provide recommendations to avoid common—but expensive—mistakes. What makes a cannabis safety testing laboratory both good and successful is not simply one or two key functions, but rather a broad list of activities and assets that need to be implemented and combined in a cogent order with exceptional employees. The foundational activities that occur in the first 6 months of planning and the level of available funding will be a major determinant of the long-term success of the laboratory.

This book refers to cannabis, marijuana, and hemp with specificity. Cannabis is a species of plant that includes both hemp and marijuana. Historically, the distinction between hemp and marijuana is the total available amount of $\Delta 9$-tetrahydrocannabinol ($\Delta 9$-THC). A cannabis plant is classified as hemp if it has a $\Delta 9$-THC of less than 0.3% of a dry weight basis and marijuana if the $\Delta 9$-THC is higher. Not surprisingly, the $\Delta 9$-THC concentration of a cannabis plant is measured in a laboratory. The legislative requirement for laboratory tested cannabis follows legalization of medical and recreational use marijuana in every US state to date as well as the production and sale of hemp.

Like a clinical blood specimen from a hospitalized patient or water aliquot from a potentially contaminated site, a cannabis plant sample is tested in an analytical laboratory. Regardless of the source of a sample requiring measurement, clinical, environmental, and cannabis laboratories share many business, compliance, operational, and technical themes that need to be addressed in relationship to the sample type. Analytical laboratories that perform cannabis safety testing do so by examining the safety and quality properties of samples of the plant and its derivatives. To address these testing needs, this book is divided into two parts. Part 1 is primarily non-technical that discusses business planning (Chap. 1), laboratory design (Chap. 2), safety, compliance, and operational requirements (Chaps. 3–7). Part 2 is technical and discusses the nuts and bolts of the primary tests that cannabis laboratories offer, the range of laboratory instrumentations and various analytes that can be tested, and in certain cases, methodology examples.

This book has some omissions. First, cannabis extraction laboratories are not represented because of tight publishing deadlines and the deliberate intent to limit the scope of this reference book to safety testing. We fully recognize that extraction laboratories are an essential component of the cannabis supply chain. Much of the information presented in the book can and does apply to them, but the book does not discuss extraction technology and instrumentation. Second, a chapter on cannabis sampling is not available because we could not identify and recruit qualified authors. Any laboratory trained staff will agree that sampling is a critical part of the analytical portion of testing and needs to be treated with the same rigor and oversight as any other laboratory process. For the same reasons as sampling, water activity and moisture content testing are not included, although we appreciate the importance of these tests as a basis of certain calculations to meet compliance standards. Chapters relevant to processing laboratories, sampling, water activity, moisture content, and others will be included in the 2nd edition or the next volume.

I accept full responsibility for all omissions and errors. I encourage you to contact me if you find an error or would like to contribute to the next edition. Future readers will be grateful for any improvements that can be offered.

Charlevoix, MI   Shaun R. Opie

Acknowledgments

Many talented and hard-working people helped to prepare this book:

I am deeply grateful to all the authors who donated their mornings, evenings, weekends, and knowledge to help prepare, review, and edit content. Writing 10,000 or so words with structure and clarity about a technically demanding subject is a large undertaking that requires a deep personal commitment. This was a massive and collaborative undertaking that could not be completed by any single person in the time span available.

Dan Falatako, my editor at Springer-Nature, who was the earliest supporter of this idea provided patient encouragement throughout the entire writing period.

The editorial team at Springer-Nature who helped clarify, organize, format, and publish.

Jen and Ella, my wife and daughter, who hold an unwavering belief that weekend mornings are for sleeping in, not writing. Without their love, support, encouragement, and personal sacrifices, this book would not have been possible.

Contents

Contributors

Tyler B. Anthony Titan Analytical Inc., Huntington Beach, United States

Susan Audino S. Audino & Associates, LLC, Wilmington, DE, USA

Patrick Bird PMB BioTek Consulting, LLC, Cincinnati, OH, USA

Kyle Boyar Tagleaf, Inc., Santa Cruz, CA, USA

Robert D. Brodnick Titan Analytical Inc., Huntington Beach, United States

Danielle Cadaret Cadaret Architecture, Royal Oak, MI, USA

Bob Clifford Shimadzu Scientific Instruments, Columbia, MD, USA

Nisha Corrigan Hygiena, Camarillo, CA, USA

Renee Engle-Goodner REG Science, LLC, Sacramento, CA, USA

William A. English Titan Analytical Inc., Huntington Beach, United States

Adam Floyd Think20 Labs, Irvine, CA, USA

Andrew Fornadel Shimadzu Scientific Instruments, Columbia, MD, USA

Krista Harlan Celtic Dragon Contracting LLC, Pinckney, MI, USA

Gary Herman Tagleaf, Inc., Santa Cruz, CA, USA

James Jaeger Laboratory Planning & Design Consultant, Carlsbad, CA, USA

Vikki Johnson Shimadzu Scientific Instruments, Columbia, MD, USA

Richard Karbowski Shimadzu Scientific Instruments, Columbia, MD, USA

Benjamin A. Katchman, Ph.D. PathogenDx, Scottsdale, AZ, USA

Maya Leonetti Titan Analytical Inc., Huntington Beach, United States

Nicole Lock Shimadzu Scientific Instruments, Columbia, MD, USA

Leyda Lugo-Morales Shimadzu Scientific Instruments, Columbia, MD, USA

Anthony Macherone, Ph.D. Agilent Technologies, Inc., Wilmington, DE, USA

The Johns Hopkins University School of Medicine, Baltimore, MD, USA

Jesse Miller, Ph.D. Neogen Corporation, Lansing, MI, USA

Shaun R. Opie, Ph.D. E4 Bioscience, Charlevoix, MI, USA

Alan Owens Shimadzu Scientific Instruments, Columbia, MD, USA

Navin Pathangay, AIA Pathangay Architects LLC, Phoenix, AZ, USA

Jon Peters Shimadzu Scientific Instruments, Columbia, MD, USA

Andrew Pham, MS ILP Scientific LLC, Westminster, CA, USA

Andy Sandy, Ph.D. Shimadzu Scientific Instruments, Columbia, MD, USA

Christine Sheehan Albany Medical Center, Albany, NY, USA

Shannon Swantek Enlightened Quality Analytics LLC, Boulder Creek, CA, USA

Mike Tunis Think20 Labs, Irvine, CA, USA

Gary Ward, Ph.D. GK Ward Associates LLC, Vancouver, WA, USA

An Introduction to Cannabis Laboratory Safety and Compliance Testing

Shaun R. Opie

Abstract The legislative requirement for laboratory tested cannabis follows legalization of medical and recreational use in every US state to date. Cannabis safety testing is a new investment opportunity within the emerging cannabis market that is separate from cultivation, processing, & distribution, allowing individuals and organizations who may have been reluctant to enter previously a new entry route to the cannabis space. However, many of the costs, operational requirements, and compliance issues are not well understood by people who have not been previously exposed to regulated, analytical laboratory testing. The purpose of this chapter is to provide a brief overview of the cannabis plant and then outline a framework for following chapters to build on by introducing some of the important business and operational considerations including legal status, business planning, laboratory design, license application, obtaining ISO/IEC 17025 accreditation, promoting a culture of quality, instrumentation purchases and methodology, and staffing.

Cannabis Classification and Phytocannabinoids

The cannabis plant belongs to the *Cannabaceae* family. While virtually identical genetically, three species as proposed by Carl Linneaus (1707–1778) are commonly recognized: *Cannabis sativa* and *Cannabis indica* (marijuana), and *Cannabis ruderalis* (hemp). This taxonomical classification is based primarily on the ratio of certain chemical compounds, phytocannabinoids, that each species produces as well as certain growth properties. In addition to taxonomical classification, at least three important differences permit separate classification of hemp and marijuana cultivars: (1) statutory definitions and regulatory oversight, (2) chemical and genetic compositions, and (3) production practices and use (Table 1).

The flowers of the cannabis plant produce two compounds of primary interest, phytocannabinoids and terpenes. When certain phytocannabinoids are consumed, they act on cannabinoid receptors (CB1 and CB2) in the nervous system human

S. R. Opie (✉)
E4 Bioscience, Charlevoix, MI, USA
e-mail: shaun@e4bioscience.com

© Springer Nature Switzerland AG 2021
S. R. Opie (ed.), *Cannabis Laboratory Fundamentals*,
https://doi.org/10.1007/978-3-030-62716-4_1

Table 1 An overview of the differences between marijuana and hemp

	Hemp	Marijuana
Botanical name	*Cannabis sativa*	*Cannabis sativa*
Statutory definition	"The plant *Cannabis sativa* L. and any part of that plant, including the seeds thereof and all derivatives, extracts, cannabinoids, isomers, acids, salts, and salts of isomers, whether growing or not, with a delta-9 tetrahydrocannabinol concentration of not more than 0.3% on a dry weight basis" (Section 297A of the Agricultural Marketing Act of 1946 (AMA)).	"All parts of the plant *Cannabis sativa* L., whether growing or not; the seeds thereof; the resin extracted from any part of such plant; and every compound, manufacture, salt, derivative, mixture, or preparation of such plant, its seeds or resin. Such term does not include the mature stalks of such plant, fiber produced from such stalks, oil or cake made from the seeds of such plant, any other compound, manufacture, salt, derivative, mixture, or preparation of such mature stalks (except the resin extracted therefrom), fiber, oil, or cake, or the sterilized seed of such plant which is incapable of germination" (21 U.S.C. §802(16)).
Current threshold for psychoactive compounds	No more than 0.3% of delta-9 THC on a dry weight basis (THIS is one of the leading psychoactive cannabinoids in cannabis)	No THC threshold specified
Other cannabinoids	Reportedly more than 60 cannabinoids (including CBD and other non-psychoactive compounds)	Reportedly more than 60 cannabinoids (including CBD and other non-psychoactive compounds)
Psychoactive properties	Nonpsychoactive	Psychoactive
Primary Federal Agencies with Regulatory Oversight	U.S. Department of Agriculture (USDA) Food and Drug Administration (FDA) (U.S. Department of Health and Human Services (HHS))	U.S. Drug Enforcement Administration (DEA) (U.S. Department of Justice (DOJ)) Food and Drug Administration (FDA) (U.S. Department of Health and Human Services (HHS))
Primary U.S. Laws	Agricultural Marketing Act of 1946 (AMA, 7 U.S.C. 1621 et seq.) Federal Food, Drug, and Cosmetic Act (FFDCA: 21 U.S.C. §§ 301 et seq.)	Controlled Substances Act (CSA, 21 U.S.C. §§801 et seq.) Federal Food, Drug, and Cosmetic Act (FFDCA: 21 U.S.C. §§ 301 et seq.)
Plant part used	Fiber, seed, and flower	Flower
Types of products	Food and food ingredient; ingredient for body products, cosmetics, dietary supplements and therapeutic products; textiles and fabrics; other manufactured and industrial products	Recreational and medicinal products
Plant height at harvest	10–15 ft (fiber), 6–9 ft (seed), 4–8 ft (flower)	4–8 ft (flower)

anatomy and generate psychoactive results associated with marijuana consumption. Terpenes are oils that produce flavor and aroma.

Phytocannabinoids, more commonly called cannabinoids, are chemical compounds synthesized primarily by the cannabis plant. Cannabis is known to produce over 100 different cannabinoids, but the three primary cannabinoids currently of interest are tetrahydrocannabinolic acid (THCA), Δ9-tetrahydrocannabinol (Δ9-THC), and cannabidiol (CBD). However, other cannabinoids are gaining consumer interest.

Δ9-THC is the cannabinoid producing psychoactive results when consumed by humans. It can be produced by all cannabis plants. However, marijuana is grown specifically for THC content, while hemp is grown for a variety of industrial properties and CBD production, currently its principle chemical product. Δ9-THC is the psychoactive, neutral form of THC, while THCA is the non-psychoactive, acidic form of tetrahydrocannabinol. The chemical structures of both molecules are very similar with THCA differing from Δ9-THC by the addition of a single carboxylic acid group (COOH) in the 1 position between the hydroxy group and the carbon chain (Fig. 1). Over time, THCA spontaneously decarboxylates by losing the carboxylic acid (COOH) group which converts the molecule into the psychoactive form Δ9-THC. Heat is a catalyst for the decarboxylation reaction and greatly increases the conversion rate of THCA to Δ9-THC.

Cannabidiol chemically resembles Δ9-THC, with the key structural difference being the addition of hydroxyl group (OH) at position 3 breaking the pyran ring (Fig. 1). Although the difference is small, CBD is non-psychoactive, and it is not a precursor to Δ9-THC. Accordingly, the level of CBD in a hemp plant has no impact of the legal status of a plant. Although environmental growth conditions impact CBD level, it is influenced primarily by plant genetics and seeds for chemovarieties that will express high yield CBD can be purchased. CBD is being investigated for a wide variety of health benefits and is popular as an additive in personal care products and nutritional supplements. Anecdotal reports suggest CBD may help chronic pain and arthritis, inflammation, anxiety, PTSD, depression, psychosis, nausea, inflammatory bowel disease, migraines, nicotine and opioid addiction, muscle spasticity, low appetite, etc. On June 25, 2018, the United States Food and Drug Administration (FDA) approved Epidiolex, a pharmaceutical grade CBD extract as an oral formulation to treat two rare and severe forms of epilepsy. It is currently the only cannabinoid approved by the FDA to treat a medical condition.

Current Legal Status of Cannabis in the US

In recent history, the distinction between hemp and marijuana originates from "The species problem in Cannabis: science and semantics", published in 1979 by Earnest Small, in which he proposed that the hemp can be distinguished from marijuana by virtue of having a Δ9-THC concentration of <0.3% on a dry weight basis [1]. This definition was widely adopted and is used internationally to classify hemp and

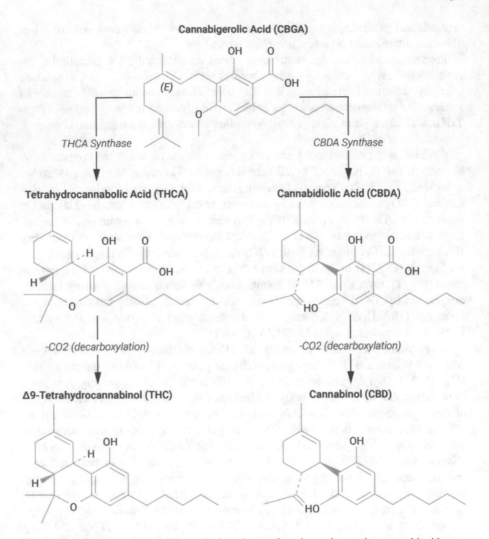

Fig. 1 Chemical structures and biosynthesis pathways for relevant hemp phytocannabinoids

marijuana today and was also used as the important cut-off in the 2018 Farm Bill. It is the critical point of having a Δ9-THC concentration of not more than 0.3% of a dry weight basis that served a basis to separate legal hemp from illegal marijuana at the federal level and is the reason that hemp growers must test industrial hemp crops. Hemp is comparatively free of Δ9-THC at the accepted concentration of 0.3% or less whereas marijuana is generally around 3–30% Δ9-THC/THCA.

Agricultural Marketing Act of 1946

Section 297B of the Agricultural Marketing Act of 1946 (AMA) provides guidance for State and Tribal plans to monitor and regulate hemp production [2]. It specifically outlines the need for "a procedure for testing, using post-decarboxylation or other similarly reliable methods, delta-9 tetrahydrocannabinol concentration levels of hemp produced in the State or territory of the Indian tribe…". The inclusion of the term "post-decarboxylation…method" was a constant source of confusion for many laboratories and hemp growers since no "post-decarboxylation method" exists and the concept of post-decarboxylation contradicts the selective assessment of the level of Δ9-THC in freshly cut plants.

Controlled Substances Act

In the modern era, the Controlled Substances Act authorizes the federal government to classify controlled substances as well as regulate the manufacture, importation, possession, use and distribution of certain narcotics, stimulants, depressant, hallucinogens, anabolic steroids, and chemicals used in the illicit production of controlled substances [3]. Consequences of a conviction for violations of the Controlled Substances Act and related laws include heavy fines and imprisonment [4].

The Controlled Substances Act classifies marijuana as a Schedule 1 drug with no medical benefit and high use potential for abuse and defines it as "all parts of the plant *Cannabis sativa* L., whether growing or not; the seeds thereof; the resin extracted from any part of such plant; and every compound, manufacture, salt, derivative, mixture, or preparation of such plant, its seeds or resin. Such term does not include the mature stalks of such plant, fiber produced from such stalks, oil or cake made from the seeds of such plant, any other compound, manufacture, salt, derivative, mixture, or preparation of such mature stalks (except the resin extracted therefrom), fiber, oil, or cake, or the sterilized seed of such plant which is incapable of germination" [3].

While opposing opinion exists about whether the classification of marijuana as a Schedule 1 drug is correct, until human clinical research proves medical benefit in controlled studies, it is likely to remain in this category.

Agriculture Improvement Act of 2018

The Agriculture Improvement Act of 2018, commonly referred to as the 2018 Farm Bill, did not change the legal status of marijuana, but it removed hemp-derived cannabidiol (CBD) from the Controlled Substance Act permitting hemp to be grown as an agricultural commodity [5]. Within Sec. 10111 (Hemp Production) of the legislation, hemp is defined as "the plant *Cannabis sativa L.* and any part of that plant, including the seeds thereof and all derivatives, extracts, cannabinoids, isomers, acids, salts, and salts of isomers, whether growing or not, with a delta-9 tetrahydro-cannabinol concentration of not more than 0.3% on a dry weight basis". The 2018 Farm Bill suggests total THC concentration—*not just Δ9-THC*—by the inclusion of "acids" in the definition, but it does not specifically state that the conversion of THCA to Δ9-THC needs to be accounted for.

Establishment of a Domestic Hemp Production Program

In October 2019, the USDA published an interim final rule, "Establishment of a Domestic Hemp Production Program" [6]. This document laid a solid foundation for a national hemp policy that allows for interstate transfer and clearly defines that the total available THC (i.e. the sum of THCA and Δ9-THC) needs to be measured.

7 CFR Part 990.1

Postdecarboxylation. In the context of testing methodologies for THC concentration levels in hemp, means a value determined after the process of decarboxylation that determines the total potential delta-9 tetrahydrocannabinol content derived from the sum of the THC and THC-A content and reported on a dry weight basis. The postdecarboxylation value of THC can be calculated by using a chromatograph technique using heat, gas chromatography, through which THCA is converted from its acid form to its neutral form, THC. Thus, this test calculates the total potential THC in a given sample [6].

7 CFR Part 990.1

Acceptable hemp THC level. When a laboratory tests a sample, it must report the delta-9 tetrahydrocannabinol content concentration level on a dry weight basis and the measure-ment of uncertainty. The acceptable hemp THC level for the purpose of compliance with the requirements of State, Tribal, or USDA hemp plans is when the application of the measure-ment of uncertainty to the reported delta-9 tetrahydrocannabinol content concentration level on a dry weight basis produces a distribution or range that includes 0.3% or less. For example, if the reported delta-9 tetrahydrocannabinol content concentration level on a dry weight basis is 0.35% and the measurement of uncertainty is +/− 0.06%, the measured delta-9 tetrahydrocannabinol content concentration level on a dry weight basis for this sample ranges from 0.29% to 0.41%. Because 0.3% is within the distribution or range, the sample is within the acceptable hemp THC level for the purpose of plan compliance [6].

Many hemp growers have expressed concern that the inclusion of THCA will make many of their crops "hot" by exceeding the 0.3% Δ9-THC threshold. This is likely to become a contested issue and hemp growers are advised to pay close atten-tion to the discussion. Commercial hemp is a highly regulated agricultural product with confounding and often contradictory legal status between the federal and laws in different states. Any hemp grower is advised to thoroughly understand the state specific regulations related to cannabinoid testing requirements. In contrast, CBD levels are not regulated by any legislation, but testing is voluntarily performed to measure the financial value of the crop.

State Regulation

In addition to national legislation, most states have enacted regulations to support industrial hemp cultivation and production. States have taken slightly different perspectives on hemp which has led to a web of confounding legal status across the nation. But many states have allowances providing for cultivation of hemp for commercial, research, or pilot programs. The National Conference of State Legislatures maintains a regularly updated database about the position of industrial hemp for each state as well as a helpful link to the formal legislation [7]. Potential hemp growers are encouraged to carefully review state specific rules.

In states that that have legalized marijuana in either medical or adult use/recreational form, much like hemp, a non-uniform and confusing set of independent regulations exist. However, from the outset, a key restriction to understand before any testing laboratory is proposed is that many states prohibit ownership in both safety testing laboratories (compliance) and cultivation/processing/dispensary businesses (manufacturing and retail). If an investor has current ownership interests in the marijuana space, it is strongly recommended to consult with an attorney to understand the legality of participating in a testing laboratory.

Business Planning and Financial Modeling

Starting a cannabis safety testing laboratory is no different than any other business where having a business plan that provides a written roadmap outlining the tasks, timelines, and costs will help the investor plan for success. While operating a cannabis safety testing laboratory is gaining popularity as an investment vehicle for entrepreneurs and larger investment groups, the infrastructure needed for laboratory testing is complex and expensive. Without prior knowledge in analytical, clinical, or environmental laboratory testing, there is a new set of terminology to learn and many hidden costs. A business plan is an essential document that should not overlooked for the sake of speed. A careful examination of the anticipated start up or pre-revenue costs, the ongoing operating costs once the lab is testing samples, and realistic revenue projections will reduce the risk of undercapitalization. By taking the time to understand the scope of the applicable laws, market, financial projections, and operating challenges, investors can better understand the legal, financial, and operational risks of entering into the cannabis laboratory space.

Market Size

Until major legislative changes at the federal level occur, the market size is determined by the state the laboratory is located in. While states are legalizing cannabis and providing certain business protections, it remains illegal to transport cannabis across state lines. This means that for marijuana flower, the theoretical market can be estimated by determining the total number of indoor and outdoor square feet permitted for growth, estimating the amount of harvested cannabis flower per year, dividing by the state legislated batch size, and multiplying by any mandated replicate testing to derive the potential number of samples to be tested on an annual basis. In addition to flower, concentrates and marijuana infused products require safety testing, but at this time there is no way to easily determine the potential number of samples that will require testing.

The potential number of annual samples multiplied by the dollar amount a laboratory believes they can charge for a test is the total theoretical revenue for any laboratory. If a laboratory can make rudimentary predictions about market share, then total laboratory revenue can be estimated.

Revenue

In August 2019, we conducted a pricing study of randomly selected laboratories in states that have legalized recreational cannabis use to determine the current average reimbursement for testing. The following terms were entered into a Google search: "Cannabis" and "Laboratory" and "Price" and "List" and "[Full state name with recreational use legalized status]".

All of the first page links were reviewed for pricing information. Laboratory name, state, broad testing category, and a rollup comprehensive state mandated compliance safety test pricing that would mirror the state of CA requirements were charted.

One observation is that many well-known labs we expected to appear on the first Google search page were absent. We suspect that some many labs could significantly improve visibility by better search engine optimization. It was difficult to compare "apples to apples" as state mandated tests, action limits, and necessary level of laboratory compliance are different leading to a wide range in testing costs. Since the state of CA currently has the most stringent testing requirements nationally, all efforts were to identify the combination of tests that meet CA guidelines to normalize for testing requirements.

Of the 31 laboratories that were identified by the constrained Google search parameters, only 15 provided pricing information. Using available pricing information, the following testing averages were calculated: flower $395.13 (range $100–$899), concentrate $389.60 (range $100–$899), edible $326.27 (range $50–$899), single test $104.57 (range $42.50–$265). One potential concern is that while testing fees are currently comparatively high and provide generous profit margins for an efficient laboratory, testing may become a commoditized service. Over time this could result in significant downward pressure on testing reimbursement favoring laboratory consolidation and laboratory automation, both leading to higher throughput with lower labor requirements providing substantial cost savings.

Start-Up Costs

The initial start-up costs for a cannabis testing laboratory can vary widely, but the primary drivers are instrumentation selection, the extent of tenant improvements that are needed, and licensing application preparation costs. A range of estimated costs are outlined in Table 2. It is not uncommon for the total cost of laboratory tenant improvements to exceed $100/ft². Another significant cost is hiring personnel, but it is not included in the initial build-out phase. Cannabis laboratories are highly specialized and require certain tenant improvements that are not considered for most commercial businesses. Additional details about laboratory planning and design are discussed in chapter "Cannabis Safety Testing Laboratory Floor Planning and Design". Cannabis testing requires multiple pieces of expensive and highly sophisticated instrumentation. Additional details equipment and methodology are discussed in chapters "Pesticide and Mycotoxin Detection and Quantitation, Cannabinoid Detection and Quantitation, Utilizing GC-MS and GC Instrumentation for Residual Solvents in Cannabis and Hemp, Elemental Analysis of Cannabis and Hemp: Regulations, Instrumentation, and Best Practices, Quantitative Terpene Profiling from Cannabis Samples, and Laboratory Safety and Compliance Testing for Microorganism Contamination in Marijuana". And finally, unless an investor has previous regulatory filing experience, an applicant may find the process to be complex, requiring multiple checklists, and discover conflicts between forms, guidance documents, and approved regulation. Additional details about licensing requirements are discussed in chapter "Preparing Cannabis Laboratory Business License Applications".

Pre-launch Costs

After laboratory construction and tenant improvements are completed and instruments are installed, there is a period of time when intense focus will be placed on testing validation and other pre-inspection activities necessary for state laboratory

Table 2 Estimated start costs for key cannabis laboratory buildout activities

Start up activity	Low estimate	High estimate
Equipment purchase	$650,000.00	$2,000,000.00
Tenant improvements	$50/ft²	$500/ft²
Consulting/legal fees	$20,000.00	$200,000.00
State and municipal license application fees	$5000.00	$75,000.00
Inspection fees	$5000.00	$15,000.00

licensing. During this time, the laboratory will need to be staffed at operational levels, consume laboratory reagents, and require all systems be tested for go-live functionality. Many owners and investors find this a frustrating time since the entire laboratory operation needs to be supported financially without generating revenue. The time between buildout and first state compliant sample is measured in months, not weeks, and investors need to ensure there is sufficient capital to cover these operational costs. The pre-launch can take between 4–12 months and is highly dependent on the quality and experience of the analytical staff that has been hired as well as the level of operational organization. Having a detailed understanding of the tasks, dependencies, estimated timing, and known costs will make the pre-launch phase much smoother.

Post-launch Costs

After the lab receives approval from the state and/or township to accept and test cannabis samples, the operating costs and projected income will ultimately determine long-term viability and profitability. Each laboratory will have a different cost structure, but ensuring a stable supply of samples, coupled with efficient, productive, compliant, and high-quality operations, should be a driving business philosophy. After payroll/labor, which can be responsible for between 30–40% of total expenses, the next largest expense is typically laboratory consumables. These include all of the solutions, solvents, plastic and glassware, quality control materials, instrument columns, etc. that are used during the testing process and is usually about 10–20% of total expense. Together these two line-items may account for 40–60% of the total laboratory operational expenses and need to be managed carefully.

Laboratory Planning and Design

Because individual commercial structures are very different, cannabis testing laboratories are always a custom build requiring more substantial tenant improvements than most other businesses. Good laboratory design creates a safe and efficient workflow that contemplates many different needs. Several excellent resources exist to help understand the specific needs of an analytical laboratory [8–10]. Some of the planning points include:

- Minimizing and optimizing staff movement.
- Providing for common and limited access areas.
- Separate areas for enclosed sample receipt, accessioning, sample storage, chemical and biological waste storage.
- Electrical needs and independent circuitry (220 and 110 V).
- Industrial gas (N2) generation, storage, and venting.
- HVAC systems for ventilation.
- Casework.

- Plumbing for sinks, showers, eye wash stations.
- Separate areas for DNA extraction, pre-amplification, and post-amplification rooms
- Unidirectional airflow recommendations for molecular testing areas.
- Comprehensive security plan providing for restricted access areas.
- Breakrooms and changing areas.
- Administrative offices.
- Sufficient parking.

If sufficient funding is available, it is worth the additional cost to improve the internal laboratory appearance. Most clients will conduct a laboratory site visit and a visually appealing, clean and organized space will help to demonstrate professionalism and sell services. Furthermore, the first appearance of a laboratory will orient regulatory auditors to degree of concern the laboratory places on safety. Additional details about laboratory safety are provided in chapters "Cannabis Safety Testing Laboratory Floor Planning and Design, and Laboratory Safety from Site Selection to Daily Operation".

Workflow

Workflow is often overlooked in spaces that are being converted from general use into a laboratory. Ideally, a good laboratory design would have a single path workflow in which samples and people move in a manner that limits retracing steps and with minimal distance between and in areas where the different pre-analytical, analytical, and post-analytical steps occur. For example, apart from complicating the limited access plan, it is not ideal for staff have to move potentially hazardous chemicals a long way to the storage area, or to have to move regularly between lab areas where aseptic and non-aseptic techniques are performed requiring additional cleaning protocols and time.

Applying for a Cannabis Testing License

While cannabis testing may provide exciting financial possibilities, the pathway to opening a laboratory begins with obtaining a license. Licensure may seem like a straightforward process, but the reality is that in most states obtaining a cannabis safety testing laboratory license is an expensive, time consuming, and challenging task. Cannabis safety testing is not a business that is "figured out as you go", rather, the application process demands a thoughtful, written plan that requires substantial upfront effort and supporting document preparation.

Preparing to submit a cannabis safety testing application is typically a lengthy process. Additional details about the application process are provided in chapter "Preparing Cannabis Laboratory Business License Applications". Even though states are highly motivated to approve applications to reduce the bottleneck in sample testing, a complete, accurate, and detailed application is a legal requirement. States and townships are working hard to provide support and guidance to applicants, however, minor modifications to application documents are common. Therefore, it is imperative to always download the most current application forms, or better, submit an electronic application when possible.

Because regulations are currently created by states, variations in application specific requirements are expected, but two general categories/phases of information are common: (1) a pre-qualification phase that includes legal and financial disclosures, business background checks, and demonstration of financial security, and (2) a physical description of the facility and a detailed operations plan. In addition to state licensing, there is also likely to be city/township/municipal licensing requirements. Fortunately, the application information tends to closely mirror state applications. The pre-qualification phase is relatively straightforward as it relies on preexisting information. The second phase is future thinking and often requires the applicant to describe their plan and include building zoning requirements, architect stamped floorplan schematics, security plan including limited access areas, hazardous waste management plan, quality management plan, marketing and advertising plan, standard operating procedures, instrumentation/equipment, staffing plan.

If accurate operational information is included, rather than generic placeholder information, the process may take 2–6 months of consistent effort to prepare and submit both the state and municipal applications. Some of the key determinants of time include: (1) the ability of business owners to produce documentation of their business history, financial strength, and litigation history and (2) prior experience and/or access to template documents that can be used as a writing roadmap. Working with a knowledgeable consulting group that has a previous record of success will significantly reduce the timeline by helping to decipher regulations and organize a prioritization plan for document procurement and preparation. One potential concern about using placeholder information is that the operational information and documents still need to be created. Since temporary applications are only granted for a limited time window, working on those essential tasks after the fact may distract employees from other essential validation and cutting into the time available for ISO/IEC 17025 accreditation activities.

Accreditation and Compliance

Cannabis safety testing laboratories are highly regulated and generally require some form of third party accreditation. Accreditation is the formal recognition from an agency or organization that provides oversight that a laboratory is able to produce accurate and defensible analytical data. An accredited laboratory is expected to demonstrate the technical proficiency to conduct an identified scope of work through standard procedures and protocols to meet defined quality standards. Accreditation requires a thorough evaluation of a laboratory's quality system, facilities and equipment, test methods, records, reports, and staff. Several organizations are aligned to facilitate cannabis laboratory accreditation including ISO/IEC 17025, AOAC International and ASTM International.

ISO/IEC 17025

The International Organization for Standardization (ISO) is an internationally recognized accreditation organization that sets standards that are meant to be applied consistently, irrespective of geographic location. In the analytical laboratory space, ISO/IEC 17025 is a widely recognized accreditation used in many industries including food safety testing and environmental testing [11]. Although several laboratory accrediting organizations exist, many states have chosen to defer to ISO/IEC 17025 accreditation as the key requirement to transition from a temporary license to a full license. Furthermore, many clients will not work with laboratories that don't have ISO accreditation out of concern about potential lapses in quality even though the lab is legally allowed to offer testing services.

ISO/IEC 17025 can be summarized as a process for a laboratory to show that a quality system exists that can reliably detect and/or quantitate analytes of interest. In a cannabis laboratory examples of analytes include the percentage of THC and/or other cannabinoids, pesticides, elemental impurities, residual solvents, microbial pathogens, moisture content, and water activity. ISO/IEC 17025 and subsequent accreditations are typically difficult to obtain, involve substantial labor, effort, and documentation, and require maintaining a high level of expertise and compliance post-accreditation [12]. Additional details about the requirement and process for ISO/IEC 17025 accreditation are discussed in chapter "Quality Assurance and the Cannabis Analytical Laboratory".

AOAC International

AOAC International, formerly the Association of Official Analytical Chemists, develops voluntary consensus standards in accordance with the U.S. National Technology Transfer and Advancement Act of 1995 (PL 104-113) and U.S. Office of Management and Budget Circular A-119. AOAC provides a helpful guidance document that effectively interprets ISO/IEC 17025 guidance, "AOAC International Guidelines for Laboratories performing Microbiological and Chemical Analyses of Food, Dietary Supplements, and Pharmaceuticals – An Aid to the Interpretation of ISO/IEC 17025" [13].

AOAC International has organized a new initiative, the Cannabis Analytical Science Program, to provide a forum where the science of hemp and cannabis, and the development and maintenance of cannabis standards and methods can be discussed. The CASP analytical community is comprised of government, academic, and contract laboratories; technology providers; private sector organizations; and allied associations. It publishes voluntary standard methods of performance requirements (SMPR) that laboratories and manufacturers are encouraged to implement.

ASTM International

ASTM International, formerly known as the American Society for Testing and Materials, is an international standards organization that develops and publishes voluntary consensus technical standards for a wide range of materials, products, systems, and services. ASTM formed Committee D37 on Cannabis to develop standards for cannabis, its products and processes. The activities are focused on meeting the needs of the cannabis industry, addressing quality and safety through the development of voluntary consensus standards. Subcommittees will focus on the development of test methods, practices and guides for cultivation, quality assurance, laboratory considerations, packaging and security.

Culture of Quality

A strong culture of quality forms a solid foundation upon which a successful, safe, and respected laboratory can function. Common quality activities include: Preparing for new accreditations, inspections, and audits; Writing, following, and documenting standard operating procedures (SOP's); Ensuring a safe working environment; Collecting, analyzing, summarizing, and sharing operational data; Performing regular proficiency testing; Investing in staff education and training. Additional detail about organizational quality and laboratory compliance are discussed in chapter "Cannabis Laboratory Management: Staffing, Training and Quality".

A successful quality program requires a daily commitment from everyone in the organization. It is essential that the executive management team and technical leaders of any laboratory are familiar with the expectations of, and fully support the time and cost of, a comprehensive quality program. Unfortunately, quality is not a revenue generating service line, so it is not uncommon, particularly during the "Pre-Launch Phase" discussed previously, for the executive team to want to take shortcuts with quality activities to reduce pre-revenue expenditures. Failure to understand basic safety and documentation rules is a key—and obvious—flag for external auditors. Most examples of cannabis license denial, suspension, or revocation can be directly linked to this major shortcoming. An audit violation becomes part of the laboratory permanent record and takes more time to fix retroactively than it does to do it correctly the first time. To help avoid these issues, a laboratory should hire an experienced and dedicated Quality Assurance/Quality Control employee and take the additional time whenever needed to prepare robust quality documentation to avoid receiving a major audit violation.

Finally, large clinical diagnostic laboratories (>$10B valuation) such as Quest Diagnostics and LabCorp, have been operating for decades and can provide valuable guidance about quality for the cannabis testing laboratory space. There is consistent overlap in their cost structure, operational requirements, and quality needs that can be directly applied to cannabis testing. Cannabis safety testing labs can learn from clinical diagnostic laboratories. These laboratories have exceptionally strong quality programs, of which an important part includes performing rigorous and detailed mock inspections using external assistance to identify areas to strengthen before a live audit occurs.

Instrumentation

Cannabis and cannabis derived products generally require several analytical tests for contaminants including: pesticides, heavy metals, residual solvents, cannabinoids (THC primarily), mycotoxin, and microbial pathogens. The instruments used to perform the first five categories of tests are predominantly based on chromatography and spectroscopy technology, while microbial pathogens can be identified by traditional microbiological culturing and/or molecular biology instrumentation/ techniques. Additional details about the key methodologies are provided in several technical chapters "Pesticide and Mycotoxin Detection and Quantitation, Cannabinoid Detection and Quantitation, Utilizing GC-MS and GC Instrumentation for Residual Solvents in Cannabis and Hemp, Elemental Analysis of Cannabis and Hemp: Regulations, Instrumentation, and Best Practices, Quantitative Terpene Profiling from Cannabis Samples, and Laboratory Safety and Compliance Testing for Microorganism Contamination in Marijuana".

Choosing the appropriate instrument vendor/make/model for a particular test can be a daunting task, but several standard considerations will recur during most conversations with laboratory equipment manufacturers including: price, sensitivity, sample throughput, consumables cost, vendor support, frequency of maintenance downtime and maintenance difficultly, anticipated time to failure, software, and ease of use. Several chromatography and spectroscopy instrument vendors are positioning themselves for the cannabis laboratory safety testing market. Current market leaders in the cannabis space include: Agilent, Perkin-Elmer, Sciex, and Shimadzu.

Because of the high financial barriers to entry, new cannabis laboratories often cite price as the primary consideration in instrument selection. As with most financial decisions, there is a performance trade-off for low or lower cost instrumentation. Lower cost invariably leads to lower sensitivity and throughput, faster time to failure, and more frequent maintenance. Higher sensitivity allows additional sample dilution, which provides for a cleaner sample being loaded onto the instrument and extends the life of critical parts. Purchasing used instruments without a manufacturer warranty is generally not advised unless the use history of the instrument is well documented. When available, manufacturer refurbished and warrantied instruments are an excellent and cost efficient way to reduce equipment expense. While the low cost may be appealing, purchasing pre-owned instruments without a manufacturer warranty from a used laboratory equipment broker is not recommended unless the technical team has substantial expertise in hardware troubleshooting.

The goal of any laboratory is to reduce maintenance to a minimum since any time repairs or maintenance is being performed, testing, and the associated revenue stream, is stopped. A cleaner and more dilute sample leaves less matrix residue, which in turn reduces instrument maintenance. However, it will increase uncertainly when analyzing results near the limit of detection (LOD) and limit of quantitation (LOQ). When limits of quantitation (LOQ's) are close to the lower end of sensitivity, it becomes more difficult to validate a new method and the likelihood of reporting a false-negative or false-positive increases. Some vendors will provide an analysis of state specific testing analytes, minimum action levels, and appropriate instrument pairing. In many states, this is not the most expensive purchase option.

An important consideration for instrument selection is the level of vendor support. Vendors are willing to provide very helpful preventative maintenance plans or perform installation qualification and operational qualification (IQ/OQ), but both are expensive and, in the current regulatory environment, IQ/OQ is not necessary. While these technical support options can be useful, any assistance in method development and validation also has an inherent value. Developing new test methods is a complex and time-consuming task even for highly trained analytical chemists and microbiologists. Therefore, a high value should be placed on any intellectual assistance a vendor can provide. Of particular interest are methods internally validated by the vendor that can be followed and used as a foundation for SOP's. Further, access to control material, educational seminars, and in-person training are desirable offerings and services.

Chromatography

Chromatography coupled to mass spectroscopy is a common laboratory method used to separate and identify the chemical compounds in a cannabis sample. During the chromatography process, samples are dissolved into a liquid or gas to create a "mobile phase" that is passed through a cylindrical column tightly packed with very small beads, the "stationary phase". Using cannabinoids as an example, because of the small differences in chemical structures of $\Delta 9$-THC, THCA, CBD and the other cannabinoids, each of these compounds pass through the stationary phase at different speeds and can be identified and quantitated as they exit the column (Fig. 2).

A variety of chromatography instrumentation can be used to determine cannabinoid levels. Gas chromatography (GC) is a common method in which the sample is vaporized under intense heat to create the mobile phase. But the heating rapidly

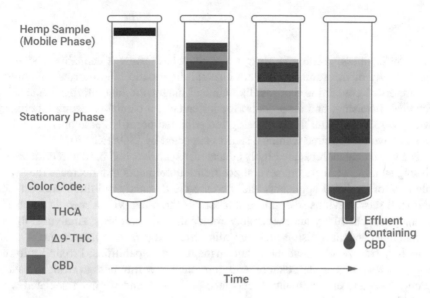

Fig. 2 Diagram of chromatographic separation of Δ9-THC, THCA, and CBD

decarboxylates any THCA present in the sample converting it to Δ9-THC. Using this method, labs report the THC value as "Total Potential THC" or "Total Available THC". While this is a valid and practical method to determine the total potential THC in recreational use samples, if provided alone it overstates the actual level of Δ9-THC in a freshly harvested hemp sample. A detailed discussion of total available THC is provided in chapter "Cannabinoid Detection and Quantitation". It is possible to mathematically convert the result into *approximate* levels of Δ9-THC and THCA by reversing the foltlowing equation:

$$\text{Total Available THC} = \left[\Delta 9 - \text{THC} + 0.877 * \text{THCA} \right]$$

High pressure liquid chromatography (HPLC) is an alternative method to identify and quantitate cannabinoids. Unlike GC, HPLC does not rely on heating, so any THCA in the sample will not be decarboxylated. Therefore, this method provides a direct and separate determination of the physical amounts of Δ9-THC and THCA present in the sample. HPLC separates the individual constituents of the mobile phase so can be measured by electromagnetic absorbance of ultraviolet and visible light or by mass spectroscopy. HPLC coupled with mass spectroscopy (HPLC/MS) is a less common method but provides two important advantages and is a best practice. First, it determines the mass to charge ratio of each compound that is separated providing a more accurate identification of a compound. Second, because of the resolving power of the mass spectrometer, it offers a more accurate and precise quantitation.

Staffing

Like any business, staffing is always a challenge. Staffing a cannabis laboratory shares many of the same issues that clinical diagnostic laboratories encounter. An important concern is that just like clinical diagnostic laboratories, because of technical, licensing, and accreditation requirements, a cannabis testing laboratory always works under the threat of being forced to temporarily cease operations if it doesn't have the required technical staff as expected by ISO/IEC 17025.

The key technical hires are highly educated, expensive, and hard to recruit. From a hiring standpoint, there are several position requirements that decrease the available pool of qualified applicants and increase the demand (and therefore cost) for their skill sets: Previous scientific and laboratory training with a specific emphasis on analytical chemistry and, separately, molecular microbiology; Analytical thinking and independent decision-making skills; Knowledge of compliance, quality, and regulatory framework; Leadership and management capabilities; Proximity to the testing facility—these staff cannot work remotely; A willingness to work in a start-up/new company environment; Acceptance of the risk and stigma associated with cannabis.

All accredited cannabis laboratories are likely to need two critical lead technologists: (1) a PhD trained (preferably) or MS trained analytical chemist with substantial experience in chromatography and spectroscopy techniques and (2) a microbiologist, preferably with molecular biology training. While it may be tempting to reduce payroll cost by merging the two positions, the likelihood of finding a person with dual training is remote. Furthermore, the testing process is time consuming and it is wholly unrealistic to think one person can perform all the testing needs and associated compliance activities alone. While it may be possible to fill additional laboratory assistant positions by someone with less laboratory experience and provide training, the technical leads need to come with an immediately deployable knowledge base or the laboratory will struggle to bring the necessary equipment and technology online. Leadership qualities in the early hiring phases cannot be overemphasized. Any critical hire must have laboratory experience and demonstrate leadership qualities since staff throughout the organization will continually look to them for guidance. Expect that qualified technologists will require higher levels of compensation and flexibility to be recruited. Additional details about staffing considerations are discussed in chapter "Cannabis Laboratory Management: Staffing, Training and Quality".

Laboratory instrumentation requires ongoing maintenance, regular troubleshooting, and an intrinsic understanding of how it works to prevent extended downtime. Therefore, independent problem solving and the authority to make critical laboratory operation decisions is an expectation for the critical technologist hires. One of the most important observations a lead technician will make is that a failed result could be the result of a process failure, rather than a sample derived problem. To confirm this suspicion, technologists will need to be able to modify instrument settings, prepare new calibration standards, conduct additional experiments, review and analyze data to identify possible causes, and, importantly, have the time to do complete these tasks. All of this is part of their scientific training; years of undergraduate and post-graduate training cannot be replaced by good intentions to learn. It should be noted that instrument manufacturers may provide technical support, but it usually comes at a significant time and financial expense should not be relied upon for regular support. Hiring clinical laboratory technologists (i.e. someone who has worked in a medical laboratory or toxicology laboratory) can offer an advantage by bringing substantial experience from working in a quality driven environment.

While not considered a technologist position, employing a full-time quality control and/or quality assurance position is highly recommended as previously noted because of the challenging regulatory environment and requirement for ISO/IEC 17025 accreditation.

Additional Hiring Challenges

Cannabis testing is a hands-on activity in which employees must be present in the laboratory. To reduce transport time, and therefore resulting turn-around time (TAT), many testing laboratories are built close to cultivators and manufacturers. While these locations may be financially appealing to laboratories, they are often rural and not close to academic or industrial centers making it harder to recruit qualified employees.

While the critical laboratory personnel need to be a major recruiting focus, general laboratory assistance including samplers, lab assistants, and administrative assistance should not be an afterthought. A laboratory that wants long term commitment, adherence to quality, and nimble action needs to pay overmarket rates, provide educational opportunities and career growth opportunities, and appreciate that knowledge workers are harder to recruit than might be expected. Unlike fast food chains, where training is relatively quick and easy, a new hire in a cannabis laboratory must go through lengthy and rigorous security, compliance, and laboratory training. Without any doubt, the importance of beginning recruiting as early as possible and continuing to develop a pipeline of qualified new hires that fits the culture of the company cannot be overemphasized. The best-looking lab with the newest equipment and most rigorous validations, cannot run a single test without competent staff.

References

1. Small E (1979) The species problem in cannabis: science & semantics. Corpus, Toronto
2. US Congress (1946) Agricultural Marketing Act of 1946. U.S. https://www.agriculture.senate.gov/imo/media/doc/Agricultural%20Marketing%20Act%20Of%201946.pdf
3. US Congress (1970) Comprehensive Drug Abuse Prevention and Control Act of 1970. PL 91-513. https://www.govinfo.gov/content/pkg/STATUTE-84/pdf/STATUTE-84-Pg1236.pdf
4. Yeh B (2015) Drug offenses: maximum fines and terms of imprisonment for violation of the federal controlled substances act and related laws

5. US Congress (2018) Agriculture Improvement Act of 2018
6. US Department of Agriculture (2019) Establishment of a domestic hemp production program
7. National Conference of State Legislatures. State Industrial Hemp Statutes. Last accessed Oct 2019. http://www.ncsl.org/research/agriculture-and-rural-development/state-industrial-hemp-statutes.aspx
8. National Research Council (2000) Laboratory design, construction, and renovation: participants, process, and product. The National Academies Press, Washington, DC. https://doi.org/10.17226/9799
9. World Health Organization (2011) Laboratory quality management system handbook. WHO, Lyon
10. Association of Public Health Laboratories (2019) Laboratory facility construction and major renovations guidelines
11. International Organization for Standards (2017) ISO/IEC 17025:2017: General requirements for the competence of testing and calibration laboratories
12. Association of Public Health Laboratories (2017) Laboratory costs of ISO/IEC 17025 accreditation: a 2017 survey report
13. AOAC International (2017) Guidelines for laboratories performing microbiological and chemical analyses of food, dietary supplements, and pharmaceuticals, an aid to interpretation of ISO/IEC 17025:2017

Cannabis Safety Testing Laboratory Floor Planning and Design

James Jaeger, Navin Pathangay, and Shaun R. Opie

Abstract Good cannabis laboratory design is the foundation for a safe and efficient testing environment. Whether an existing lab is being adapted to accommodate cannabis testing or being built from the ground up, it is essential to assemble a team of experts to help guide the process. In addition to providing guidance about adequate mechanical ventilation, electrical, and plumbing, it is important to consider the size, space utilization, workflow, security, and safety requirements for a complete and code-compliant infrastructure. To solve these important issues, it is strongly advised to work with an architecture firm, laboratory consultants, instrument manufacturers, and lab furniture manufacturers who have previous experience designing and building cannabis testing laboratories.

Introduction

The key to the success of a cannabis laboratory is being familiar with the infrastructure and being able to create a design that accommodates large laboratory equipment, working areas, storage, and safety. Several publications provide an excellent overview of construction and renovation guidelines [1, 2]. The cannabis industry is a thriving business in the world, and many researchers and scientists are developing and designing high-quality testing facilities to meet the growing demand. These laboratory facilities are more in demand than in the past, however, cannabis laboratories have a unique set of design requirements, regulations, and equipment that must be considered.

Labs are crucial in the ever-growing marijuana industry for analysis and ensuring the development of quality cannabis products. While many people believe that a cannabis testing lab should be designed the same as any other lab, it's essential to understand that they have a set of unique requirements. The purpose of this chapter

J. Jaeger (✉)
Laboratory Planning & Design Consultant, Carlsbad, CA, USA

N. Pathangay
Pathangay Architects LLC, Phoenix, AZ, USA

S. R. Opie
E4 Bioscience, Charlevoix, MI, USA

© Springer Nature Switzerland AG 2021
S. R. Opie (ed.), *Cannabis Laboratory Fundamentals*,
https://doi.org/10.1007/978-3-030-62716-4_2

is to offer guidance about important considerations to understand before designing and building a cannabis testing laboratory.

Regulations

Before building a cannabis testing laboratory, it is important to have a thorough understanding of state and local regulations. If the lab is found not to comply with the regulations during an inspection, it will be permitted to fix the issues found, however it is exceptionally interruptive and expensive after construction is finished.

Different states have different regulations when it comes to cannabis labs. There is no one established standard for these facilities. Both state and local jurisdictions have their own regulations that labs must comply with. In addition, all labs have electrical, ventilation, air filtration, and cleanliness regulations they must adhere to, making it challenging to understand which regulations apply are and how to comply. The National Fire Protection Agency (NFPA) Fire Code 1, Chapter 38 covers several aspects of marijuana facility design/build, such as permits, fire protection systems, means of egress, and processing or extraction—including LP gas extraction, flammable and combustible liquids extraction, CO_2 extraction, and transfilling [3]. While NFPA Chapter 38 does not cover safety testing laboratories specifically, most of the guidance is directly applicable and will be considered by the local fire marshal.

Size

One of the first steps in designing any laboratory is to develop a list of all laboratory instruments and equipment planned for the facility, including dimensions and installation/utility requirements. This list will help determine the types of spaces needed, the size of these spaces, the workflow and layout of the laboratory, and the specific functional requirements of the space itself.

Cannabis testing laboratories are typically sized to accommodate an expected throughput or number of 'suites' of instruments. Throughput is the number of samples the laboratory is able to process in a given period and is largely dependent on the number of instruments, or 'suites' of instruments in the laboratory. One 'suite' of instruments is the minimum grouping of instruments needed to perform all compliance tests mandated by the governing municipality. It is common for cannabis testing laboratories to be planned and designed for multiple suites of instruments. Most cannabis testing laboratories in the US have one to four suites of instruments, which process a throughput of approximately 20–200 compliance samples per day, and are sized from 2500–15,000 ft^2.

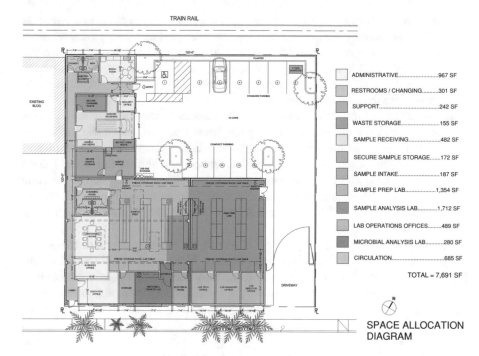

Fig. 1 Cannabis testing laboratory—space allocation diagram

The laboratory in Fig. 1 was designed to accommodate up to three suites of instruments with an expected throughput of 150 compliance samples per day and is sized at approximately 7700 ft².

Recommended Spaces

The following spaces are recommended for all cannabis testing laboratories. It is important to note that not all of the spaces listed need to be individual rooms; some of them can simply be dedicated areas in the facility, however, please refer to local cannabis regulations for further clarification on this matter.

Sample receiving—This is an area where samples are unloaded from sampling vehicles and transferred to sample intake. It is recommended this area be secure, but it is not required that the area be a garage (as designed into the laboratory shown in Fig. 1). If this space is outdoors in a parking lot or loading dock, provide a security fence and control access to this area.

Sample intake—This is where samples are received and entered into Laboratory Information Management System. Typically, this includes photography, bar coding, and data entry processes to be accommodated by moveable laboratory tables or office desks.

Secure sample storage—This is a secure space where samples are stored. Size this space appropriately to accommodate duplicate samples that are typically required to be stored for certain period after initial testing occurs. Include space for numerous racks/cabinets, refrigerators, and freezers. Ensure appropriate power and HVAC for refrigerators and freezers. Refer to local cannabis regulations for further requirements of this space.

Sample preparation laboratory—This is a standard wetlab space where samples are prepared for either microbial or chemical analysis. Moveable laboratory benches are recommended with power and utility drops from the ceiling to accommodate benchtop instruments. This space should have at least one large laboratory sink with an emergency eyewash, one safety shower, and at least one fume hood with chemical/flammable storage base cabinets. Typical laboratory equipment found in this area are refrigerators, freezers, centrifuges, grinders, sonicators, and digestors.

Microbial analysis laboratory—This is where microbial testing and analysis is performed and is typically designed to meet Biosafety Level 2 (BSL-2) standards. To meet BSL-2 standards, the space must be separated from the other labs completely, must have a sink, eyewash, automatic door closer, and must directly exhaust all air from the space. In addition, the microbial analysis laboratory must be negatively balanced to the adjacent sample preparation and main analysis laboratories regarding airflow/pressure. Typical laboratory equipment found in this area are PCRs, incubators, and biosafety cabinets.

Main analysis laboratory—This is where cannabinoid, terpenoid, pesticide, heavy metal, mycotoxin, residual solvent, and process chemical testing is performed. Moveable lab benches are recommended with power and utility drops from the ceiling to accommodate instruments. This space should have at least one large laboratory sink with an emergency eyewash and one safety shower. The main analysis laboratory will be the largest space in the facility, and will require the most mechanical, electrical, and plumbing infrastructure. Typical laboratory equipment in this area are liquid chromatography instruments, gas chromatography instruments and various types of mass spectrometers. The chemicals commonly used in a variety of cannabis-testing procedures can corrode or damage furnishings, so it is recommend to select furniture and surfaces that are designed to withstand harsh chemicals and that are easily cleanable, such as epoxy or phenolic resin.

Laboratory operations offices—These are offices for laboratory operations staff. Preferably both private offices and open offices with multiple workstations are included. It is important these offices be located directly adjacent to the laboratory space. Most local cannabis regulations require a space for secure records storage, which is typically accomplished by placing a locking file cabinet in a private laboratory operations office.

Locker/changing room—This is a space where laboratory operations staff can change out their street clothes and into proper clean gowning attire/PPE (Personal Protective Equipment). The laboratory spaces in the facility should be treated as 'clean' space and a proper changing space at the entry of laboratory is recommended. The space should include lockers for secure storage of staff belongings, space for racks with gowning attire/PPE, benches, private areas for changing, restrooms, and (ideally) showers.

Security office—This is a space where security personnel can control access to the building and monitor security cameras. Refer to local cannabis regulations for further requirements of this space.

Conference room—While not required, it is recommended to have a meeting/gathering space within the facility. In addition to internal meetings, seminars and parties that can occur in this space, it is important to have space to meet with clients. In the facility detailed here, the conference room is designed with a large viewing window into the laboratory, which is an opportunity to show your clients your laboratory without giving them access into the laboratory.

Break room—This is another space that is not required but is highly recommended. It is preferable to have a dedicated space in the facility for staff to take breaks and eat meals. It is not preferable for staff to have to leave the property to get food or to take a break.

Storage—This is a space where laboratory consumables and supplies can be stored until needed. This space often gets overlooked as it is not critical for operations, but problems will arise if there is no dedicated storage space in a laboratory.

Electrical room—There will need to be space in your facility dedicated to electrical service panels and distribution equipment. Ideally, this space should not have to be accessed via the 'clean' laboratory space, but as was the case in the example provided, designers often need to work with as-built building conditions when renovating an existing facility. Relocating electrical service equipment is quite an expensive, lengthy, and disruptive task. Typically, this space will also accommodate all telecommunications, security system, fire alarm system, and IT/IS server equipment.

Secure cannabis waste—This is a secure space where cannabis waste is stored until officially destroyed. Typically, this is a small room with a couple tables and receptacles where waste can be weighed, documented, and officially destroyed. It is preferable to locate this near the service entry/exit, where it can easily be picked up by an outside vendor without them needing to enter the clean laboratory space, or so it can easily be transferred to an outside dumpster (if allowable). Refer to local cannabis regulations for further requirements of this space, and the cannabis waste process.

Hazardous waste—This is a space dedicated to collecting and storing biological and chemical waste material until it leaves the facility. It is preferable to locate this near the secure cannabis waste area, adjacent to the service entry/exit, where it can easily picked up by an outside vendor without them needing to enter the clean laboratory space.

Janitor's closet—This is small space to store all cleaning equipment. It should have just enough space for a cleaning cart and racks for cleaning supplies. Ideally this space will also include a mop sink in the floor.

Administrative Offices—In addition to laboratory operations offices, it is important to have administrative space that is separate from the laboratory for non-laboratory operations functions, such as compliance, accounting, information technology, safety, sales, and support. It is not recommended that access to these offices is through 'clean' laboratory space.

Workflow

A cannabis testing laboratory can be classified as a 'process' facility that should be planned and designed to support the efficient flow of product, work, and people throughout the laboratory. Cannabis testing laboratories are largely operated based on a set of specifically developed Standard Operating Procedures. The layout of the laboratory needs to support these operating procedures to make each process as efficient as possible. It is imperative to consider all aspects of how products (samples) flow through the laboratory, from when they are received until they are removed as part of the waste stream. It is also important to understand how people and waste flow through the facility as well, as those processes will also affect the design of the space. Figure 2 illustrates the way products (samples), personnel, and waste flow through the facility.

If it often helpful to consider the daily life of someone working in the lab and what they will need to do and where they will go. All items within the space should

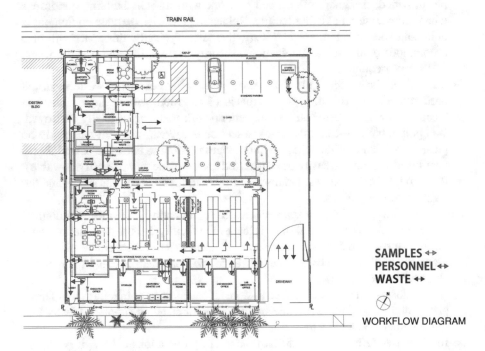

Fig. 2 Cannabis testing laboratory—workflow

be laid out for workflow efficiency, including lab tables. Staff should not have to walk any further than necessary to access everything that they need.

Laboratory cabinets and other storage solutions will also optimize workspaces, providing workers with areas to put items that they don't immediately need at easy access. Having laboratory casework and cabinets will allow staff to continue to focus on the work instead of searching for an item that they can't find when they need it.

Security

Security is an important aspect to consider when planning and designing a cannabis testing laboratory. Besides from the obvious reason to secure the facility, local cannabis regulations typically require some level of security. The following are different types of security measures often included in cannabis testing laboratory design:

Access controls—This is typically accomplished via card access and readers at doors to all spaces needing access control. Biometric access controls are also starting to be implemented, typically for the most secure spaces such as sample storage and records storage. Refer to local cannabis regulations to determine access control requirements for different spaces in the facility.

Surveillance cameras—Surveillance cameras are used to provide continuous surveillance of the entire facility and to record that surveillance over time. Every angle of the entire facility must be captured, and numerous cameras are required to accomplish this. Refer to local cannabis regulations to determine specific requirements for surveillance cameras.

Security station/office—Most local municipalities require at least one security guard in addition to access controls and surveillance cameras. If a security guard is required, it is important to design in a security office or space for this person to monitor surveillance cameras and control access to the facility.

Safety

Laboratory safety is an important topic as laboratories can be dangerous spaces to work if not designed and operated properly. However, with proper design and appropriate operating procedures, laboratories can be a safe place to work. First, it is important to include a sufficient number of sinks throughout the laboratory, all of which should include emergency eyewash units. Safety showers need to be provided as well, in appropriate locations. Space and supporting infrastructure for fume hoods, chemical and flammable storage cabinets, biosafety cabinets, and other laboratory safety equipment needs to be included in the design. In addition, layout of laboratory benches and equipment must not create safety issues by creating dead ends or limiting path of travel. Lastly, emergency exiting must be considered. It is preferable, and sometimes required, for all laboratory spaces to have at least two exits.

Fire code limits the amount of hazardous solvents or materials in a given space or 'control zone'. The lab manager should know how much solvent they plan to store on-site and how much future storage they will need if the design plan includes a multistage buildout. Every state has differing local storage and handling regulations of hazardous material.

Electrical

Laboratories consume large amounts of power and designing for power efficiency is highly recommended. This is because both the instruments inside the laboratory, and the building infrastructure equipment (needed to serve the laboratory) consume large amounts of power. Most laboratories also use a few different types of power—normal power, standby (emergency) power, and UPS (Uninterruptable Power System) power.

Normal power—This is the standard building power that is provided by the local utility company. The amount and type of power supplied to the laboratory is one of the most important infrastructure items to consider when evaluating properties for your lab, or when building from the ground up. As a general guideline, it is recommended that your facility have a minimum of 200 amps per suite of analysis instruments. For example, the laboratory pictured in this chapter has 600 amp of power, for three suites of instruments. While not required, three-phase power is recommended over single-phase power for most laboratories.

Standby power (emergency generator)—This is the power supplied by the emergency generator. In laboratories, ideally the generator provides standby power to temperature sensitive equipment such as freezers, refrigerators, incubators, and critical analysis instruments. Having an emergency generator and standby power system will allow your laboratory to continue operating and producing revenue during power disruptions. It is important to note that there is a time lag between when the power disruption occurs and when the emergency generator starts. In addition to note is the spike in power when the emergency generator starts. This is acceptable for freezers, refrigerators, and incubators, but the spike of power when the emergency generator starts up can damage your laboratory's most sensitive instruments.

UPS (uninterruptable power system)—UPS units provide short amounts of continuous power for the laboratory's most sensitive instruments. In the event of a power disruption, the UPS unit provides enough back-up battery time to either properly shut down these instruments, or to keep them running until the emergency generator is started and generating steady power.

Mechanical/Heating, Ventilation and Air Conditioning (HVAC)

Laboratories require more robust and complex mechanical systems than most other types of facilities. Laboratories are filled with heat-generating equipment and require a large amount of cooling to maintain appropriate ambient temperature. If laboratory spaces become too hot, these instruments will not function properly. In addition to heating and cooling, a laboratory must have an appropriately sized exhaust system. It is not recommended to recirculate laboratory air, which means all laboratory space air should be directly exhausted. Fume hoods and analysis instruments also have exhaust requirements that need to be maintained for laboratory safety. It is recommended to consider filtering the exhaust air to prevent nuisance cannabis odors from affecting neighboring businesses. Airflow should also be considered when designing laboratory mechanical systems. In general, air should flow from cleanest spaces outward via pressure gradient created by air supply and exhaust balance. In particular, the sample preparation laboratory and the main analysis laboratory should be positively balanced to adjacent spaces (air should flow out of these spaces). The microbial analysis laboratory should be balanced negative to adjacent spaces (air should flow into the space), as this space is considered Biosafety Level 2 (BSL-2) space, and negative airflow is necessary to achieve proper biocontainment.

Plumbing/Piping

All laboratories incorporate standard plumbing systems such as domestic/industrial cold and hot water and sewer/sanitary waste, which are connected to external (municipal) plumbing systems. In addition, most laboratories also utilize specialized internal plumbing and piping systems for various forms of purified water, laboratory vacuum, compressed air, and laboratory gasses such as nitrogen, argon, helium, hydrogen, and carbon dioxide. Piping systems are needed to carry these gases and liquids from their source to their point-of-use. It is recommended to locate these source gas cylinders and water purification systems in a space outside the 'clean' laboratory that is easily accessible by outside vendors and gas delivery vehicles. If a laboratory gas piping system is not utilized, gas cylinders will need to be located at each point-of-use in the 'clean' laboratory space, taking up valuable floor space and bringing outside contaminants into the laboratory. In this case, consider onsite gas generation options.

References

1. Association of Public Health Laboratories (2019) Laboratory facility construction and major renovations guidelines
2. National Research Council (2000) Laboratory design, construction, and renovation: participants, process, and product. The National Academies Press, Washington, DC. https://doi.org/10.17226/9799
3. National Fire Protection Agency (2018) Fire Code 1, Chapter 38

Laboratory Safety from Site Selection to Daily Operation

Krista Harlan and Danielle Cadaret

Abstract The chapter on lab safety is intended to instruct people interested or actively engaged in opening a cannabis safety and compliance testing laboratory. The content includes information from location selection and lab layout to development of operational safety procedures and how to establish a culture of safety. This information is structured for compliance to national Occupational Safety and Health Administration (OSHA) requirements and includes advice for identifying local or state requirements. The resource list includes links to online examples of checklists, chemical hygiene plan (CHP) instructions, and regulatory guidelines, as well as some of the chapter sources that contain valuable content to the reader but too lengthy to include in a short chapter focusing on lab safety in a book covering all aspects of cannabis lab start-up fundamentals.

Introduction

What is your first reaction when you think of "lab safety"? Do you think of lab coats and protective eyewear? Do you inwardly brace yourself for important but boring information? Or, is safety enticing because doing things the right way and knowing people are protected is energizing and rewarding? Regardless of personal feelings about safety, it is important. This chapter provides a strong foundation for creating an intentional mindset about safety using ideas and questions relevant to cannabis compliance test labs. The scope is defined for commercial labs in the United States, but the concepts are applicable to all labs performing analysis. We will examine safety concepts for investors, architects, builders, and managers as well as realtors and employees. Each person will benefit from considering the ideas and questions provided in this chapter. If you are responsible for Environmental Health and Safety (EHS), or the Chemical Hygiene Plan (CHP), this chapter will help define the

K. Harlan (✉)
Celtic Dragon Consulting LLC, Pinckney, MI, USA
e-mail: Krista@celticdragonconsulting.com; https://www.celticdragonconsulting.com

D. Cadaret
Cadaret Architecture, Royal Oak, MI, USA
e-mail: Danielle@cadaretarch.com

© Springer Nature Switzerland AG 2021
S. R. Opie (ed.), *Cannabis Laboratory Fundamentals*,
https://doi.org/10.1007/978-3-030-62716-4_3

parameters you should be considering and provide links to helpful resources [1]. If you want to get into the cannabis testing business as an analyst or technician, this chapter will have more information than you need, and because it is safety, you still need to read it.

Cannabis is regulated differently in different parts of the country. Fortunately, safety is universal, and the differences in regulation are limited to how each region measures and enforces the regulations. Since this book is not region specific, it does not provide specific codes or laws below the federal level, but it does provide guidance on how to find your state and local requirements.

Federal Occupational Safety and Health Administration (OSHA) created the standards everyone must follow. Because of the authority granted to states and territories, some regions have created their own safety standards. At the time of publication, OSHA has authorized 28 states or territories to operate their own occupational safety and health agencies and these locations are called 'state-plan' states. In all cases, the state plans must be as good as or better than the federal plan which means some states, such as California, have chosen to adopt stricter standards. Six areas (Connecticut, Illinois, New Jersey, New York, Maine, and the Virgin Islands) have plans that cover only public sector employees and therefore working in the private sector is subject to federal OSHA standards. In the remaining state plan states, the state has jurisdiction over both public and private employers [2].

Complying with federal OSHA regulations may not be enough; it is necessary to verify if your state/territory is subject to a state plan. Although the federal OSHA regulations set a minimum standard for all state plans, some states have additional requirements or have adopted different approaches to enforcement and fines. Whenever possible, gather all the laws and codes that are applicable to your location. This process can be done through various efforts and/or expense. The easiest route, but most expensive, is to hire a code compliance specialist. These consultants put together a list of all the local, municipal, state, and federal laws, statutes, and codes that must be followed. If you want to do the research on your own, most of the information is available to the general public (certain professional standards such as ISO 9001:2015 must be purchased). Since we are talking about safety, the best place to start is with the local Fire Marshal and OSHA consultant. There are programs available for smaller companies that provide an OSHA consultant at no charge with no risk of penalty so long as the findings are corrected by the time you have the official OSHA inspection.

Why do we need to be safety conscious? Most people answer with ideas about protecting workers, assets, or consumers (products). Some acknowledge that the law requires safe operations, and some will claim that a safe environment sets the tone for a quality operation. And everyone is right. Practicing safety is what protects people, property, and product (a.k.a. revenue) as well as maintains compliance with regulatory requirements AND, it supports quality.

Let us address the question 'Why do we need to be safety conscious?' from a new perspective. In blunt, perhaps even cynical terms, we answer 'because humans are both intelligent and creatures of habit, and because our society loves a good lawsuit.' Getting emotionally charged, about anything, is a growing part of our social media culture and though it is a great distraction from the day to day, it feeds into instant gratification the practice of placing blame. This cultural conditioning is the opposite of what is needed for a culture of safety and so we must work to provide the elements of balance in the workplace. In more statistical or traditional terms, we need to be safety conscious because 5250 workers died on the job in 2018 (3.5 per 100,000 full-time equivalent workers)—on average, more than 100 a week or more than 14 deaths every day [3]. Because safety is such a buzz topic, people often forget to stop and question what it means to be safe. The numbers shown here are deaths, not injury or exposure. These statistics sound bad, but when compared to historical data it is an impressive improvement. "In more than four decades, OSHA and our state partners, coupled with the efforts of employers, safety and health professionals, unions and advocates, have had a dramatic effect on workplace safety. Worker deaths in America are down-on average, from about 38 worker deaths a day in 1970 to 14 a day in 2017." [4] If we want to impress the importance of safety, it is interesting to note that in the scientific category, OSHA verified 70,500 work related injury or illness cases. That is 1356 incidents per week, and if we assume a five-day work week, 271 incidents per day. If we average across all 55 states and territories, the number is five every day in every state or territory [5].

Now that you have an idea why you should practice safety, what does that look like? Here is the definition of safety used in this chapter: Use safe practices in all areas of work and business, create a culture where safe practices and habits are valued and supported, and most importantly, have an intentional mindset. This last statement means that you promote safe practices on purpose with deliberate intent. It means NOT operating on auto pilot or assuming more safety is better safety. It means caring enough about safety to set aside habit and repetitive exercises and thinking and acting with intention. This focus shifts the goal from 'doing safety' to 'I want to be a part of something that is safe'. It is the first step to having a workforce, management team, investors, and lawyers all on the same page thinking critically about safety because each member of the team knows how he/she benefits from having a safety conscious culture and intentional mindset [6].

The next step in creating a safe workplace is understanding the relevance of safety vs risk to business success. The most obvious risk is the hazardous materials used to perform routine work. Equally common is the risk of cuts or abrasions, but what about equipment related injuries, or unexpected chemicals (unexpected samples or materials) coming in the form of research and development (R&D) activities? What about fatigue related accidents because the lab is too warm? What about a fire that becomes uncontrollable because the fire extinguisher was blocked in a closet? Not all labs will offer R&D services, but it is still possible to receive samples that are not expected, or to receive samples contaminated with something that 'couldn't' be there. When products require analysis prior to release, there are subtle, and sometimes not so subtle, pressures to push material through to release. There is a responsibility for the results produced, and there are varying levels of familiarity with the tools available to accomplish the work. The pressure and complexity of the work are good ways to introduce mistakes, shortcuts, or angry customers (when you must report a failed product). Knowing the pressures ahead of time can help the organization build in support systems and strategies for the hazards and risks not stereotypically considered when thinking about safety.

At this point, the hope is that you are interested enough in safety to walk through the different components of safety. There are four categories used to talk about safety: architectural/facility, physical/personal, chemical, and procedural. We start with the structure, building, or facility space since many readers are using this book as a guide for how to start a lab. Investors, builders, and designers, this topic is particularly relevant to you. What kind of space do you want? It is not enough to have the square footage. You need to know at the very beginning what types of things will be in the space. And, knowing what types of things will be in the space, you need to know the restrictions on those things for the specific location under consideration. Sometimes, existing operations have been grandfathered for their hazardous activities and that can impact a new business going into the same space. Are there other businesses nearby or newer ordinances that create any issues? This scenario is one of the reasons why having a list of hazardous materials and building a relationship with the local fire marshal or fire safety bureau is useful up front. Will you have the appropriate storage available? Will the heating, ventilation, and cooling (HVAC) and fire systems support storage in the building, or will you need an outdoor/annex space? Will you have to update the electrical to support instrumentation? Do you know if hoods and showers will be easy to install in an existing space or will you be better off building from scratch? These questions are surface level safety because you will need all those things to operate safely; but let us also look at these questions from an intentional mindset. If my structure is difficult to remodel, or costly, or prohibited from certain expansions or features, how does that impact safety? Immediately, it can influence whether the ideal tools and resources are available to operate safely. How? By impacting the budget and timeline. Construction and renovations are often hard to keep on target for the budget and timeline because there are so many factors that are hard to anticipate. Choosing an experienced construction contractor and consulting experts for every aspect of the build and as much information about the hazards at this stage is time and effort well worth the investment. The construction/renovation stage is often under immense pressure to be completed but rushing the start of construction opens the door to problems. Did you know the construction industry has the most workplace incidents according to OSHA and has a special mention in their statistical reporting? It is the number one industry for workplace incidents as of 2018 (This is the most recent data available on the national level) [7]. There is a proverb that comes to mind for this scenario: an ounce of prevention is worth a pound of cure. The specific questions to ask and concepts to include in the pre-construction planning are addressed in the facility segment of this chapter. This phase is easily the most important because it sets the precedent for the operation goals and priorities. Experience shows that as a lab goes through the four phases of operation, each phase is impacted by the foundation on which it's built. As labs become functional, some discover that the physical layout causes inefficiencies and can lead to some of the common pitfalls such as rushed work and higher costs [8]. These factors are one of the most common root causes for safety shortcuts.

The second category of safety is physical (or personal) safety. This topic includes a person's interaction with the workplace as well as concepts that are person specific. Following the current format, we begin with the obvious questions. Are measures in place to provide appropriate access to materials and equipment? This includes things like lockout/tagout procedures for equipment, and specialized storage cabinets or rooms for hazardous materials. How well does the workplace incorporate ergonomic considerations? Is there reasonable space to do the work, or are things congested and hard to organize? Do you need to provide accessible features for persons with disabilities? What personal protective equipment (PPE) is required for your set-up and how do all these things translate to our deeper look at safety from the intentional mindset? Simply stated, everything comes back to human nature; if daily work requires extra thought or effort unrelated to the task at hand, the opportunity for disregard and shortcuts becomes possible. Here are some examples of what this issue looks like. If the proper place to hold samples is in a sample storage cooler or controlled temp refrigerator, is it near where the work is done? Do you risk employees holding on to samples for a 'little while' because it is too much hassle to take them back to the approved storage location only to retrieve them 30 min later?

Is it going to feel like a waste of time (and thus a source of resentment or unwillingness to follow procedures) if the people are constantly getting in each other's way to do assigned work? These might seem like exaggerated considerations but how long does it take for someone to ignore a rule that is 'annoying'?

Chemical safety is probably the first thing most people think of because of the obvious risks to dangerous or unfamiliar materials. A good chemical hygiene program (CHP) is going to be the foundation and truly the heart of all chemical safety considerations. The purpose of a chemical hygiene program is to create a safe workspace with hazardous materials. A large part of a CHP is the standard operating procedures (SOPs)—the instructions on how to do things. SOPs include policies and procedures for procurement, storage, handling, and disposal. Once a good chemical hygiene program is in place, implementation and support (including enforcement), are key components to having a safe workspace in real life, instead of concepts living only on paper.

The lab standard was developed to address the unique conditions and scale of a laboratory environment, which is often different from larger industrial operations [9]. While both HazCom and the Lab standard require a plan, there are differences between them. An article published by EHS Daily Advisor from the Bureau of Labor Relations (BLR) discusses the requirements and differences:

Lab Standard vs. HazCom
There are similarities and differences between the Laboratory Standard and the HazCom.

- The Hazard Communication Standard was developed to inform employee who work with hazardous chemicals of the risks associated with those substances.
- The Laboratory Standard was specifically developed for laboratory operations because these environments often differ from industrial environments in their use and handling of hazardous chemical.

- A chemical hygiene plan (CHP) essentially combines both these requirements in order to establish a standardized framework for chemical hygiene practices, information dissemination, and training.

Both HazCom and the Lab Standard mandate that the employer make a "plan":

- HazCom mandates a Hazard Communication Program (HCP).
- The Laboratory Standard mandates a CHP.

Many employers/workplaces require both a HCP and a CHP. [10]

Successful plan implementations include targeted training and having a solid foundation of engaged management and employees. If information is easily accessible (both in understanding and physical access) and visually efficient, it can mean the difference between someone choosing the right action over the wrong one. The need for clarity and nearly automatic reactions or responses is important for routine tasks and exponentially grows in emergency situations. Emergency responders are trained to handle dangerous situations and they do it well. It is not suggested or reasonable to expect employees to have the knowledge and training of an emergency responder; but everyone should be as confident as an emergency responder for their expected actions such as notifying the appropriate responsible person and following the outlined protocols.

The last category of safety we will explore is procedural. Training is mentioned repeatedly through several elements of this chapter and this iteration of training is again from a different perspective. This is procedural training. This idea is about how you conduct the training and the procedures built into the processes of your facility. Do you approach procedures as a one and done, or are they constantly evolving instructions? Is it part of your procedures to have routine safety updates or is it only through the constructs of mandated compliance? And shifting slightly, again, do your procedures point to clear and concise policies when it comes to safety? Does the work force and management team know where to find safety information or who to ask? The easier it is for people to get the information, the easier it is for them to participate in the culture of safety. Bold statements leave little room for misinterpreting expectations. Be careful about the language you use. Must and shall are words that leave no room to guess whether something is mandatory or not. Should and could are words that leave room for questioning whether the procedure or rule is mandatory. Ironically, this chapter is full of 'should' statements because this book is a guide, not a mandate.

During start-ups and then again when a new hire joins the team, and throughout daily operations, consistent enforcement of safety will set the tone of your safety culture. People are creatures of habit; make safety a habit by actively soliciting safety conscious behaviors and thought processes. Make it self-reinforcing by simplifying the safe way of doing things. Inevitably, shortcuts will happen; instead of immediate disciplinary actions, evaluate the scene to see if there is an impediment to the safer process. Perhaps, the shortcut identifies an area of inefficiency; perhaps, the shortcut reveals an unintended message of throughput over safety. It is possible the shortcut is as safe as the original directive, but more commonly, it identifies someone who is not fully convinced of the risks present or is not fully committed to the safety culture. Sometimes, the shortcut is valid and can be made safe with a small adjustment. This approach helps managers and business owners see where employees are feeling overtasked or not understanding the risk vs benefit of doing the action safely. Keep the process fluid and have an open dialog with personnel to foster growth. This approach leads to procedure improvements that are both documented and evaluated for safety considerations. When personnel can actively participate in developing easy and efficient safety procedures, they will take ownership of the process that helps keep them safe. At the same time, if the shortcut is found to be unsafe and cannot be modified for safety, it is an opportunity to educate the employee on the risks and reinforce the correct way of doing the procedure. This fluid environment also helps keep things relevant and good procedural documentation is useful for compliance with many quality systems.

Architectural and Facility Safety

Now that you have an overview of how and why safety is relevant to success, let us discuss some things in greater detail. Going back to the beginning, we start with design and layout. Construction and renovation are the perfect times to quite literally 'build' safety into the space. Starting with the right design ensures a safe environment that meets the users' needs and has the lovely side effect of supporting productivity. Understanding what needs to go into lab design can be overwhelming. If you ask your operations or lab manager what they need, you will get a list of items. If you ask them what they will be doing, you will start to get a picture of where those things need to go and a few other ideas [7]. It is critical you select experienced consultants and contractors who know what questions to ask to address every need. Renovations require one additional step vs new construction. Any space that is going to be renovated needs to be evaluated for existing hazardous substances. Asbestos and lead paint are common in older construction, and some older lights and thermostats contain mercury [11]. If these are removed, you will need appropriate remediation solutions. If the space was previously used for any work that used chemicals, you will need to ensure the space is free of contaminants from the previous occupants. Time and expense are the two biggest issues when incorporating hazard removal in a renovation project.

As the first responder and department most likely to have jurisdiction over many of the safety aspects of the lab, fire marshals are one of those consultants you should connect with early on. It is prudent to include and even invite them to do a preliminary inspection or provide informal comments on the design plans for the lab. OSHA offers consulting services and for many small or medium size businesses, these services are of charge. Even if you have a great architect and builder, each locality has the authority to determine what must be there to pass the final inspection. When you include the known chemicals (with estimated quantities) up front, you get feedback on any special requirements you might need to build in or add to your design. Things like fire rated walls, explosion proof rooms, HVAC capacities, fume hood recommendations, or even odor control if you have material coming onsite in sufficient quantities to merit such precautions. (Odor control is a common requirement for grow and extraction facilities and may not be required for safety compliance labs, but design is the best place to discover if your local fire marshal wants it addressed in any way.)

As you work with experts, it is helpful to have some idea of what is needed. This chapter is not specifically about construction or design but for the safety aspects, we must discuss the process. The following segment will touch on the form and function of a lab layout and design so that you can effectively include safety in every aspect [11]. Having a well-designed layout will not guarantee you never have accidents, but it will ensure none of those accidents are caused by a poor design. Being able to respond quickly to spills or having plenty of room to move past other workers does reduce the risk factors that cause accidents and injuries [7].

Layout

When designing a lab space, consider non-lab spaces the building will need as well. Lab space needs to be separate from office space, which is separate from the kitchen or eating space, which is separate from meeting space, and these are again separate from loading docks or storage spaces [12]. Easy access and convenience go a long way to encouraging employees to adhere to safety policies and activities that are permitted in each space. Placement of doors, furniture such as benches and cabinets, vents, and even the type of flooring and transition from one space to the next can have a big impact on the safety of the lab. Will any of the doors be fire rated and what does that entail? What type of special features do you need? Will you need any distribution systems for gases or DI water? Some areas will need signage and designing space and provisions to do so in the beginning will make things easier down the road. (Think brackets, magnets, and sleeve holders.) [12] Unless restricted by code or security measures, it is a good idea to have some visual access between hallways and lab spaces. This feature facilitates assessment and response in the case of an accident or injury and prevents the possible collision of opening a door into someone on the other side. Place large equipment so that it is accessible for maintenance. If you have items such as an autoclave, include provisions for adequate drainage and to prevent floor damage [12]. Build space for pre and post material staging; security of the lab and materials is easier to manage when the original design is sufficient from the start. Plan forward to have the option to expand or change things by including multiple electrical circuits and know ahead of time if any equipment requires special voltage or grounding measures. Extension cords and plug adaptors are strongly discouraged and may be against code, so this consideration is particularly relevant to renovated spaces that might have fewer outlets. If you need to add a subpanel or move the main electrical panel, make sure it is accessible and labeled in detail.

Furniture and Fixtures

Once the layout of the lab is established, you can begin to assign materials for construction and surface treatments. Employees will appreciate smooth, easy to clean surfaces on work benches, counters, and tables. Consider any chemical resistant features you may need [12]. What kind of weights will the structures need to support? Analytical equipment can be very heavy and be sensitive to, or cause, vibrations. One of the most common mistakes people make when building a lab is selecting the wrong benches, tables, and casework. It is not uncommon for equipment to fail because it was too heavy for the table causing the legs to warp and the instrument to be tilted or to be too long with multiple instruments causing vibration issues. These problems can be easily solved with a lab grade custom bench placed in the ideal location in the beginning stages. Selecting ergonomic features such adjustability, user friendly configurations, and healthy lighting will ensure you get the best return on your investments long term. Tables, benches, and workstations should have knee space for employees to use chairs when any prolonged operations or procedures could take place. Sink space should be enough for the tasks intended such as hand washing or filling containers. Determine if any of them need foot, elbow, or sensor controls. Place them in line with workflow patterns so that one isn't ignored and becomes a dry trap odor source or take up valuable counter or storage space.

Storage

Storage is a significant topic for lab safety. Access, security, and special requirements all need to be included as a part of the storage design. Choose sturdy and durable constructions made of materials compatible with the items stored and identify when a generic cabinet or shelf is suitable for things like electronics or non-chemical items. Choose designs that meet special needs such as ventilation, fire rated, or corrosion resistant. Consider the economic impact of using liners or coated fasteners vs potentially more expensive options of whole construction using special materials. Consider bookcases and stand-alone cabinets instead of wall mounted shelves or built in cupboards [12]. These are future friendly features and in the case of cabinets, can be specifically designed for things like flammable storage (both liquid and gaseous) which must be considered for code compliance and quantity limitations. Keep in mind any materials that need monitoring, venting, or is incompatible with other materials in the lab. Certain materials have rules about where they can be stored with relation to egress, HVAC, or electrical circuits. All labs will have supplies employees need to perform the job and ensuring space for boxes of gloves and other disposable PPE, tissue, glassware, paper, and daily tools like pens, notebooks, tape as well as the reagents and consumables will keep the space uncluttered and user friendly [11]. Any time supplies are hard to find or out of the way, people will take short cuts. Provisions might be needed for spill kits, waste collection, and other items needing special care. Choosing how and where these items are placed will encourage easy access and proper handling. Storage is often the first cost saving measure examined but appropriate laboratory storage is one key component often deleted from the construction budget to gain cost savings. If adequate storage isn't provided then equipment and materials occupy lab benches, lab bench knee-spaces, and lab aisles. All of these reduce valuable area to work efficiently and safely [7].

Laboratory Ventilation

There are different requirements for ventilation/air exchanges in office space vs lab space and often different rates in the lab space if it is occupied or unoccupied. These issues are something you will need to verify for your facility during the design stage. To further define ventilation requirements, you will need to look at fume hoods and whether your operation requires them, if so, how many, and what type (bypass vs auxiliary) [12]. The location of hoods, supply and intake vents, operable windows, and doors should be carefully planned with safety in mind. Some materials and equipment cannot be placed near intake vents, some door and window placement can compromise the efficacy of a hood, and some layouts can inhibit the number of air exchanges you might otherwise achieve. This is why choosing your consultants and contractors is so important. When selecting equipment such as a hood, you will want to understand what kind, if any, functional verifications are required. There are models and devices available that monitor face flow rate and tests that can be done to determine laminar flow patterns in the room. In general, labs should be negative pressure compared to adjoining halls or non-laboratory rooms. You need to consider any cabinets or equipment that requires venting during these early stages so that you don't run into complications or require additional duct work after construction is completed. A small but valuable consideration you should include is air filtration. Instruments can be susceptible to excessive dust and implementing an easy to maintain filtration system in the original lab design will cut down on maintenance, service calls, and wear and tear on the analyzers. It also may benefit employees who have allergies or asthma.

Emergency Equipment

Emergency equipment such as eye wash stations, safety showers, spill kits, emergency shut offs, and monitoring equipment is the next topic to include in the design and layout of the lab [12]. Depending on the requirements for type and quantity, you will want to decide early on where drains and electrical lines need to be and if you need non-slip flooring. Fire suppression systems can have a direct impact on the types and quantities of flammable materials you are allowed to store in the lab and this issue is an area where an architect is invaluable. For a new construction, it is easy to include things like sprinkler systems and multiple exits, but for a leased location or remodel, it is more difficult to meet the requirements unique to laboratories. An architect that is familiar with laboratories can address things like 'common path of travel' where the max distance to an exit is determined by the presence, or absence, of a sprinkler system. These factors also impact things like whether the hallway needs to be fire rated, but it does NOT impact how many and where the fire extinguishers need to go [13]. The requirements for fire extinguishers are based on the class of combustible materials present, the size of the building, and occupancy. Creating intentional storage alcoves and allowing space for wall mounting is smart and efficient. Understanding this impact will help you design the lab for the most efficient operation. Another kind of emergency equipment is detection and alarm equipment. It is possible to have systems for gas detection, HVAC, centralized fire detection, security, and even lighting controls. Consider a building integration system if you have multiple detection or control systems. Alternately, include identifiers for each light or audible alarm to clearly indicate what has been detected or activated. Emergency lighting for any space without windows is an important feature.

Materials Handling

When it comes to docks or material handling, you will appreciate the foresight to allow enough space to accommodate the different types of trucks and containers. Some trucks require a forklift to unload pallets, and some trucks need a dock plate. Equipment and materials can be excessively heavy such as a gas chromatagraph/ mass spectrometer (GCMS) or cryogenic liquid can, and receiving them safely and efficiently is a worthwhile investment [12]. We discuss doors in the layout section, but they must also be chosen with material handling in mind. Understand where carts and people will travel and whether automatic sensor or extra wide doors are necessary [11].

Physical and Personal Safety

In the segment on facility safety, we touched on ergonomics and considered the layout of equipment, adjustable workstations, and accommodating areas that might need chairs or extra counter space. When we look at these same considerations from the physical perspective, we look with the intent of meeting the needs of the employee. Employees will have more thought resources available for safety when they do not have to worry about physical stresses like back or neck strain, headaches from the lighting, overcrowding, misplaced materials, or general clutter. Sitting on a chair at a station that has no leg room is annoying and makes reaching over to a bench awkward. It is impossible to predict the nuances of any lab set-up until it is functioning but knowing that you can make a reasonable attempt at providing a comfortable workspace with some specific features and spacing is a good start. People will come to work in every state of rest or stress and there is no way to prevent this. However, it is possible to minimize the stress and fatigue caused by the workplace. In northern climates, where there is reduced sunlight in the winter months, consider putting in a full spectrum light in the break areas to supplement the lack of sunlight. In southern climates where it is often very bright and hot, provide reasonable lighting that is soothing to the eyes and consider lab appropriate window treatments. Some employees might be overwhelmed by constant noise or loud sounds; consider placement of loud equipment such as a cryogenic liquid can, vacuum pumps, or noisy compressors with relation to employee desks or workstations. Provide reasonable distances, noise dampening walls, or noise reducers such as whisper vents or pump casings when you choose the materials and equipment for the lab [9].

The less people must think about whether they are in a safe environment, the easier it is to support employee participation in the safety culture [7]. Limiting exposure to hazardous areas, cabinets, or equipment provides a sense of security. Providing easy access to the areas, cabinets, or equipment someone needs to do their job, demonstrates trust and empowerment (presuming necessary training has been provided). Access to information is included here in the discussion of access because everything hinges on knowing what to do when, where, and how. Employees expect access to procedures and policy and giving access to the reasons why those procedures and policies exist is even better. Providing ongoing access to information through varied training is one of the most effective ways to ensure the information is retained and accessible.

Most direct physical and personal safety measures come in the form of PPE. Laboratories usually have standard PPE such as safety glasses, gloves, lab coats or aprons, and face shields. When procuring these items, you should choose them for specific features. Safety glasses can be sized to fit over glasses, which is doubly important if you restrict the use of contact lenses. Over glass styles can be bulky so consider the budgetary effect of one size fits all vs the comfort of a style for people who do not wear glasses. Perhaps, you will offer a prescription safety glass option instead of layering eyewear. Gloves must be rated for the materials in use. Will you need any special protection beyond standard chemical resistance? Do you have any cryogenic or heated processes that require insulated gloves from either heat or cold exposure? Do you need any cut resistant gloves for receiving or handling broken glassware? How many sizes do you need? It may seem like a small consideration but if gloves make it harder to do the required tasks, will employees be tempted to skip them? How about ill-fitting gloves that introduce additional pinch injuries? During employee orientation include PPE fitting that captures eyewear needs, glove size, lab coat, and any other safety items the employees will be required to wear and then make them available. In lieu of standard uniforms, it is prudent to have a dress code. Requiring lab employees to cover their skin minimizes the risk of chemical splashes or spills. Do not allow open toe shoes and require non-slip soles such as athletic shoes or work boots.

The flow through the lab is another area that touches on physical safety. When people are working in a shared space, it can be synergistic or disruptive. Physical safety is often tied to having resources for the task at hand and knowledge to effectively use those resources. The simple concept of people being able to perform a task without running into each other, waiting for someone to get out of the way, or putting things in the wrong spot because there is not enough room seems logical [7]. Unfortunately, it must be mentioned because the assumed parameters can easily be left out of the equation because they are assumed. No one is spending time thinking about the things everyone assumes is obvious. Intentionally thinking about physical proximity, walk patterns, and personnel flow will ensure this concept of simplified workflow.

The risk of musculoskeletal injury is significant enough for OSHA to publish a fact sheet about ergonomic training for lab employees. Train employees to think about posture, train employees to think about chair position, train employees to understand and use proper screen distance, and train employees about footrests. The fact sheet is linked in the reference sections and because so much of it relies on the training of personnel, it seemed worth the time to mention some of the key ideas. Small choices such as using an electronic pipette instead of a manual pipette can reduce the repetitive motion that leads to pain in the hands and fingers or thumb [14]. This example is a small detail, but it represents the entire way of thinking about personal safety and how much can be accomplished during training. It is important to teach employees both about proper posture and the tools they choose to complete a task. It is not unrealistic to have 20, 30, or even 50 opportunities in a shift to adjust posture, position, and choice of tools. These choices, and ways to reinforce the thought process, are presented and encouraged through a robust training program that is dedicated to personal safety. Employees will feel like they are important and cared for when their personal safety (including comfort designed to minimize strain) is an obvious priority. When employees feel their safety is important, they are much more willing to participate in the safety culture, even when there is the temptation to skip some of the less popular aspects of safety compliance.

Chemical Safety

A chemical hygiene program is the heart and soul of chemical safety considerations. There are abundant resources available to support someone creating a program from scratch and equally as many services to provide a completed program specific to your needs. This chapter does not provide instruction on creating a CHP, but it does define key concepts and resource links, so you have a starting point [15]. The primary structure of a CHP includes four broad topics: Procurement, Storage, Handling, and Disposal [16].

Procurement is the task of obtaining materials and/or services. Depending on the size of the company, procurement responsibilities may reside with one person, a department, or split between several steps and approval requirements. Whatever structure is in place, safety must be included in the process. The person selecting the materials must be aware of hazards, know how to handle them so that anything required for safe handling can be procured simultaneously or ahead of time, and someone needs to understand cost efficiency vs risks associated with the material. This approach often results in a team with shared responsibilities and is why SOPs for procurement are included in a CHP. Say an employee needs a new chemical; a manager authorizes the purchase through a safety review, a purchasing agent orders the material, and finally funds are allocated or released. It is a rare individual who possesses the skills, knowledge, and authority in an organization for all the safety considerations of all the materials; this eventuality is where the CHP comes into play. Chemicals fall into predefined hazard classes. If the specific materials requested are not easily defined by the default categories, using SOPs to set the rules for safe chemical procurement eliminates an entire body of knowledge that a single person would have to know. The SOPs should include who is responsible for each step, starting with a review of the SDS (Safety Data Sheet) to provide the hazard class. Then the CHP will dictate the handling rules for ordering and whether additional precautions or approvals are needed to bring the material safely into the facility. It can also flag certain known substances that are critical to the operations and ensure that the precise material is on a pre-approved list [17]. Sometimes, there are less expensive options that might work less efficiently or that would introduce undesired impurities. A person tasked only with a responsibility to spend wisely might choose an economic option not realizing the additional costs a substandard material can cause. Including specific requirements for chemical procurement processes will reduce the chance of bringing in an incorrect material or more hazardous substance. Minimizing hazardous quantities onsite directly affects procurement procedures and whenever possible, safe quantities vs bulk cost savings should be included in the process. Receiving chemicals should be a two-part process as well; procurement can document receipt dates in the chemical safety inventory and verify expiration dates against projected use rates. This process naturally flows into the storage aspects of the CHP and enhances both inventory management and quantity limits.

Storage is more than putting materials on a shelf or cabinet. It also means having the right kind and number of storage units and that can be anything from a locked cabinet to a dedicated room or a vented cabinet near where the material will be used. Chemical storage in this chapter includes wet chemicals, dry chemicals, and gaseous chemicals (compressed gases). The safety issues with storage are many levels deep and the surface issues are easily described as having a place for everything and everything in its place. What happens when you run out of space? That scenario is a good reason to review the disposal section of your plan which we discuss later and it is a good time to look at overflow habits. How easy is it to follow a protocol when you need more space? It might be perfectly fine to have benign materials out on counters but that sets the stage for clutter, which is its own issue and can leave a bad impression on an inspector walking through. The purpose of correct storage is to keep people safe from inappropriate exposure and protect the integrity of the materials. Looping back to procurement and quantity limits, it is a good check and balance to have procurement monitor quantity on hand before approving new orders. If there is excess material in storage, it is a good indicator that the procurement and inventory processes have missed something.

Handling encompasses the rules for how employees must interact with hazardous materials. Handling protocols are where PPE, procedures for access, and interacting with the materials, are defined. Some materials require using a hood while working with them and others must be maintained in a specific temperature range so the time it can be handled is closely restricted. Some materials are reactive with the air and can only be used in an inert glove box while others are used safely with a simple pair of gloves and eye protection. The chemical hygiene plan defines the procedures for common materials onsite and designates authority to assess new or unknown materials.

Some materials have both a physical and health hazard and it can be easy to dismiss or overlook one of them. Carbon dioxide (CO_2) is used for fire suppression, some equipment, cultivation, and for transporting perishable materials (in the form of dry ice). This common material has multiple hazards and it can be easy to overlook one of them. As a fire suppression agent, it is compressed and when released displaces oxygen. This item is an example of both a physical hazard (high pressure) and health hazard (asphyxiant). CO_2 is also toxic above 400 ppm, and when CO_2 is compressed in a cylinder, it liquefies and has the cryogenic hazard of extreme cold (which is why it can be used as a backup for −80 C freezers)? It is easy to remember the cold associated with dry ice, but if CO_2 is not being used for fire suppression or refrigerated transport, most people are not going to focus on the cryogenic aspect. The point here is that many materials can present multiple hazards and having a thorough CHP is critical to helping employees identify and take proper precautions for any material that might come onsite. Another 'hidden' hazard is familiarity. Using a formal hazard analysis is both efficient and prudent. It allows people to examine an area through an auditor's criteria [17].

One of the easiest ways to ensure safe handling is the employees' ability to recognize when something is not normal. If a label warning has changed or a piece of equipment sounds funny, these are abnormal conditions. The intentional mindset discussed in other sections is a key component to noticing when things are not as they should be, or not as expected [18]. If the safety culture is healthy and employees are engaged, it is easy to report the anomalies. If employees feel pressure to produce results over working safely, they might gossip about issues during lunch, but they will not report them. Early detection of changes, even if those changes are intentional, is a great skill to have and contributes to mitigating potential issues.

The last of the four primary components of the CHP, is disposal. Just as procuring chemicals requires SOPs, training, and appropriate storage space, so does the disposal process. There are some materials that can be poured down a drain or thrown in general trash, but chemicals typically require more care [19]. Disposal safety includes procedures for identifying what kind of disposal is required, where the appropriate vessels are located, where those vessels reside once filled, how long the filled (or partially filled) vessels can remain onsite, and both how the vessels are removed from the site and who is responsible for removing them [19]. The details of this process rely heavily on the materials brought into the lab and can vary dramatically from one facility to another due to quantities, external disposal services hired, or specific hazards of the materials in use. Because things like flammable liquids have such strict quantity limits, and waste products always count toward total facility limits, it is a good thing to compare the cost of internal storage and final disposal vs hiring a chemical disposal service. It is not uncommon to have some materials defined for internal final disposal and some materials, such as flammable, highly toxic, or reproductive hazards to be handled under a service contract.

Training is a theme repeated in many segments of this chapter. This is because training takes place in so many areas of safety from so many different aspects throughout the entirety of employment, that it cannot be referenced once and assumed that is enough. There is much that goes into safety training and there are numerous books dedicated to laboratory safety training. In addition to subject matter content, there are books about how to have effective training programs for laboratories, what topics and how often they should be included in training, as well as how safety training can promote quality and profitability. Training is so important across all industries, that people need to be taught, so they know how to train, i.e. 'train the trainer'. It might be less common for a laboratory manager or safety manager to attend a train the trainer program but having the awareness of the importance of *effective* training is potentially the most critical aspect of a successful safety program. Chemical safety is a traditional safety topic. Everyone expects to attend chemical safety training in a laboratory. One can hope that everyone hired to work in the laboratory has some experience and prior safety knowledge, but assuming everyone, especially experienced employees, has the same foundation is a mistake. People naturally key into the information needed to perform a job or set of responsibilities and file away the things that are not used regularly. Chemical safety training should cover all aspects of chemical safety from fundamentals to job specific hazards.

Training provides the mechanism for repeated reinforcement and brings the topic of safety to the forefront. This consistent reintroduction of safety material helps cultivate the intentional mindset. When safety content is broken up into individual topics and presented frequently with the encouragement to keep thinking about safety, and employees are regularly engaged in this process, it helps minimize the risk of familiarity and complacency. The intentional mindset is a personal investment from the employees and is a strong sign of their buy-in to the safety culture. Using multiple styles and formats for training shows the trainer's investment and is helpful for maintaining interest and avoiding rote repetition.

Training is also when attention is called to the specific hazards onsite such as issues or concerns found during a safety audit or walk through [17]. Each department should receive the findings in their area and an overview of general facility safety health. If too much detail is presented to people who have no control over the situation or no interactions with the risks presented, there is a risk of losing the personal investment. Safety managers, or whoever is responsible for safety, should be acutely aware of their responsibility to the intentional mindset so relevant information is included in the ongoing training [4]. This event is also when you can present mitigation efforts, which is a valuable tool for demonstrating the commitment to safety and dynamic response. When employees become educated on what to look for and how to address it, the safety partnership is strengthened.

Some materials have obvious hazards, and some have multiple or less obvious hazards. Ongoing training is a great time to pick out one material or one concept that bolsters the confidence of employees who must work with the hazardous material. The more something is understood, the less assumption has to be made about why precautions are in place. Sometimes, the hazard is not to the people working with it, but to the material itself. Keeping the integrity of reference materials and reagents or consumables is as important to safety as every other aspect. This concept crosses over into quality but many experts agree that safety and quality cannot be separated [20].

Using the immediate environment for ongoing safety training is much more personal and engaging to employees. Generic safety training topics can be made personal to your lab with a few select references or photo examples. When it is difficult to convey a hazard related to the materials in use, real world examples are very effective. For example, some labs find it difficult to get employee cooperation to wear lab coats every day. Even though lab coats clearly protect employees from minor chemical spills and offer a professional image, sometimes the risk is not clearly understood. An example that is useful for multiple points is the hydrogen leak example. Pretend you use compressed hydrogen in the lab. Now, let us create a worst-case scenario. The cylinder is located next to the analytical instrument and the employee workstation is also nearby. Assume the cylinder has a small leak. The leak is not big enough or fast enough to trigger the sensor near the ceiling of the lab, but it is enough to start getting into the fibers of the employee's outer clothing. Because we are creating a worst-case situation, this employee is about ready to take a smoke break. The employee steps outside and right as the cigarette is lit, the lab alarm goes off indicating a flammable gas leak. The alarm startles the employee, and they drop the cigarette onto their shirt, which still has hydrogen trapped in the fibers. Next, the shirt ignites, and our employee gets burned. If the employee had a lab coat on while working and took it off before going on break, the result would have been a small burn in the shirt and nothing more. Is this example a likely scenario? Probably not. But what if? 'What if' are two excellent words in the intentional mindset. Using this kind of real-world example in safety training along with the information that most safety procedures are developed for the unlikely 'what if' scenarios is an effective way to communicate the value of proactive safety. Any time you can find and include real world examples, it makes the reasons for following safety procedures that much more personal.

So far, you have read about the components of safety and the ideas behind them. For the safety systems to work, it is critical to get personnel on all levels to participate. Because people are most invested in themselves, and appropriately so, it is useful to show the employees how this participation benefits them. Creating, and inviting input, for rules that are clear and easy to follow will eliminate the resistance tied to feeling it is too hard or irrelevant. If the rules are part of the natural process, it flows much easier than an add on practice. Consistently enforcing the rules is an effort that is well worth the time and energy. With enforcement, you empower others and prevent the tendency to micromanage. It is a process of passing on values as much as it is teaching rules. For some, this process comes naturally; for others, it can be a lot of trial and error to find what works. The key is dedication to the process and persistence over time. The HR field has developed and implemented wonderful programs about change management. Those same principles are required for any company culture shift. If those ideas and methods are applied from the very beginning, the safety culture will grow and thrive with every passing benchmark.

Procedural Safety

Procedure and work culture are closely related for this segment. The expected routine, attitude, and unspoken expectations are critical to how things function long term in the lab. There needs to be strong built in rules for how things are safely handled. Here is a real-world example of something that occurred. During a plasma donation pre-screening, a finger prick bled more than expected and the nurse in training who was performing the screening got some blood on her glove. The trainer instructed the nurse to switch out the single glove and continue. After the screening was completed, the trainer continued to instruct the nurse in an alternative procedure if an auditor was to be present which included changing both gloves and disinfecting the table. The person giving the plasma donation asked why there are two sets of instructions and the trainer replied that the more elaborate steps are policy, but it does not really need to be done every time. This event is a perfect example of someone who did not buy into the safety culture. Either the risk was deemed too insignificant to merit the effort, or the person thought the policy was inappropriate. Perhaps, the individual was being pressured for throughput or the office was understaffed, but something in the operation norms of that day had caused a supervisor to instruct an employee to take safety shortcuts. Is it a guarantee that those actions presented a safety risk? No. Is it a problem that a procedure existed and wasn't followed? Yes. It is likely the supervisor felt the policy required more time than was available for the level of risk she felt existed. This response is a breakdown of the system. If the procedure is unnecessary, it should be possible to change. The employees should be able to present concerns about time consuming procedures that hold no benefit. Or, the management needs to educate the employee of possible risk, discover what the impediment to compliance is, and enforce the policy.

Policies about how things run, the safety goals of the company, and what to do when things go wrong, all need to be clearly communicated and this is part of the safety training. There is much more to safety than knowing how to run a test correctly. Lab personnel are often assigned preventive maintenance tasks, supply/inventory tracking, and housekeeping duties. What does that have to do with safety? A lot. These tasks are where training is critical to the safety culture of the company and procedural safety is most visible. When equipment gets dirty, it can malfunction; malfunctioning equipment can be dangerous. When PPE runs out, everyone recognizes this as a safety issue. But what if something else runs out; what if the reagents run out? How can safety be the main focus when basic materials are unavailable? How can employees engage in a safety culture when they struggle to do their job without the proper materials?

So much of safety is focused on safe practices, policies, features, and attitudes, that it can be easy to overlook the safety of the products being tested. The purpose of marijuana compliance testing labs is to ensure released products are safe. This is where the relevance of lab methods, ISO accreditation(s), functional equipment, personnel with a sense of accountability, and last but not least, a management team absolutely committed to quality and thereby safety comes together. It would be easy to get off topic at this point and go into all the ways to ensure quality and how important it is to manage time and volume of work, but that is better suited to the topic of lab management. As far as safety goes, here are the components to keep in mind from the product safety perspective. Test the product for all required contaminants. Strongly consider testing for the 'should' components (things not yet required but probably will be required someday). In the early stages of planning, make sure you have a budget appropriate to the instruments and resources you will need. Getting a false test result due to inadequate equipment or reagents is worse than not testing. Talk to a professional who understands testing. You will quickly understand that not all analysis is equal. It is possible for some industry practices to be insufficient or outdated and you will only benefit from being on the forefront of doing it right. Spend the extra time and funds to get an experienced analyst in the lab. If you can only afford one, get someone who is an experienced trainer to get the rest of your staff up to speed. Reporting systems, ISO accreditations, security, quality, and business requirements all come into play for the safe testing of cannabis products. This book is an example of the voluminous tasks one embraces to achieve a functional safety compliance lab. While each chapter outlines a critical component of the process, it must always be with the intent to provide a quality operation that provides consumers and manufacturers with a safe product.

Safety and Common Sense

It would not be prudent to close a safety chapter without talking about common sense. It seems like an easy topic but what is common sense? Common sense includes all the things obvious and logical when deliberately problem solving or thinking about a situation (but not always thought of in the moment!). It is speaking up when something is unclear or safety prohibitive. For this to happen, it means providing a culture where personnel are comfortable speaking up. Sustaining that culture means creating a system where management is empowered to address, enforce, and respond to input received from personnel. To empower management, the system should not be burdensome. Proactive measures, which often mean more effort up front, allow things to run smoothly and efficiently. Having PPE available and accessible means it is easy to use. Inventory done regularly prevents the panic cycle of rushing an order of gloves because the lab just ran out. Having a clear procedure for what items to keep in inventory, who is responsible for maintaining those items on a predefined schedule, and building in time allotments to perform the inventory count eliminates the problem of someone feeling like they have to do extra work that 'isn't part of the job'. Preventing possible resentment and including all employees in safety ensures that everyone takes safety seriously. The intentional mindset prevents the process from becoming common, invisible, or too hard. The intentional mindset helps each person seek fulfillment in the task of safety. It gives responsibility and inclusion in a process that is a foundation of laboratory success. The intentional mindset is a way to be honest with oneself about the level of commitment to safety. It can be the difference between following the rules and engaging in safety.

Safety is a simple idea that is complex in practice and requires a workplace culture dedicated to safety for it to be fully effective. It is possible to write entire books on some of the topics briefly mentioned in this chapter and there are plenty of those books already published. This guideline is intended to demonstrate how integration, an intentional mindset, and knowing what questions to ask the experts will ultimately come together to build a safe laboratory. Safety should be a part of every process, decision, action, and structure. The balance between the no blame approach and the accountability approach is called the "Just Culture". A just culture takes safety culture one step further. It allows for the distinction between human error, at-risk behavior, and recklessness while providing a mechanism for individual accountability. It is most effective when the entire team is focused on safety improvement first, evaluation and retraining, and finally discipline for egregious disregard [21]. Individual personalities play a big part in building a culture of safety, so the process needs to start with both stuff and people. Dissonance is disruptive to a workplace and while it cannot be prevented 100%, it can be minimized by setting expectations. Adding criteria to job descriptions that reflect safety culture goals and commitments to safety lets people know of the company's dedication to safety. Teaching critical thinking as part of professional development enhances the intentional mindset as well as bolsters the skill set of your employees.

Summary

Employees, policy, leadership, commitment, and enforcement are all important to safety. The single most important factor to safety is desire. Wanting something makes it so much easier to expend effort to get it. The hardest part of building a safe operation is igniting that desire in someone who has little to no interest. Find people who already desire a safe workplace, people who are committed to doing things right instead of fast, people that bring a sense of personal accountability. If you are unable to find people with these qualities, invest more heavily in education and incentivize safe practices. Sometimes, safety culture can be reinforced through habit; if someone repeats an action or idea enough times, it becomes the norm. The music industry uses repetition to shift what the public likes. Sometimes, a song is played so often that even if you do not like it at first, it becomes familiar. Human nature is to gravitate to the familiar. The song that is familiar eventually becomes tolerable, and then sometimes, even likeable. The same process applies to safety. When a person does not gravitate to safety concepts, present the concepts over and over again. Make the ideas and mindset familiar. Operate as if safety is the most direct route to long term ease because it is, and then create a business where employees want to stay long term.

Safety can be overwhelming. That is why this chapter approaches safety from many perspectives but emphasizes the need for the intentional mindset. If safety is a part of the process from start to finish, it becomes less its own task and more of an aspect of how you think. Yes, you will still need to write and/or obtain safety specific items like a CHP and training programs, but the rest is intentionally including safety in your thoughts when going about your daily business.

Additional Resources

The following list of resources is meant to provide access to regulatory documents, checklists, and valuable articles not used as source material for this chapter. There are several articles related to the 'How' of ensuring your training program is effective.

OSHA Standard (All regulations related to labor safety). https://www.osha.gov/laws-regs/regulations/standardnumber/1910

Lab safety audit and inspection Safety Audit/Inspection Manual [PDF]. https://ACS.org/content/acs/en/about/governance/committees/chemicalsafety/publications.html (scroll to Industry & Small Business)

Identifying and evaluationg Hazards in research laboratories. https://www.acs.org/content/dam/acsorg/about/governance/committees/chemicalsafety/publications/identifying-and-evaluating-hazards-in-research-laboratories.pdf

US Accreditation Bodies. https://www.standardsportal.org/usa_en/resources/USaccreditation_bodies.aspx

8 workplace safety tips every employee should know. eSafety Training. https://www.esafety.com/8-workplace-safety-tips-employees-should-know/

Five tips for providing effective safety training. ISHN RSS. https://www.ishn.com/articles/109678-five-tips-for-providing-effective-safety-training

Culture of safety. PSNet. https://psnet.ahrq.gov/primer/culture-safety

Instruction and training - types of workplace health and safety training. Healthy Working Lives. https://www.healthyworkinglives.scot/workplace-guidance/managing-health-and-safety/instruction-and-training/Pages/types-of-workplace-training.aspx

Learn the basics of hazardous waste. EPA. https://www.epa.gov/hw/learn-basics-hazardous-waste

Make safety training more effective. StackPath. https://www.ehstoday.com/training-and-engagement/article/21914833/make-safety-training-more-effective

7 steps to create an effective workplace safety training program. https://blog.nsc.org/7-steps-create-effective-workplace-safety-training-program

Education and training. United States Department of Labor. https://www.osha.gov/shpguidelines/education-training.html

Safety leadership: making a culture of safety the foundation. StackPath. https://www.ehstoday.com/safety-leadership/article/21919022/safety-leadership-making-a-culture-of-safety-the-foundation

References

1. Occupational Safety and Health Administration (2020) 1910.1450 - Occupational exposure to hazardous chemicals in laboratories. United States Department of Labor. https://www.osha.gov/laws-regs/regulations/standardnumber/1910/1910.1450. Accessed 4 May 2020
2. Occupational Safety and Health Administration (2020) State plans. United States Department of Labor. https://www.osha.gov/stateplans. Accessed 4 May 2020
3. U.S. Bureau of Labor Statistics (2019) Census of fatal occupational injuries. U.S. Bureau of Labor Statistics. https://www.bls.gov/news.release/cfoi.toc.htm. Accessed 1 May 2020
4. Occupational Safety and Health Administration (2020) Commonly used statistics. United States Department of Labor. https://www.osha.gov/data/commonstats. Accessed 4 May 2020
5. U.S. Bureau of Labor Statistics (2019) Employer-reported workplace injury and illness, 2018. U.S. Bureau of Labor Statistics. https://www.bls.gov/news.release/osh.nr0.htm. Accessed 2 May 2020
6. Occupational Safety and Health Administration (2020) Safety and health topics. Laboratories. United States Department of Labor. https://www.osha.gov/SLTC/laboratories/index.html. Accessed 4 May 2020
7. Hersh E (2017) Five ways effective laboratory design impacts health, safety, and productivity. Executive and Continuing Professional Education. https://www.hsph.harvard.edu/ecpe/five-ways-effective-laboratory-design-impacts-health-safety-and-productivity/. Accessed 5 May 2020
8. Kaul S (2020) Managing a cannabis lab. Clinical Lab Manager. https://www.clinicallabmanager.com/business-management/managing-a-cannabis-lab-22822. Accessed 3 Jul 2020
9. Occupational Safety and Health Administration (2020) Laboratory safety guidance - occupational safety and …. OSHA. https://www.osha.gov/Publications/laboratory/OSHA3404laboratory-safety-guidance.pdf. Accessed 4 May 2020
10. Lawton P (2014) Facts about OSHA's lab standard and chemical hygiene plans. EHS Daily Advisor. https://ehsdailyadvisor.blr.com/2014/05/facts-about-oshas-lab-standard-and-chemical-hygiene-plans/. Accessed 6 May 2020
11. Yale (2020) Guidelines for safe laboratory design. Yale. https://ehs.yale.edu/sites/default/files/files/laboratory-design-guidelines.pdf. Accessed 5 May 2020
12. Lab Manager (2020) Health & safety design considerations for laboratories. Lab Manager. https://www.labmanager.com/lab-health-and-safety/health-safety-design-considerations-for-laboratories-20387. Accessed 5 May 2020
13. NFPA Codes and Standards (2020) Free access. NFPA. https://www.nfpa.org/Codes-and-Standards/All-Codes-and-Standards/Free-access. Accessed 5 May 2020

14. Occupational Safety and Health Administration (2020) OSHA fact sheet: laboratory safety ergonomics for the prevention of musculoskeletal disorders. United States Department of Labor. https://www.osha.gov/Publications/laboratory/OSHAfactsheet-laboratory-safety-ergonomics.tml. Accessed 4 May 2020
15. Occupational Safety and Health Administration (2020) OSHA fact sheet: laboratory safety chemical hygiene plan (CHP). United States Department of Labor. https://www.osha.gov/Publications/laboratory/OSHAfactsheet-laboratory-safety-chemical-hygiene-plan.html. Accessed 4 May 2020
16. Occupational Safety and Health Administration (2020) 1910.1450 App A - National research council recommendations concerning chemical hygiene in laboratories (non-mandatory). United States Department of Labor. https://www.osha.gov/laws-regs/regulations/standardnumber/1910/1910.1450AppA. Accessed 4 May 2020
17. OSHA (2002) Job hazard analysis (OSHA 3071). United States Department of Labor. https://www.osha.gov/Publications/osha3071.pdf. Accessed 21 May 2020
18. EPA (2010) Lab safety - use common sense! EPA. https://blog.epa.gov/tag/laboratory-safety/. Accessed 25 May 2020
19. Hazardous Waste Experts (2019) How to dispose of laboratory and pharmaceutical chemical waste. Hazardous Waste Experts. https://www.hazardouswasteexperts.com/how-to-dispose-of-laboratory-and-pharmaceutical-chemical-waste/. Accessed 5 May 2020
20. Maxfield D (2010) Workplace safety is the leading edge of a culture of accountability. StackPath. https://www.ehstoday.com/safety/article/21904308/workplace-safety-is-the-leading-edge-of-a-culture-of-accountability. Accessed 5 May 2020
21. Boysen PG (2013) Just culture: a foundation for balanced accountability and patient safety. Ochsner J 13:400–406. https://www.ncbi.nlm.nih.gov/pmc/articles/PMC3776518/. Accessed 28 May 2020

Preparing Cannabis Laboratory Business License Applications

Shaun R. Opie

Abstract In the United States, obtaining a cannabis business license is a require-ment to open a cannabis safety testing laboratory in most states that have legalized medical and/or recreational marijuana consumption. The licensing process begins by preparing an application. Submitting a comprehensive application is time consuming and benefits from the input of an experienced and multidisciplinary team. The team should include people with areas of knowledge in finance, architec-ture, construction, security, laboratory operations, laboratory compliance, and labo-ratory safety. Building a strong team and gathering the information and documentation is a costly, time consuming, and challenging task. However, the team will be able to help assemble a large amount of supporting documentation including legal and financial disclosures and a prospective facility and operations plan. Having a comprehensive understanding of the timing and content of an appli-cation will help an investor understand the time and financial commitment before the decision to apply for a cannabis laboratory business license is made.

Application Timelines

Because there is no uniform federal oversight, application preparation requirements and review times vary at the state and local government levels. This point cannot be stressed enough, and it is essential to understand that until federal oversight occurs, differences in application requirements should be expected between the indepen-dent state and local governing organizations. Accordingly, this chapter will focus on the generally accepted common elements on the application process and try to high-light major differences when appropriate.

Preparing to submit a cannabis safety testing application is a lengthy process and applicants should generally plan on a minimum of 2–4 months of focused effort to prepare, submit, and receive feedback for both the state and township/city applica-tions. This time range can easily be extended if the team has competing priorities or distractions. Some of the key determinants of time include: (1) the ability of applicants

S. R. Opie (✉)
E4 Bioscience, Charlevoix, MI, USA
e-mail: shaun@e4bioscience.com

to stay organized and focused, (2) the specific state application requirements, (3) the ability of applicants to produce documentation of their business, financial, and litigation history, (4) whether applicants have prior analytical laboratory experience and/or access to existing policies and SOP's.

Working with people who have successfully completed cannabis laboratory applications before substantially increases the likelihood of meeting the anticipated timelines. Investors and business owners may choose to complete applications without previous experience, but it is more common for applicants to hire knowledgeable consultants. Consultants often include attorneys, certified public accountants, cannabis business license application specialists, and/or laboratory experts. Consultants will be familiar with the regulations, understand the nuances of application requirements, provide an organized and prioritized checklist for document procurement and preparation, and know which sections are going to be reviewed with more scrutiny. The benefit of using a consultant is shortening the submission timeline and reducing the possibility of the state or local government review committee requesting follow-up documentation.

Many state and local governments limit the number of business licenses that are available making the submission process competitive. In some locations, licenses are awarded in the order a fully completed application is received and the site inspection is passed, while in other locations a point system is used to score and rank applications which ultimately determine whether an applicant qualifies for a license. In competitive license locations, the speed at which an application can be completed must be a high priority. In these cases, it is common for applicants to hire a consulting team, rather than an individual, in addition to their own team to ensure rapid completion since time is critically important.

Fees and Costs

Application Fees

All cannabis business license applications have a fee due at the time of submission. The application fees can vary significantly depending on location. Fees are intended to cover the costs of the regulatory agency documentation review as well as an inspection fee for the on-site assessment plus travel and per diem expenses for the auditor or audit team. See Table 1 for an overview of fees. The fee may or may not be refundable. Fees are typically only a few thousand dollars, but in some states

Table 1 Range of fees associated with submitting a cannabis business license application

Fee	Approximate cost
Application fees	$1000–$200,000
License renewal fees	Up to $75,000
Inspection fees	$1000–$10,000

they can be much higher. For example, in New York the non-refundable application fee and refundable registration fee are $10,000 and $200,000, respectively, while in Utah the application fee is $500. In some states it is necessary to submit separate applications for both medical and recreational marijuana. Although the content of these applications may be virtually identical, the process requires appropriate licensing agency personnel review, inspection, and administrative support ostensibly supporting a duplicate submission fees, but also requiring additional time and effort from the applicant. While not part of the application process, it worth noting that most states have an annual renewal fee that, like the application fee, can vary substantially by location.

Application Preparation Costs

Because the specific requirements for each state application are different, expenses required to prepare the application can vary. Some requirements can be fulfilled with little to no expense while others require third party assistance and will drive cost upwards. Some examples: certain markets don't require real estate to be secured, stamped architectural blueprints of your facility, or zoning approval; a complicated business structure will increase attorney fees much more than if you are a sole individual; some states require proof of ISO/IEC 17025 accreditation documentation which is a massive cost increase in both time and money.

Minimizing costs and maximizing future revenue is always a primary goal for investors. The time it takes to prepare and submit a cannabis business license application impacts investor cost and downstream revenue. Calculating the loss of revenue from delays that prevent a laboratory from operating may help to focus many application preparation decisions. Assuming the average revenue per safety test is $500 and a lab can process 50 tests per day with one set of instrumentation, then every day that an application is delayed defers $400 \times 50 = $20,000—**per day**—of downstream revenue. With this in mind, it is not surprising that many applicants chose to work with consultants who have previous experience organizing, preparing, and submitting the necessary documentation. Consultants who state that an application can be prepared for a comparatively low price ($5000) on an hourly basis should be questioned about their previous experience and successes. If an experienced laboratory consultant charges $250/h (far less than most attorneys) that provides for only 20 h of work. Applications regularly exceed 100 pages and it is virtually impossible for a customized cannabis laboratory license application to be prepared in 20 h.

Application Overview

States and local governments have the authority to create their own cannabis business license application requirements. Variations between them are common, but two general categories of information are expected: (1) Business and personal disclosures including legal and financial disclosures, business background checks, and demonstration of financial security to qualify applicants for the right to submit a complete business application, and (2) Laboratory information including a physical description of the facility and a comprehensive operations plan. In addition to state licensing, local government licensing is normally required. Fortunately, this application information tends to closely mirror the state application. The business and personal disclosure phase is relatively straightforward as it relies on gathering preexisting information. However, individuals with the ability to invest in a cannabis laboratory often have a substantial number of disclosures that take time to acquire. The laboratory information phase is like a comprehensive business plan without detailed financial statements and requires the applicant to provide a large amount of original descriptive detail.

States and townships work hard to provide support and guidance to applicants, but because the cannabis industry is still in its infancy, changes to application documents and application requirements are common. Something as simple as a missing form or using an outdated form can stall an application review. Therefore, it is imperative to use the most current application forms and understand the state licensing structure. See Table 2 for website links to currently available state application forms. Applicants should download the most current information from the state and township and understand all application requirements. This packet of information generally includes five sets of documents: (1) state government application forms, (2) local government application forms, (3) any application guidance documents and/or technical bulletins, (4) a copy of the state legislation approving marijuana testing facilities, and (5) a copy of the local government ordinances approving medical and/or recreational marijuana. Although regulatory agencies often provide summary checklists, it is highly recommended that these are converted into a single consolidated spreadsheet that lists action item, priority, responsible person/party, due date, and comments so task progress can be tracked carefully. Because of the volume of information, without a robust organization system it is easy to overlook required application elements and experience unnecessary and frustrating delays.

Laboratory licenses have typically not been as competitive as cultivation and dispensary licenses so the need for differentiation is decreased. However, many of the principles other license types use to elevate themselves from the competition can still be incorporated into the application. While there is a shift in public perception toward increased acceptance of marijuana, investors who to enter the industry or want to apply for a new license should consider making community impact and perception a priority. Being able to describe any additional value to the state and local communities is a tremendous differentiator. This may include sections explaining how the laboratory will engage with the local community, provide community

Table 2 URL's for state cannabis laboratory business license application forms

Alabama	No application at this time
Alaska	https://www.commerce.alaska.gov/web/amco/MarijuanaLicenseApplication.aspx
Arizona	https://www.azdhs.gov/licensing/medical-marijuana/index.php#labs
Arkansas	https://www.healthy.arkansas.gov/images/uploads/pdf/Lab_Form_20170706.pdf
California	https://bcc.ca.gov/clear/forms.html
Colorado	https://www.colorado.gov/pacific/enforcement/med-new-regulated-marijuana-business-license-application
Connecticut	https://portal.ct.gov/DCP/Medical-Marijuana-Program/Medical-Marijuana-Program
Delaware	https://dhss.delaware.gov/dhss/dph/hsp/medmarhome.html
Florida	https://knowthefactsmmj.com/rules-and-regulations/
Georgia	No application at this time
Hawaii	https://health.hawaii.gov/statelab/wp-content/blogs.dir/10/files/2013/07/2016_Medical_Marijuana_Testing_Facility_Application.pdf
Idaho	No application at this time
Illinois	https://www2.illinois.gov/sites/mcpp/Documents/Lab-Application-Approval-Form.pdf
Indiana	https://www.oisc.purdue.edu/hemp/index.html
Iowa	https://idph.iowa.gov/omc/For-Manufacturers-and-Dispensaries
Kansas	No application at this time
Kentucky	No application at this time
Louisiana	No application at this time
Maine	https://www.maine.gov/dafs/omp/adult-use/applications-forms
Maryland	https://mmcc.maryland.gov/Pages/testinglabs.aspx
Massachusetts	https://www.mass.gov/how-to/apply-for-an-independent-testing-laboratory-certificate-of-registration
Michigan	https://www.michigan.gov/lara/0,4601,7-154-89334_79571_87302%2D%2D-,00.html
Minnesota	https://www.health.state.mn.us/people/cannabis/manufacture/lab/index.html
Mississippi	No application at this time
Missouri	https://health.mo.gov/safety/medical-marijuana/how-to-apply-fi.php
Montana	https://dphhs.mt.gov/marijuana/laboratories
Nebraska	No application at this time
Nevada	https://tax.nv.gov/FAQs/Marijuana_License_Applications/
New Hampshire	No application at this time
New Jersey	https://www.nj.gov/health/medicalmarijuana/alt-treatment-centers/applications.shtml
New Mexico	https://nmhealth.org/about/mcp/svcs/
New York	https://www.wadsworth.org/regulatory/elap/medical-marijuana
North Carolina	No application at this time
North Dakota	https://www.health.nd.gov/mm

(continued)

Table 2 (continued)

Ohio	https://medicalmarijuana.ohio.gov/testing
Oklahoma	http://omma.ok.gov/laboratory-application-information
Oregon	https://www.oregon.gov/olcc/marijuana/Pages/Recreational-Marijuana-Licensing.aspx#Labs
Pennsylvania	https://www.health.pa.gov/topics/programs/Medical%20Marijuana/Pages/Laboratories.aspx
Rhode Island	https://health.ri.gov/applications/MMTestingLabPreliminary.pdf
South Carolina	No application at this time
South Dakota	No application at this time
Tennessee	No application at this time
Texas	No application at this time
Utah	https://medicalcannabis.utah.gov/production/labs/
Vermont	No application at this time
Virginia	https://www.dhp.virginia.gov/pharmacy/PharmaceuticalProcessing/default.htm
Washington	https://lcb.wa.gov/mj2015/testing-facility-criteria
West Virginia	https://dhhr.wv.gov/bph/Pages/Medical-Cannabis-Program.aspx
Wisconsin	https://datcp.wi.gov/Pages/Programs_Services/IHLicRegFee.aspx
Wyoming	No application at this time

benefit, and offer educational support. Laboratories have an advantage over retailers and growers since in addition to creating high paying jobs, they can also claim that they are providing the community a safety benefit by ensuring that cannabis products are free of harmful contaminants. Developing an educational community outreach program to share information about the potential safety concerns of consuming cannabis and how laboratory services reduces those risks is a highly recommended activity. Think about working with local not-for-profit groups and establishing relationships with local and state officials by attending meetings.

Some states offer a research use license application that allows the holder to engage in research and development testing. Where available and allowable, this can be used an opportunity for safety compliance facilities to begin method validation on cannabis and cannabis products without having to meet the infrastructure requirements of a regulated compliance testing laboratory.

Application Sections/Information

Ownership Disclosure

States and townships frequently use a "pre-qualification" application to understand who is in control of any cannabis business, the level and source of financial resources backing them, and to obtain information about their legal history. These legal and financial disclosures, business background checks, and demonstrations of financial security are used to qualify applicants for the right to submit a complete business application. Table 3 provides a list of common required application ownership dis-

Table 3 Common business and personal application disclosures

Business disclosures
Business formation documents
Corporate governance documents
Operating or partnership agreements
Agreements relating to profit/loss sharing
Ownership structure including an organizational chart
Business affidavits
Business liability insurance
Permission to lease/purchase facility space
Adequate capitalization and source
Proof of meeting labor requirements
Personal disclosures
Criminal history record check including misdemeanors, felonies, arrests/citations
Fingerprint cards
Social security number
Full disclosure of past bankruptcies and loan defaults
Child support and alimony payments
Individual and personal tax returns
Personal affidavits/attestations
Bank statements
Tax check authorization
Investigation authorization

closure documents. It is recommended to consult an attorney and/or CPA for advice for guidance about these documents and disclosures, so they align with application requirements.

It is common requirement for any business unit applying for a marijuana license to be formed in the state in which the laboratory will be located. Frequently, members of the business must all have a residency requirement and must maintain residency if a license is issued.

Facility and Operations Plan

The facility and operations plans are critical and lengthy sections of the application much like a comprehensive business plan without financial statements. Depending on the state and local government, minor differences in requirements for these sections are expected, but practically all require a core set of common information. Table 4 provides an overview of documentation that should be anticipated in most facility and operations plans.

Table 4 Common facility
and operations plan
application sections

Facility
Building zoning requirements showing distance restrictions
Evidence of legal right to occupy and use the proposed premises location
A boundary sketch and floor plans for the proposed premises
Operation
Quality management plan
Standard operating procedures
Security plan including limited access areas
Sample transport and transfer procedures
Inventory control plan
Marijuana storage plans
Marketing and advertising plan.
Staffing plan
Fire safety plan
Hazardous waste management plan.
Less frequently requested
Architect stamped floorplan schematics.
Instrumentation/equipment
ISO/IEC accreditation

Facility Plan

Most states have enacted "Green Zone" legislation that allow local governments to opt-in or opt-out of permitting cannabis related businesses. Assuming an applicant has identified a green zone location that allows safety testing laboratories, specific zoning restrictions need to be considered. Zoning distance buffers are common and include being a minimum prescribed distance from any elementary school, middle school, high school, college or university, either public or private, including child care or day care facility, to ensure community compliance with Federal "Drug-Free School Zone" requirements; any church, house of worship or other religious facility or institution; any residential zoning district or existing residential dwelling; any halfway house or correctional facility; and any private park. As a result, finding properly zoned real estate can be one of the most difficult—yet important—aspects of starting a cannabis laboratory. During property identification and selection period is a good time to create boundary sketches can be made by obtaining using satellite images from google maps. Ensure a distance legend is included.

Even though a building may be in a green zone and meet all zoning restrictions, few buildings are well suited for a laboratory. Any building is very unlikely in its unmodified state to meet the infrastructure requirements of a laboratory and the applicant should expect to do more tenant improvements than a typical office environment. Including someone experienced in laboratory design and workflow can be very helpful to understand potential building infrastructure constraints before a lease is signed or a purchase is made.

Once property access has been secured, the applicant needs to demonstrate a lease or proof of building ownership. The intent to operate a cannabis safety testing laboratory should be discussed well in advance of any lease agreement as many landlords are opposed to being associated with marijuana related activities. A letter that acknowledges and approves use of the facility for should be included whether the structure is leased or owned.

The level of detail in floor plans and the trueness to the final physical space required varies widely among states and local governments. In a few, no floor plans are required, but it is more common to expect identification all doors (or other points of ingress or egress), walls, partitions, counters, and windows; fire walls; the location of all cameras, assigning each camera a number or other unique identifier; A clear designation of the limited access or restricted area boundary within and outside the building. If the laboratory has multiple floors, it is necessary to submit a floor plan for each floor. While placeholder floorplans can often be submitted, a better strategy is to work with an architect with previous laboratory design experience and spend the time to design the final laboratory layout. Chapter "Cannabis Safety Testing Laboratory Floor Planning and Design" provides a comprehensive overview of laboratory design considerations.

Because commercial structures are widely different, cannabis testing laboratories are always a custom build. Good laboratory design creates a safe and efficient workflow that contemplates many different needs. Some of the planning points include: minimizing staff movement; providing for common and limited access areas; separate areas for enclosed sample receipt, accessioning, sample storage, chemical and biological waste storage; security surveillance; laboratory instrument infrastructure including power (220 and 110 V); industrial gas generation; HVAC systems for venting; laboratory bench arrangement; water for sinks, safety showers, and eye wash stations.

Operations Plan

The major categories of information in the operations plan include quality, safety, and security. Quality and safety are foundations of all analytical laboratories, i.e. research and development, clinical diagnostics, environmental testing, cannabis safety, etc. Without them, the accuracy and reliability of testing data is suspect and there is the potential to expose employees to dangerous working conditions. In addition to these two key initiatives, cannabis safety testing laboratories must also prevent sample diversion, so applications must include a thorough security section.

Quality Management System

Applicants should emphasize and value quality above all since providing accurate and reliable data—a byproduct of quality—is the core function of a laboratory. Therefore, a comprehensive quality management system (QMS) is a critically important for overall laboratory operations. Chapter "Cannabis Laboratory Management: Staffing, Training and Quality" of this book will take a detailed review of laboratory quality but selected areas of the QMS are noted below since they are generally required by most applications. A QMS is a set of documents created over time that describe: how the laboratory will commit to producing quality processes and data; writing, following, and documenting standard operating procedures (SOP's); collecting, analyzing, summarizing, and sharing operational data; preparing for new accreditations, inspections, and audits; Investing in staff education and training; and performing regular proficiency testing. It is highly recommended that applicants have someone with laboratory compliance and regulatory experience assist with developing the quality plan since ISO/IEC 17025 accreditation is dependent on these documents.

Standard Operating Procedures (SOP's) are required as part of the QMS for any analytical laboratory. SOP's are written to describe the step by step process for completing a task with sufficient detail so that anyone with the appropriate training and skills should be able to reasonably complete without assistance. These documents help to provide a consistent approach to performing laboratory work, streamline laboratory activities, help reduce variation in testing results, and mitigate loss of knowledge when staff turnover. While laboratory staff tend to focus on the analytical method SOP's—those directly related to the process of testing, pre- and post-analytical activities (e.g. sampling, transport, accessioning, ordering, reporting, etc.) should not be ignored as poor performance in those areas will impact the accuracy of results. The most difficult SOP's to prepare are typically the analytical method and reporting SOP's. If these are required for an application, an early strategy is to rely on protocols provided by instrument and LIS vendors. However, it should be expected that before the laboratory can be operational, they will need to be modified to reflect actual instead of proposed activities.

Security

Security is carefully scrutinized in *all* cannabis business applications, not just safety testing laboratories, because controlled substances will be transported, handled, stored, and disposed of at the facility. The security plan is a detailed document which describes the physical and electronic safeguards that will be used throughout the facility. Physical barriers include exterior fencing, internal walls and doors that demarcate unrestricted and controlled zone access, key card operated magnetic door locks, doors with standard key locks, visitor sign in logs, and the possibility of an on-site security guard. Electronic security include the location of high definition video cameras that can create time stamped images and capture 100% of the area in the facility where cannabis may be handled, motion detectors, glass break sensors, and the alarm system. It is good business practice to hire a licensed security firm, who are at all times responsible for on-site monitoring of all intrusion equipment, authorizing all entrance by staff, visitors and delivery vehicles. Most facilities will need a dedicated I/T room with a server that will store all continuously recorded and monitored video for a minimum of 30 days and provide real time on and off-site access to regulators and law enforcement.

Safety

Because of the potentially hazardous nature of analytical chemistry, microbial pathogens, industrial gases, and waste storage—both chemical and biological— license applications need to have a robust safety plan. The purpose of the safety plan is to protect the welfare anyone who may enter the facility. Chapter "Laboratory Safety from Site Selection to Daily Operation" provides a comprehensive overview of laboratory safety that extends beyond the application needs. For the application, a safety plan for chemicals should include a list of chemicals in the facility and the storage locations and conditions for: acids, bases, and water reactive chemicals by class; flammables by National Fire Protection Agency (NFPA) standards; peroxides; and oxidizers. The safety plan for industrial gases and volatile chemicals should address local exhaust ventilation (air exchanges per hour) to ensure that dangerous gases do not accumulate in the facility as well as a plan for gas detection and automatic shutoff. The safety plan for biohazardous materials should include descriptions of biosafety hoods, biological waste containment, and personal protective equipment. Standard laboratory safety items such as locations of showers, eye wash stations, chemical spill clean-up kits, should be discussed and cross referenced to the floor plan.

Laboratory operations generate hazardous waste that requires proper storage and disposal. This waste should be stored in an independent space separate from other consumables and workflow. The application should include a description of how hazardous waste will be properly contained, sealed, and picked up, and disposed of by a licensed hazardous waste transporter for treatment in compliance with all federal, state, and local regulations. Air quality is becoming increasingly important in applications, since the testing of marijuana is a source of both odors and volatile organic compounds (VOCs) which can impact air quality and cause offsite nuisance. When needed, air quality requirements include, but are not limited to, air permits, registration program and fees. Burning marijuana and the waste associated with the plant is not allowed, so include a description for how residual sample not used in the testing process will be disposed of in a safe and proper manner.

Additional Application Items

Although quality, safety, and security are the primary focus of the application, additional smaller elements may be required. A local government notice affidavit is often required for marijuana establishment license applications when a proposed premise that is located within a local government. As soon as practical after initiating a marijuana license application, an applicant should provide notice of the application to the public by submitting a copy of the application to each local government and any community council in the area of the proposed licensed premises. The local government usually has a reasonable window (30 days is common) to respond with an approval, objection, or no response to your license application.

In some states, providing evidence of additional community support is a desirable addition. These may include being a certified woman- or minority-owned business, describing plans for environmentally friendly operations, facilitating medical cannabis research, plans to positively impact your local community, plans to combat substance abuse, and employee-friendly policies.

Tracking of marijuana from seed to sale to prevent diversion and to facilitate product recalls is a standard state requirement. Two marijuana traceability software systems are the industry standards, Metrc or BioTrackTHC. Before a full license is granted, marijuana testing laboratory employees within a will need to train on how to use the software and provide documentation of successful completion.

Application Submission and Review

Once the application packet is complete, it must be submitted to the state and township. Submission may be paper or electronic, but it is highly recommended to use an online submission portal instead of paper whenever possible since everything is time stamped and it is much harder for to lose electronic copies. It is common for the state to require a review and acknowledgement from the local government before it will be accepted creating a staggered submission process.

After the necessary documents are provided to the appropriate governing bodies, a marijuana facility license investigator is assigned and contacts the applicant to verify information and help answer questions pertaining to the application process. The license investigator may communicate any additional requirements for the application including revisions or additional information related to documents that were submitted, additional documents related to persons or entities with a financial interest in the business, additional information related to the criminal history of certain persons with a financial interest in the business, additional information related to the facility and/or the operations plan. Failure to provide any requested information may result in denial or revocation of a license.

The laboratory licensing process is long and complex and designed to eliminate unqualified applicants. Common issues include: missing information; failure to meet basic licensing requirements; objection from the local authority; premise is located within 1000′ of a restricted entity; questionable source and/or amount of funding; indicators of hidden ownership; criminal history, i.e. conviction of certain felonies, conviction of a gross or simple misdemeanor involving liquor or drugs, any series of violations that show a disregard for the law; misrepresentation of fact.

Once the application review is finished and deemed complete, the marijuana facility licensing investigator will arrange a time to visit the laboratory to perform a visual inspection by confirming the items described in the application. Frequently, the investigator will bring a small team with specific expertise to help distribute the workload. The team will review a long list of items including at a minimum: a physical inspection of the premises to evaluate the orderliness of the physical layout; review of environmental monitoring for industrial gas safety; implementation of controlled access areas for storage of marijuana test samples, waste, and reference standards; sufficient space allocation for each testing area; a review of personnel records for the laboratory director, testing personnel, and ancillary staff; a review of quality assurance protocols; procedures for the transport and disposal of unused marijuana, marijuana products and waste; the use of a record system that allows for readily retrievable test results; verifying that complete testing SOP's, with current approval by the laboratory director, are readily available to staff; chain of custody documentation from receipt to disposal.

Because of the complexity of both the application and laboratory operations, it is not uncommon for minor delinquencies to be discovered. Laboratories are allowed sufficient time to correct any problems. Assuming delinquencies are resolved, a short follow-up inspection may occur after which the laboratory will be granted a temporary or full license, depending on whether ISO/IEC 17025 accreditation is part of the application. If the application requirement is to show only that the laboratory is in the process of working on accreditation or if the state sets a deadline for accreditation (usually 1 year from granting of the temporary license), then the laboratory will quickly shift its focus to method validation activities.

Some states use a single step application process allowing the laboratory to show they are in the process of working on accreditation or if the state sets a deadline for accreditation (usually 1 year from granting of the temporary license), then the laboratory will quickly shift its focus to method validation activities. However, many states use a two-step licensing process. In the first step, a laboratory that has passed a security and safety inspection may be granted a provisional or temporary license that allows them to obtain and handle cannabis samples. This allows a laboratory to develop laboratory specific methods and corresponding SOP's that reflect actual activities rather than what the propose will work. The second step, done in collaboration with the state, is to have the state prospectively review and signoff method validation protocols and summary data so that the laboratory has some affirmation that their work will meet the state inspection requirements.

Hemp

In October 2019 the USDA published the interim final rule, "Establishment of a Domestic Hemp Production Program" [1]. This document laid a strong foundation for a national hemp policy by allowing for interstate transfer, removing any uncertainty about the requirement to test for the **total potential** delta-9 tetrahydrocannabinol concentration, and requiring that laboratories register with the U.S. Drug Enforcement Agency (DEA). While a licensed marijuana testing facility can provide all the necessary testing required for hemp, it may not bypass the step of registering with DEA.

Interestingly, soon after the legislation was passed, the requirement was temporarily delayed [2]. This was because of the overwhelming number of industry stakeholders and consumers who expressed a concern about the limited number of DEA certified hemp testing laboratories and how that would impact the supply of cannabidiol (CBD). Under this guidance, the testing can be conducted by labs that are not yet registered with DEA until the final rule is published, or October 31, 2021, whichever comes first. This is intended to allow more time to increase DEA registered analytical lab capacity. It is strongly recommended that laboratories who intend to offer hemp testing services do not wait to begin the registration process. The registration application is 2 pages long and requests general information [3]. It is not nearly as complicated as a state cannabis business license application but—importantly—it requires the state laboratory license number. Obtaining a state laboratory license number requires a separate state application unrelated to marijuana testing. The intent of requiring the state license is to show that the laboratory is authorized to prescribe, distribute, dispense, conduct research, or otherwise handle the controlled substances in the schedules for which are being applied under the laws of the state or jurisdiction in which it is operating or proposes to operate. The legal landscape for hemp and marijuana are very different and anyone wishing to test both cannabis plants within the same facility should consult an attorney early in the planning process for guidance.

Conclusion

Cannabis testing laboratory business applications are long, complex, and require a large amount of highly detailed information. It is highly recommended to work with an experienced team who have previous experience completing successful applications since they will understand required disclosures and be able to successfully navigate the facility and operations plan requirements.

References

1. Department of Agriculture (2019) Establishment of a domestic hemp production program
2. https://www.ams.usda.gov/rules-regulations/hemp/enforcement
3. https://www.deadiversion.usdoj.gov/drugreg/reg_apps/225/225_form.pdf#search=DEA%20 Form%20225

Quality Assurance and the Cannabis Analytical Laboratory

Susan Audino

Abstract Testing laboratories rely on a broad array of standards from sources such as regulatory bodies, analytical test methods, and materials to ensure the reliability and trueness of their test results. In the cannabis testing industry, one of the most common standards is ISO/IEC 17025. This chapter will provide a general overview of the most common laboratory standard, intent of several clauses within the standard, and discuss some of the more egregious risks to laboratories. Included will be more detailed description of a laboratory's Quality Management System and its major components, and discussion on some of the more challenging issues common to laboratories testing cannabis.

Introduction

A standard practice or process is one by which an authoritative body has established a set of conditions, principles, or measures to drive specified responses. In science, "standards" reference those documents that contain specific information or instructions to perform defined activities. These are essential tools that facilitate clear understanding of processes, products, and services to ensure consistent applications and to ultimately promote marketplace trade. There are basic elements in every process that reflect to what everyone learns in high school such as hypothesis, experiment, and conclusion. It is the necessary complexity of the process and its standardization that creates a universal practice so that a product maintains its quality. The replication of the process in its entirety and the maintenance of all the factors in its habitual performance sets up the essential element of quality assurance. Equipment for example, must be calibrated to the same strict guidelines so that the same product maintains its viability and strength wherever it is manufactured.

In most of the United States, cannabis has been accepted for medical use and, in many states, it has also been decriminalized. Cannabis products' availability is as close to some consumers as the corner strip mall. Customers are driving these goods into becoming a strong and easily obtainable market item. The industry must work

S. Audino (✉)
S. Audino & Associates, LLC, Wilmington, DE, USA

© Springer Nature Switzerland AG 2021
S. R. Opie (ed.), *Cannabis Laboratory Fundamentals*,
https://doi.org/10.1007/978-3-030-62716-4_5

toward a reliable product base so that sellers can trust that the products they supply are all of the same, whether to know the cannabinoid concentration (or "potency"), presence or absence of contaminants, or any other desired constituent.

Testing laboratories rely on a broad array of standards from sources including regulatory bodies, analytical test methods, and materials to ensure the reliability and trueness of their test results. In the cannabis testing industry, one of the most common standards is ISO/IEC 17025.

This chapter will provide a general overview of the most common laboratory standard, intent of several clauses within the standard, and discuss some of the more egregious risks to laboratories. Included will be more detailed description of a laboratory's Quality Management System and its major components, and discussion on some of the more challenging issues common to laboratories testing cannabis. This chapter is not intended to provide a full description of ISO/IEC 17025 or any laboratory quality system nor is it intended to offer explicit guidance on any subject.

The International Organization for Standardization

The International Federation of the National Standardizing Association (ISA) was founded in 1926 to develop standards for mechanical engineering and later disbanded in 1942. However, in 1946, delegates from 25 countries met in London and ultimately founded the International Organization for Standardization (ISO) and established headquarters in Geneva, Switzerland in 1947. ISO is a global network of currently 164 countries, each represented by one member. The American National Standards Institute (ANSI), for example, represents the United States in ISO.

While the acronym in English also somewhat represents the name of the organization, the shortened form of its name is also standardized. The organization wanted to have consistency in the recognition of the organization within all languages and selected ISO for this purpose. ISO is traced back to the Greek word "isos" which means "equal," the goal of internationalizing approved standards of testing [1].

As an independent and non-governmental organization, ISO provides a collaborative platform that brings together global experts who work cooperatively to develop standards in response to stakeholder needs. There are three major benefits to this approach:

- Provision of a competitive advantage to industry by developing a framework for consistent quality assurance, thereby increasing profits and decreasing loss;
- Global recognition of systematic/unified regulations, thereby ensuring credibility and improving global trade; and
- Provision of assurance to consumers that a given product or process was competently developed.

As a brand, ISO is a recognized leader within a global network carrying with it a long history of impartiality, integrity and profound commitment to both consensus and fair trade. ISO standards cover a broad range of services and products, for example camera film speed (ISO 6), medical devices (ISO 13485), and images for computer files (ISO 9660). Smaller families of standards encapsulate specific areas and include environmental management (ISO 14000 series), food safety (ISO 22000 series), and quality management (ISO 9000). In partnership with the International Electrotechnical Commission (IEC) several standards were developed for conformity assessment bodies, such as testing and calibration laboratories (ISO/IEC 17025). Since its inception, ISO has published more than 23,000 standards and has nearly 250 technical committees [1].

ISO defines a "standard" as "Documents that provide requirements, specifications, guidelines or characteristics that can be used consistently to ensure that materials, products, processes and services are fit for their purpose" [1]. Conformity assessment is the attestation by an external third party that a given organization meets the requirements of a given standard. The main forms of conformity assessment are certification, testing, and inspection. Certification is written assurance that a product or service meets specified requirements. ISO 9001, for example is a form of certification often sought by organizations to demonstrate they have a quality management system that meets the standard requirements. Testing refers to activities in a controlled environment typically to describe the characteristics and/or properties of a product. Inspections involve routine checking to assure that a product or service meets specified criteria. Both testing (ISO/IEC 17025) and inspection (ISO/IEC 17020) bodies are accreditation activities which require a third party Accreditation Body to issue a declaration of competence to fulfill the respective standard requirements.

Although ISO facilitates the development and housing of standards, the organization neither certifies nor provides conformity assessments. However, the ISO Council Committee on Conformity Assessment (CASCO) is tasked with developing policies and standards to address these needs. Among other functions, CASCO prepares international guides and standards relating to certification, testing and inspection of products and services, as well as to assessment of management systems, testing laboratories, inspection bodies, certification bodies, and accreditation bodies. Through CASCO and in partnership with IEC several of these are familiar to cannabis testing laboratories and include ISO/IEC 17025, ISO/IEC 17011 (Accreditation Bodies), ISO/IEC 17020 (Inspection Bodies), ISO/IEC 17043 (Proficiency Test Providers), and ISO/IEC 17034 (Reference Material Providers). Through these processes, CASCO promotes mutual recognition and acceptance of assessment systems and the appropriate uses of the international standards [1, 2].

International Laboratory Accreditation Cooperation

In 1977, at a conference in Copenhagen, Denmark, scientists met to discuss the international acceptance of accredited tests and calibration results to facilitate international trade. As a result of these discussions, in 1996 the International Laboratory Accreditation Cooperation (ILAC) became a formal cooperative. A charter within ILAC was established in 2000 for a global network of mutual recognition arrangements (MRA) among accreditation bodies [2].

The MRA is the formal recognition of equivalence of accreditation provided by MRA partners, and it maintains an open marketplace for free and unencumbered trade between countries while providing regulatory bodies with the confidence in the credibility of the processes and products produced under a specific standard. ILAC is the over-arching international organization responsible for laboratory accreditation bodies operating within the strictures of ISO/IEC 17011 [2]. By 2020, the ILAC-MRA extended to ISO/IEC 17025, ISO/IEC 17020, ISO/IEC 17043, and ISO/IEC 17034.

Accreditation Hierarchy

ILAC establishes the standards and requirements for accreditation bodies and recognizes five regional accreditation groups to manage this work (Fig. 1). Regional groups recognize the accreditation bodies through the Mutual Recognition Arrangement and provide peer-evaluation for each accreditation body to the ISO/IEC 17011 standard.

The Accreditation Bodies subsequently assess conformity assessment bodies (e.g., testing laboratories) to the appropriate standard, thereby accrediting them to perform work in accordance within a defined scope. Figure 2 describes the hierarchy of governance of the standards. Figure 3 summarizes the most common standards applicable to the cannabis testing laboratories.

Quality Management System

There are many definitions of "quality." However, most simply put, quality refers to conformance to some specified requirement or set of requirements. Very often these requirements are primarily driven by customers and then specified by regulatory

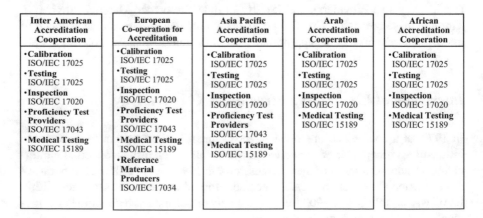

Inter American Accreditation Cooperation	European Co-operation for Accreditation	Asia Pacific Accreditation Cooperation	Arab Accreditation Cooperation	African Accreditation Cooperation
• Calibration ISO/IEC 17025	• Calibration ISO/IEC 17025	• Calibration ISO/IEC 17025	• Calibration ISO/IEC 17025	• Calibration ISO/IEC 17025
• Testing ISO/IEC 17025	• Testing ISO/IEC 17025	• Testing ISO/IEC 17025	• Testing ISO/IEC 17025	• Testing ISO/IEC 17025
• Inspection ISO/IEC 17020	• Inspection ISO/IEC 17020	• Inspection ISO/IEC 17020	• Inspection ISO/IEC 17020	• Inspection ISO/IEC 17020
• Proficiency Test Providers ISO/IEC 17043	• Proficiency Test Providers ISO/IEC 17043	• Proficiency Test Providers ISO/IEC 17043	• Medical Testing ISO/IEC 15189	• Medical Testing ISO/IEC 15189
• Medical Testing ISO/IEC 15189	• Medical Testing ISO/IEC 15189	• Medical Testing ISO/IEC 15189		
	• Reference Material Producers ISO/IEC 17034			

Fig. 1 Recognized regional cooperation bodies

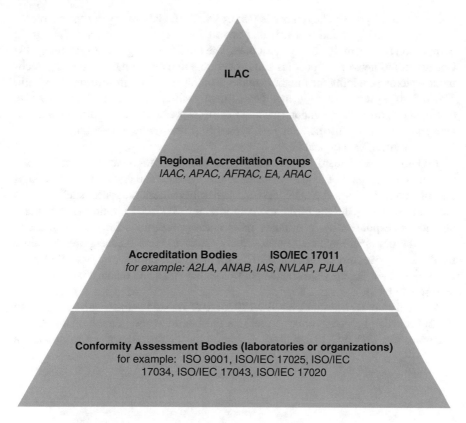

Fig. 2 Accreditation hierarchy

Standard	Application
ISO/IEC 17025:2017	Testing and Calibration Laboratories
ISO/IEC 17020:2012	Inspection Bodies
ISO/IEC 17050:2004	Product Certification
ISO/IEC 17065:2012	Product Certification
ISO/IEC 17043:2010	Proficiency Test Providers
ISO/IEC 17034:2016	Reference Material Producers
ISO 9001:2015	Quality Management System

Fig. 3 Popular standards applicable to cannabis industry

bodies. A quality-based laboratory is one in which the laboratory conforms to customer needs using sound methods that are fit for purpose and carried through to completion transparently and without bias. The quality management system (QMS) is a structured network of policies and procedures that encompass quality and technical systems providing the umbrella under which the quality infrastructure is built. The quality system generally includes subject matters such as customer service, document control, and vendors whereas sample preparation, instrument maintenance and analytical methodology are typically found within the technical system. Figure 4 represents a typical QMS.

ISO 9001 is a standard for which any type of organization can seek certification for the establishment of a quality management system. The cumulation of policies and procedures within the QMS address general business practices, such as good customer service [3]. One of the more significant ISO 9001 requirements is management's responsibility to evaluate risks and opportunities for improvement to ensure the organization's quality system continues to meet business and customer needs. The most current ISO/IEC 17025 imported this philosophy from ISO 9001 and requires a laboratory's management to identify risks and manage those risks on an ongoing basis.

While it may appear that ISO/IEC 17025:2017 relaxed some requirements from the 2005 revision, having done so presents the laboratory with additional risk. For example, the ISO/IEC 17025:2005 requirement for the laboratory to appoint a

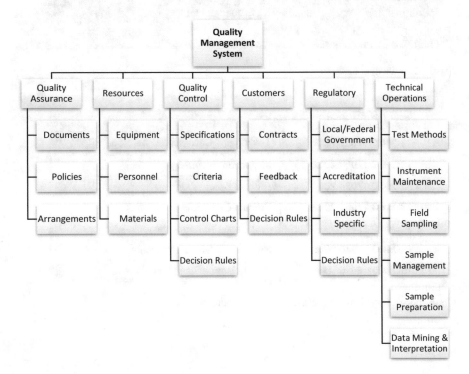

Fig. 4 Quality management system

Technical Manager and a Quality Manager was removed from the 2017 revision.¨ With this greater flexibility comes greater responsibility for the laboratory to know and mitigate a host of risks.

ISO/IEC 17025

As stated earlier, ISO/IEC 17025 is the international standard for the general requirements for the competence of testing and calibration laboratories. Building on the strengths of ISO 9001, the ISO/IEC 17025:2017 revision more closely aligns with processes than with requirements for procedures. In doing so, laboratories are afforded greater opportunity to develop a quality management system.

Quality Assurance

Quality assurance refers to the detailed processes and systems that build the foundation of the quality management system. This foundation must ensure that operations throughout the laboratory are defined and implemented at all turns and encompass the laboratory's technical competence, access to necessary equipment, materials, effective providers for services and supplies, empirically sound test methods, and transparent data analysis.

Quality assurance is part of every industry from toymaking to rocket science. It is well known that companies such as Hasbro and Mattel use a group of target-aged children to play with their new designs to demonstrate the effectiveness of a new product.

The failure of quality assurance can be aggravating and annoying, but it can also be life-threatening. One example can be found in forensic science when the investigation of several murders kept directing detectives in one major city to believe that a single serial killer was responsible for all the deaths. However, the detectives were unable to find any other links among the victims who were from vastly different demographics and educational backgrounds, for example. The finding that finally broke the case was that a worker at the manufacturer of cotton swabs used for DNA collection was uncomfortable using the required gloves when handling the cotton. So, she worked without them. Thus, it was her DNA, not the killer's, found at every crime scene. The police spent thousands of hours chasing a ghost. There was no serial killer [4].

Another example of a life-threatening failure is the tragedy of the space shuttle Challenger. On January 28, 1986, Challenger exploded 75 s after liftoff killing all seven astronauts on board. After a comprehensive investigation, it was concluded that a simple O-ring was to blame for the explosion. In their findings, NASA found that the O-ring in the joint of the right solid rocket booster failed at liftoff [5, 6]. The O-rings were not designed to withstand the unusually cold conditions on that

January morning. The Rogers Commission, formed by President Reagan to independently investigate the Challenger failure, found that NASA managers knew since 1977 that the manufacturer of the solid rocket boosters was aware of a potentially catastrophic flaw in the O-rings but failed to resolve it [7]. Scientists had warned that the weather was too cold for a safe launch, but the NASA managers failed to report this to their superiors. Both the manufacturer and NASA were found at fault for this unprecedented failure of quality assurance.

What the cotton swab manufacturer and the Challenger disaster teach everyone is that the small items are just as important, if not more so, as the big ones. Quality refers to everything in a laboratory, even the personnel, both scientists and non-specialized workers. A very small mistake can cause disastrous results.

Because organizations designate one or more persons serve as QA Manager(s), staff and managers alike often assume that the designated QA manager(s) are solely responsible for all quality operations. Indeed, the QA Manager is responsible for ensuring all quality elements are implemented. However, the responsibility falls to all personnel for the consistent implementation of all elements of the quality management system at all times. To be effective, the laboratory must embrace the ideology that quality is the responsibility of every individual in a laboratory. There are no exceptions.

Quality Control

Quality Control refers to the specific measures the laboratory takes to ensure their measurements or test results are both accurate and precise. Where QA speaks to the foundation of laboratory and testing quality, quality control speaks to the veracity of test results and the confidence a laboratory takes to release each test result to the respective customer. When the QC performs as expected, the laboratory can confidently release customer test results knowing that the analytical process was efficiently and effectively implemented. However, when the QC does not pass or performs poorly the laboratory is able to examine and where possible, remedy any defect before releasing customer test results.

One example is in the automobile industry. In February 2020, *Fortune* magazine [8] reported that Hyundai announced a 430,000 car recall because of water getting "into the antilock brake computer, [causing] an electrical short and possibly an engine fire." This computer is manufactured for Hyundai and its sister company Kia by an outside provider. The QC for this item failed to identify and correct this potentially life-threatening deficiency. The recall and the replacement of the antilock brake computer will cost Hyundai huge penalties in finances and in consumer confidence. Word of mouth alone will cause the cost of this failure to multiply exponentially as potential buyers look elsewhere for a new car. Brand success and consumer confidence rest firmly in the QC of the testing facility. Cannabis testing laboratories are no exception where business is very much 'word of mouth'.

There are many ways to implement quality control that are usually incorporated as part of sample batches. QC data can be obtained from initial or continuing calibration verification (ICV/CCV), internal standards, reagent blank, matrix blank, fortified matrix blank, matrix duplicate, fortified matrix duplicate, or some other well characterized laboratory control sample. The laboratory must determine which quality controls to use, where each one has distinct benefit and risk.

ISO/IEC 17025:2017 and Cannabis Laboratories

The ISO/IEC 17025:2017 standard is organized in eight sections [9]: Scope, Normative References, Terms and Definitions, General Requirements, Structural Requirements, Resource Requirements, Process Requirements, and Management System Requirements. Sections 4–8 specify the conformity clauses. However, the respective Accreditation Body will audit the laboratory to every applicable clause in the standard. Throughout the standard, several clauses have "notes" attached, the purpose of which is to provide suggestion or clarification for the reader. Where the Accreditation Body will assess or audit the testing or calibration laboratory to every applicable clause within the standard, "notes" are excluded. Very often, however, these provide exceptional guidance where the laboratory would be well served to integrate into their quality management system.

Section 4: General Requirements

Although a relatively brief section, impartiality and confidentiality are imperative for transparent and ethical laboratory practice. ISO/IEC 17025:2017 [9] defines impartiality as the "presence of objectivity," and this applies to all persons who may have business with the laboratory. Risks to impartiality, for example, may arise from laboratory owners or investors, customers, laboratory personnel, lobbyists, regulatory bodies, contractors, volunteer activities, or previous employers. Identifying these risks occurs on an on-going basis may present at any time and with any person. Once identified, the laboratory is tasked with understanding its severity and then mitigate, if not completely remove, the risk. For example, laboratory owners or investors are often concerned about seeing a fiscal return on their capital investments and may not understand the need for increased consumables for quality control samples, certified reference materials, and viable working standards to establish calibration curves. To accommodate the fiscal needs, laboratory personnel are at risk of not working objectively and instead bias the analytical process.

Laboratories that perform analytical work impartially treat all customers and laboratory personnel objectively and fairly, without bias or prejudice. Nearly every quality policy or mission statement this writer has reviewed includes a direct or implied statement of the laboratory's intent to perform work impartially, objectively and without conflict of interest. Rarely are the risks to these principles identified and transparently managed. There are many places in the laboratory where objectivity and conflict of interest can be compromised or lost altogether.

Although risks to impartiality exist in any laboratory, cannabis testing laboratories are especially prone to influence by customers who may take their business elsewhere if they feel the laboratory did not provide test results that are to their liking and meet their needs. Consequently, a laboratory may offer to re-test a sample or to change the scope of testing from "compliance" which has strict specifications that must be reported to a regulatory body, to "quality" or "research and development" where results may not be required by the regulatory body. The realistic fear of losing business can turn an organization's business model from being science-driven to customer-driven.

Laboratory personnel may also present a risk to impartiality if they have worked in other competing laboratories, particularly if they established personal relationships with former co-workers. Laboratory owners themselves may pose risks to objectivity. If they don't understand the science of laboratory work, owners may make promises for laboratory testing that are not feasible. Similarly, investors with limited knowledge of laboratory science anticipate a return on investment that may be unreasonable in terms of the science driving the laboratory services. Short-cuts to method development and putting forth analytical test methods that are not fully qualified, are commonplace in the interest of cost and time savings. And laboratory personnel accommodate short-cuts and similar compromises in the interest of job security.

As a new industry, regulations around cannabis are dynamic and vary on anticipated final use of the product. Occasionally, regulatory specifications are based on arbitrary test method values which inadvertently reward substandard test methodologies.

Another significant risk of concern among cannabis laboratories is the risk to confidentiality. These risks may go unnoticed by all parties because of the requirement that laboratories follow state regulations. Many state cannabis/marijuana regulatory bodies have tight product specifications and require customers to request laboratory services through a state-managed database system. Once tested, the lab delivers the test results to the customer using the same database. When this happens, the laboratory is releasing confidential customer data to a third-party data base over which the laboratory has no control. Ensuring the customer is aware of the laboratory's release of confidential information to a party other than the customer, strengthens the laboratory's position of defense in the event of a data-breach and release to the public.

Perhaps one of the most egregious compromises to confidentiality occurs when a laboratory markets itself as being the "laboratory of choice" by one of their large customers. This is a violation of that customer's expectation of confidentiality unless the laboratory has explicitly sought specific approval by the customer. However if the customer takes initiative to advertise the laboratory they use for their products then the laboratory is free to disclose the relationship.

Section 5: Structural Requirements

Another relatively brief section of the standard, the laboratory must be a legal entity and be committed to carrying out laboratory activities to comply with relevant standard clauses and any regulatory specifications.

The laboratory is required to establish the basic structure of their operations including identification of personnel with responsibility and authority to access resources to improve the management system. The management must ensure that the integrity of the management system is always maintained, even when changes are planned or implemented. Essentially, even though upper management or owners may not be involved in the daily operations, they should have solid understanding of the quality management system. The laboratory's reputation and the reputation of its management and ownership depends upon the quality and integrity of the laboratory's only product: the data they generate and report to customers.

Section 6: Resource Requirements

Laboratory Personnel

Even under the best of circumstances and intentions, a laboratory cannot function well unless it has adequate resources such as competent personnel, equipment, and materials. ISO/IEC 17025:2017 expanded the requirements for personnel and especially focuses on the competence of any person whose actions may influence the validity of test results. The ISO/IEC 17025:2005 requirement to specify personnel qualification has been replaced with competence requirements and monitoring competence in ISO/IEC 17025:2017 (Fig. 5).

As stated earlier, test results are directly influenced by many actions and multiple people. The testing process begins with sample receipt and includes every subsequent step through sample storage, sample preparation, reagent preparation, equipment maintenance, instrument performance, data mining and interpretation, and reporting. Authorization to perform each of these processes, as well as a host of additional processes, as well as a host of additional processes not included here, requires assurance that a person can do so correctly and efficiently. The laboratory must determine specific competence requirements that reflects knowledge, skills, and education for each activity that has influence on the validity of a test result.

Equipment

Laboratories require appropriate equipment and instruments capable of performing the intended work. This is often referred to as "fit for use" and should be implemented at various points:

Fig. 5 Technical
competence

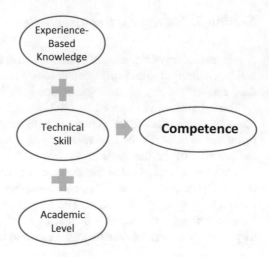

- Installation Qualification (IQ) ensures the equipment has been installed correctly and is supported by a report that justifies acceptance criteria;
- Operational Qualification (OQ) ensures the equipment's operational compliance within stated specifications in its specified location/environment;
- Performance Qualification (PQ) ensures the equipment is continuously meeting performance criteria.

In all cases, the ISO/IEC 17025 assessor will look for records to demonstrate *Initial Qualification* and *Operational Qualification* meet pre-determined specifications to demonstrate fitness for purpose before any new instrument has been used for customer samples. Further, Operational Qualification is expected whenever the equipment has undergone primary service or significant maintenance, or if it has been moved to a different location. Finally, it is recommended laboratories establish ongoing *Performance Qualification* specifications, which may include routine system performance checks or other system-quality control check, as one mechanism of quality assurance.

Metrological Traceability

Metrological traceability is defined as the "property of a measurement result whereby the result can be related to a reference through a documented unbroken chain of calibrations, each contributing to the measurement uncertainty" [10]. There are several ways to demonstrate compliance to metrological traceability, the most common is the use of ISO/IEC 17034 Certified Reference Materials (CRM) which are formulated under stringent procedures to characterize a given property or properties with measurement uncertainty. A CRM is defined as a "reference material characterized by a metrologically valid procedure for one or more specified properties, accompanied by a reference material certificate that provides the value of the specified property, its associated uncertainty, and a statement of metrological traceability" [10]. Often with no statement of associated measurement uncertainty, Reference Materials (RMs) are defined as "material, sufficiently homogeneous and stable with respect to one or more specified properties, which has been established to be for its intended use in a measurement process" [11] (Fig. 6).

An ISO/IEC 17035:2016 requirement is that the Certified Reference Material Producer shall "take measures to ensure that the integrity of each individual [Reference Material] unit is maintained until the seal, if any, has been broken or up to the point when first used" [10]. This means that once the ampoule or container has been opened, the stated expiration date is invalid and the integrity of the CRM may be compromised during future use. CRMs have limitations and it is the laboratory's responsibility to ensure each CRM is valid once the container has been breached, even if the stated expiration is months or years later [11]. Regrettably, most laboratories do not establish the shelf-life of the CRMs upon opening and instead rely on the expiration date printed on the label or Certificate of Analysis, thereby compromising the validity of test results. Similarly, it is acceptable to use an RM beyond its stated expiration date provided the laboratory produces records that empirically demonstrate it is fit for use. Unfortunately, CRMs and RMs are not typically given the attention and care necessary to ensure their role(s) in the validity of test results.

Externally Provided Products and Services

Similar to the strict requirements that laboratory personnel be explicitly qualified and authorized before performing work, laboratories are also required to ensure that materials and services obtained from providers outside the laboratory are qualified

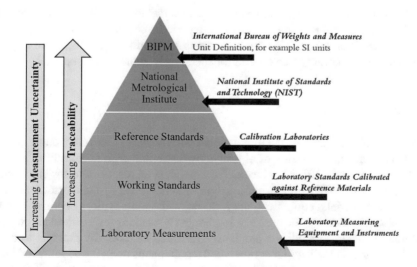

Fig. 6 Metrological traceability

and appropriate for intended use. The laboratory is tasked with identifying the requirements to qualify vendors, consultants, and sub-contractors, and ensuring no services or products are used unless they meet laboratory specifications. In many cases, qualifications could be as simple as ISO/IEC 17025 where appropriate (e.g., calibration providers, sub-contractors), or ISO/IEC 17034 for certified reference materials, ISO/IEC 17043 for accredited proficiency test providers, when available, or perhaps ISO/IEC 17011 for Accreditation Bodies. Additionally, ISO 9001 could be considered a qualifier for chemical distribution companies. The laboratory is required to retain records of qualification for each external provider.

Section 7: Process Requirements

Selecting Appropriate Test Methods

Test methods can be qualitative, quantitative or semi-quantitative. The analytical method must always be specified prior to beginning work on customer samples. Understanding the customer's needs is paramount in the selection of test methods, which must be fit for intended purpose. Test methods can be sourced from any number of places such as, but not limited to:

- Voluntary consensus methods, such as AOAC International, ASTM
- Standard test methods, such as EPA, ISO, or FDA
- Published methods, such as Journal of Analytical Chemistry, Journal of AOAC International, or Journal of Chromatography, etc.
- Methods developed or specified by equipment/instrument manufacturer
- Laboratory developed test methods

When a laboratory chooses to develop its own method(s), they must first have an established protocol to implement and have periodic checks to ensure the intended purpose is still on target. This represents a specialized task and therefore requires well-established competence criteria for specifically authorized personnel.

Method Verification

ISO/IEC Guide 99:2007 is the international vocabulary of metrology and it defines verification as the "provision of objective evidence that a given item fulfills specified requirements" [12]. This means the laboratory is responsible for empirically demonstrating that an established method is suitable using their resources (e.g., personnel, materials, equipment, etc.). Some of the most typical opportunities for method verification are the desire to implement a standard test method and when a laboratory wishes to modify a previously validated method, for example to determine if the method will still meet specified requirements upon introduction of a new or alternate sample matrix. Changes in concentration range and changes in chemicals or reagents may also trigger method verification. Records of all verifications must be retained by the laboratory for the life of the method.

Method Validation

ISO/IEC Guide 99:2007 defines validation as the "verification where the specified requirements are adequate for intended use" [12]. ISO/IEC 17025:2017 requires laboratories to validate "non-standard methods, laboratory-developed methods and standard methods used outside their intended scope or otherwise modified."

The validation process must begin with a pre-defined protocol that clearly specifies performance parameters and decision rules on the method's fitness for purpose. There are many performance parameters that must be well characterized during the validation process. Several of the most common are listed below [13–15] (Fig. 7):

1. Selectivity: The demonstration and assurance that a method can unconditionally detect or discriminate an analyte of interest in a mixture without interference from other compounds.
2. *Limit of detection: The lowest analyte concentration that can be detected but not necessarily quantitated. Typically, a signal to noise (S/N) ratio between 3:1 and 2:1 is generally acceptable, and can be calculated using the following formula:

$$\text{Detection Limit} = 3.3 * \text{standard error } (\sigma) / \text{slope} \qquad (1)$$

3. Ruggedness: The measure of a method's capacity to withstand small changes in deliberate changes to conditions.

Parameter	Quantitative Analysis	Qualitative Analysis
Selectivity	✓	✓
Limit of Detection	✓	✓
Ruggedness	✓	✓
Bias/Trueness	✓	
Limit of Quantitation	✓	
Precision	✓	
• Repeatability	✓	
• Reproducibility	✓	
• Intermediate Precision	✓	
Linearity/Dynamic Range	✓	

Fig. 7 Common performance parameters

4. Bias/trueness: The difference between the mean test results and a known or accepted reference value (e.g., CRM). Bias may be caused by both random and system errors and should be evaluated during the method development process.
5. *Limit of quantitation: The lowest analyte concentration that can be confidently detected and quantitated, within an acceptable level of uncertainty. The LOQ can be calculated by using the following formula:

$$LOQ = 10 * standard error (\sigma) / slope \tag{2}$$

6. Precision: The degree of agreement among independent test results under pre-determined analytical conditions.
7. Repeatability: The measure of short-term variability of the analytical method's ability to generate the same results under the same conditions.
8. Reproducibility: The measure of performance when the method is implemented under different conditions and environments, typically during inter-laboratory studies.
9. Intermediate precision: Intra-laboratory reproducibility when monitored over time.
10. Linearity/dynamic range: The statistical relationship between known analyte concentration and a response factor such as area count. The strength of relationship is calculated by a correlation coefficient, where the valid range is specified as the range between the lowest and highest concentrations. The equation for the line is used to calculate the concentrations of unknown samples using the following formula:

$$Response (y) = slope (m)*concentration (x) + y-intercept (b) \tag{3}$$

When a series of working standards are prepared and analyzed, it is commonly referred to as a "calibration curve" where the equation is determined by ordinary least squares regression.

Method validation must be performed to the extent necessary to establish that the method is adequate for intended use [9]. Similarly, method verification can often lead to partial validation studies, or in some instances full method validation. It is difficult to predict the parameters that will need to be addressed, as it depends upon the extensions or modifications introduced. Events or conditions that may trigger re-validation include:

- QC monitoring demonstrates trends that cannot be explained through routine investigations
- Proficiency test performance
- Consistent customer requests to re-test samples
- Current literature
- New or updated validation data provided by compendial or standard method
- Longevity of method

Figure 8 provides a brief and general summary of parameters to address during method verification and method validation processes.

Evaluating Measurement Uncertainty and Measurement Error

Measurement Uncertainty and error are commonly used inter-changeably however they have different meanings. The former, measurement uncertainty, refers to a range of values within which the true value lies and does not estimate error. The latter, measurement error, is the difference between a single test result and its true value, which ironically can't be known exactly.

Measurement Uncertainty, as defined by the Guide to Uncertainty of Measurement (GUM) [16] is "non-negative parameter characterizing the dispersion of the quantity values being attributed to the measurand, based on the information used." These are generally classified as Type A or Type B uncertainties [14, 15]. Type A uncertainties are obtained from repeat measurements under the same conditions where the analyst will calculate and take into consideration response means and standard deviation, and other measures of dispersion. Type B uncertainties are deterministic come from sources such as Certificates of Analysis, specifications, tolerances, and calibrations.

Measurement errors come in two basic categories, systematic and random. Systematic errors are reproducible inaccuracies that remain consistent or constant in one direction or another. There is some degree of predictability to systematic errors that can be somewhat minimized by using offset factors and correction factors, for

Test Method	Required Action	Minimal Parameters to Evaluate and Report Before Method Deployment
Laboratory must use an already established "fully validated method" (e.g., compendial, official, etc) that includes analytical characteristics.	Verification	• Precision • Bias (including matrix variations) • Possibly linearity
Laboratory uses an established "fully validated method" but using a new/different matrix.	Verification/Matrix Extension Possible Partial or Full Validation	• Precision • Bias (including matrix variations) Objective is to ensure the new matrix has not introduced new sources of error.
Laboratory uses a published but no compendial/official method (e.g., journal) that may or may not include analytical characteristics.	Partial Validation	• Precision • Bias (including matrix variations) • Ruggedness • Linearity
Laboratory develops an in-house method.	Validation	• Selectivity/ Specificity • Linearity • Accuracy • Precision • Limit of Detection • Limit of Quantitation • Robustness

Fig. 8 Method performance characteristics

example. Random errors are unpredictable, preventing the laboratory from acquiring an exact measurement under exact conditions such that even repeated values differ from previous values.

There are three broad approaches to measurement uncertainty. The first is Target Measurement Uncertainty in which the maximum uncertainty is defined for specific measurand [17] which is growing in popularity amongst regulators. The second method is to specify and quantify the individual contributions of the overall uncertainty and combine using the root sum squares model [11]. The Kragten spreadsheet approach [18] is a helpful aid to facilitate accounting all contributions. Finally, the most basic approach is to establish good estimates of the long-term precision of the analytical process, generally found through within-laboratory reproducibility data (Fig. 9).

Samples and Sampling

The term "sample" refers to a manageable unit from larger population of units. With respect to the cannabis and marijuana industry the sample is complex. Regardless of the person or organization choosing cannabis samples from the larger facility, usually a cultivation facility, the laboratory should have information as to the source of the sample. For example, is the sample that arrives at the laboratory a composite? Is it a sample of leaves, or trim, or buds? Is this a finished product and if so, was cannabis infused into the manufacture process or was it applied as a topical? Understanding the nature/composition of the sample is critical for the laboratory to ensure it has the proper resources to provide appropriate testing.

When the laboratory is going to sample from the field (e.g., cultivation facility, dispensary, finished product, raw product, etc.), a sampling plan and method are required, and, whenever possible, the plans shall be based on appropriate statistical methods. The conditions under which sampling occurs could strongly bias final test results.

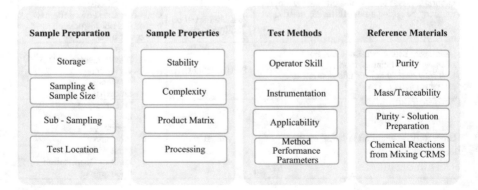

Fig. 9 Potential sources of uncertainty

The first step to address the sample and sampling process is to understand the types of samples. Representative samples selected from the population are classified as static (stable over time), or as dynamic (where the material changes over time or over conditions). Samples can also be selected with sampler bias and subjective judgment.

Once the type of samples is identified, determining the sampling plan is paramount. It is fascinating to observe laboratory personnel who claim to perform "random" and "representative" sampling when their method involves using forceps to move around material until something "looks" random and is subsequently selected. Although there are many sampling methods, probability and non-probability in nature, the cannabis industry is generally well served by the plans listed in this section.

Especially with the dry flower (or any biomass), the easiest and most appropriate sampling method is probability sampling. In this case, every unit in the container has an equal chance of being selected, making each selection representative from the entire population. A probability sampling can be [13, 15]:

1. Simple random where incremental samples are selected from bulk in such a way that each incremental sample has an equal probability of being selected;
2. Stratified where the container is either physically or visually subdivided into strata and incremental samples are randomly selected from select strata;
3. Cluster where the bulk has been divided into groups (e.g., trim, bud, etc.) after which incremental samples may be randomly selected.

Discrete, finished or packaged products are also amenable to probability sampling by using a systematic approach. In this case the sampler identifies a random starting point (say for example the third item) and selects every nth product thereafter, for example if $n = 4$, then products in positions 3, 7, 11, 15, ... Are selected for testing.

The number of samples to be selected is sometimes specified by regulatory bodies and when they are not, the laboratory needs to predetermine the number of samples to be gathered and the size of each of those samples. A typical approach is to specify a percentage of sample mass to be sampled from the bulk or harvest. Given the economic value of cannabis material, it is unusual to sample more than 0.5% of the harvested batch and 0.35% is typical. Most sample sizes are 1.5–2.0 g.

Composite samples are combinations of previous selected samples, regardless of the sampling strategy. The benefit of this approach is that composite represents the average of all collected samples.

One of the most overlooked sources of error is that which arises from sampling, yet personnel in nearly every cannabis testing laboratory will voice concern about the heterogeneity of the sample and pay little attention to the importance of implementing a sound and statistically based sampling plan. Sampling error is exacerbated when test results are based on samples that do not represent the population or have under-represented the sample. Consequently, the bias can be serious and skew errors in one direction or another, distorting the normal distribution [13, 15].

Establishing a Quality Control Program

Ensuring the validity or soundness of test results, requires the laboratory to establish an effective quality control program prior to performing any work on customer samples. As mentioned earlier, this is an integral component of the quality management system. Where quality assurance is the foundation and pathway to analytical measurements, quality control essentially evaluates the performance of the quality management system through the identification and control of measurement errors. An effective QC program applies a strict schedule typically by defining the number of samples in a batch and including one or more QC materials. The choice of QC materials should be relevant to the task and in consideration of their individual limitations. Choosing to use a continuing calibration verification (CCV), for example, is a check against itself and does not provide information that that analytical method was properly performed. Further, matrix fortifications or "spikes" can be challenging because it is difficult to know with certainty the baseline concentrations in the sample, or to know with certainty the efficiency of sample fortification.

Quality Control Materials

Any number of materials can be used to monitor sample preparation and system suitability. The most common are listed in Fig. 10.

Quality Control Considerations

The laboratory should consider the following as they develop their QC system:

- Analyte of interest: be certain to identify and quantitate before performing work on customer samples
- Specify QC material: Pre-determine which QC material or materials make the most sense for your application.
- Batch or Analytical "Run" size: specify the maximum number of samples to analyze with any given QC materials before analyzing additional QC to ensure system suitability. Although the batch size is at the discretion of the laboratory, common practice is to limit batches to 20–30 samples and then add appropriate QC materials before introducing an additional sample batch.
- Acceptance Criteria: Pre-determine acceptance and or failure criteria, including precision of replicates
- Plot the QC and pre-determine actions to take when either a trend is present or when there is an outlier.

Keep in mind that the QC system is to provide a check for the entire analytical process, beginning with sub-sampling and sample extractions. Adding a spike at the end of extraction provides no insight about sample preparation effectiveness.

Quality Control Material	Preparation	Objective
Initial Calibration Verification (ICV)	Prepared from second source; independent preparation from calibration standards.	Verify validity of calibration standards.
Continuous Calibration Verification (CCV)	Midrange standard	Verify that the calibration is still acceptable as the batch is in progress.
Laboratory Control Sample (LCS)	Well characterized material with known/established analyte parameters.	Demonstrate the analytical method covers the target analytes of interest.
Matrix Blank (MB)	Analyte-free matrix that is carried through every step of the assay.	Demonstrate the analytical method does not introduce contamination.
Matrix Spike (MS)	Known amount of analyte is added to the sample prior to any work performed on the sample.	Provides measures of method accuracy and precision.
Internal Standard (IS)	Addition of a known amount of a compound similar to the analyte of interest that is added to every sample and calibrator.	Calibration is based on response ratio of calibrator to internal standard.
Reagent Blank (RB)	Reagents only; no matrix.	Determine contributions, if any, the reagents may have on the measurement results.
Replicates: Matrix Blank, Reagent Blank, Matrix Spike, etc.	At least duplicates of a sample or QC material.	Provide measures of precision.

Fig. 10 Common quality controls

Quality Control Charts

Control charts are visual representations of data that demonstrate the amount and nature of variation of some attribute over time. When properly and routinely implemented, control charts provide three discrete values to the laboratory:

1. offer insight into operational variance and help identify assignable causes of observed variations;
2. estimate specific parameters of a given assay; and
3. reduce the variability of a given process or assay [16]

ISO/IEC 17025:2017 does not dictate the use of control charts. The standard simply requires the laboratory to "have a procedure for monitoring the validity of results" and that the "data shall be recorded in such a way that trends are detectable." The most common control chart used for chemical assays is the Shewhart Control Chart, named for Walter A. Shewhart who, in 1924 designed a chart to indicate whether or not observed variations in measurements were acceptable [19]. Plotting measurements over time and comparing to historical data to establish upper and lower limits, Shewhart control charts provide the user with immediate and clear course of action. Limits are established first by calculating a mean of (typically) at least 30 data points, then calculating the dispersion around the mean, which is the standard deviation.

Typically, upper and lower warning limits are established at ±2 times the standard deviation whereas the control limits are established at ±3 times the standard deviation. These limits are not arbitrary and in fact reflect the confidence intervals of approximately 95% and 99.7% of the normal distribution. A laboratory should take action when trends are evident, not just when the data are beyond stated limits. The number of points that constitute a "trend" can vary from laboratory to laboratory, however there are guidelines that a laboratory should consult when predetermining the actions to take when data are trending or extend beyond the control limits [19].

When new methods are developed, the 30 initial points are often taken from the method validation data then updated once 100 points have been collected or after some period of time has elapsed. Good practice dictates routinely updating the warning and control limits. Occasionally a laboratory will choose to calculate the limits with each new data point, typically encompassing the most recent n-number of points. This moving average approach captures the dynamic changes of the system where this is a weakness in a Shewhart control chart, however because the limits are constantly changing, trends may be difficult to detect and Type I and Type II acceptance/failure of control data are more likely to occur.

Section 8: Management System Requirements

Risk Based Thought Process

The risk-based thinking approach of ISO/IEC 17025:2017 has compelled laboratory managers to apply decision rules when statements of conformity are required. Very simply, decision rules provide the framework around accepting or rejecting a given outcome within the management system. Decisions could be to pass or fail a test sample, or to accept or decline a customer's request, or even to judge the acceptability of test results taking uncertainty into consideration [20, 21]. When statements of conformity are required, the laboratory must weigh the risks of Type I error (false positive) and Type II error (false negative) in their decisions, so measurement uncertainty should not be ignored or trivialized.

Figure 11 represents four test results in relation to the regulatory limit; calculated concentrations that exceed the regulatory limit 'fail' and those below 'pass'. In this example, A and B clearly pass, however C and D are questionable. Specifically, the value of C is right on the line, however the error bands around it suggest the true value could exceed the limit and fail. The value of D is clearly above the line and appears to fail, however the error bands around it suggest the true value could be below the limit and pass. Both C and D in this example neither clearly pass nor clearly fail, but are considered indeterminant.

There are in fact six clauses in ISO/IEC 17025:2017 that call specific attention to decision rules and statements of conformity [9]:

- *Clause 6.2.6.b: requires that the laboratory shall authorize personnel to perform "analysis of results, including statements of conformity or opinions and interpretations".*

Fig. 11 Determining conformity

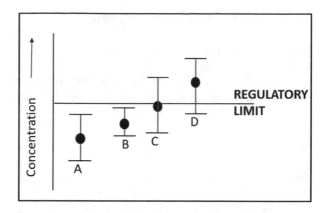

- *Clause 7.1.3: requires that "When the customer requests a statement of conformity to a specification or standard for the test or calibration (e.g. pass/fail, in-tolerance/out-of-tolerance), the specification or standard and the decision rule shall be clearly defined. Unless inherent in the requested specification or standard, the decision rule selected shall be communicated to, and agreed with, the customer."*
- *Clause 7.8.3.1.b: states "where relevant, a statement of conformity with requirements or specifications" and clause 7.8.3.1.c states "where applicable, the measurement uncertainty presented in the same unit as that of the measurand or in a term relative to the measurand (e.g. percent), when it is relevant to the validity or application of the test results, when a customer's instruction so requires, or when the measurement uncertainty affects conformity to a specification limit".*
- *Clause 8.8.4.1.a: states "the measurement uncertainty of the measurement result presented in the same unit as that of the measurand or in a term relative to the measurand (e.g. percent)." Clause 7.8.4.1e also states "where relevant, a statement of conformity with requirements or specifications".*
- *Clause 7.8.6.1: states "When a statement of conformity to a specification or standard for test or calibration is provided, the laboratory shall document the decision rule employed, taking into account the level of risk (such as false accept and false reject and statistical assumptions) associated with the decision rule employed and apply the decision rule."*
- *Clause 7.8.6.2: requires that "the laboratory shall report on the statement of conformity, such that the statement clearly identifies: (a) to which results the statement of conformity applies; (b) which specifications, standard or parts thereof are met or not met; (c) the decision rule applied (unless it is inherent in the requested specification or standard)."*

Actions to Address Risks and Opportunities

Risk assessment and management is not a one-size-fits-all. The laboratory is responsible for considering the actual and perceived risks and opportunities that exist across all laboratory activities. There are a number of process control systems that could help a laboratory identify and mitigate risks on an on-going basis.

Internal Audit

The purpose of internal audits is to provide the laboratory to delve into their own operations to learn which processes and systems are working well, and which processes and systems offer opportunities for improvements. Laboratory personnel who simply use a checklist to ensure they have procedures and policies in place, are not using the internal audit as it is intended. The internal audit is perhaps a laboratory's greatest tool to critically evaluate the totality of its operations. Unlike external audits which are limited to small sampling exercises over a very short period of time, qualified laboratory personnel can develop internal audit programs over days, weeks, or even months to ensure all elements of the quality management system are captured.

The internal audit begins with an experienced and well-qualified auditor who is able to conduct the audit independently or lead a team of qualified auditors. The scope of the audit must be pre-determined and must encompass all of the laboratory's operations: management operations, quality systems, and technical processes. At no point should the auditor(s) give a 'pass' for any non-conformance and should instead, initiate the process for non-conforming work/corrective action or other mechanism that will permit an a deeper examination of the area. Through this disciplined process, the laboratory management will gain a practical understanding of the effectiveness of the quality assurance, the effectiveness of their risk identification and management processes, and a critical evaluation of their technical operations. It should be pointed out that identifying areas/processes/systems that are working well is just as important as identifying weak or deficit areas. Although the ISO/IEC 17025:2017 does not specify the interval for internal audits, not performing annually (under most circumstances) would be a disservice to the laboratory. It is acceptable for laboratories to access qualified external consultants to perform internal audits, or to train laboratory staff by mentorship. In addition to the ISO/IEC 17025:2017 standard, ISO/IEC 19011 [22] provides specific guidance.

Common Misconceptions of ISO/IEC 17025

New laboratory owners and investors often approach ISO/IEC 17025 with trepidation for three main reasons: unfamiliarity with the standard, regulatory bodies require laboratory accreditation to ISO/IEC 17025, and misbelief that the standard

ISO/IEC 17025 Standard	Inaccurate Assumptions about ISO/IEC 17025
Provides general requirements for all types of testing/calibration labs	Specific instruments or instrument manufacturers are dictated
Requires laboratories to define the scope of their work	The laboratory must operate with a specific number of personnel.
Requires implementation of appropriate test methods	The laboratory's square footage and organization must meet the standard
Requires policies and procedures necessary to fulfill the scope of activities	Suppliers of services and supplies must meet the standard's requirements.
Requires competent staff	Test methods are issued by ISO
Requires a functional and appropriate management system	The scope of laboratory testing is determined by ISO.
Requires transparent communication with customers	Safety and Hygiene protocols must meet ISO requirements.
Requires the laboratory to provide objective evidence they do what they said they would do	All technical operations must be best practice.
Requires laboratory to identify risks and manage those risks	ISO requires a LIMS or cloud-based electronic storage mechanism.
Requires constant attention to the dynamic quality management system	All procedures and policies are in place only when then assessment occurs.

Fig. 12 ISO/IEC 17025:2017 requirements and inaccurate assumptions

dictates operational and fiscal regimens to the laboratory. As discussed earlier, ISO/IEC 17025 was not created just for the cannabis industry and has been embraced by private, state, and federal laboratories world-wide. Figure 12 offers a quick reference to the practical functions of ISO/IEC 17025 and what is does not require.

Challenges to Implementing a Quality-Driven Laboratory in the Cannabis Space

Testing cannabis is no different than laboratory testing in any other agricultural commodity. The instrumentation is the same; test method development and method validation follow the same protocols, and data analysis is no different. The most significant challenge to testing cannabis is its recent acceptance into contemporary society with murky recognition by state, federal, and even international governing bodies. As with other nascent industries, cannabis testing laboratories is evolving unto itself. Where there were no standard or consensus test methods, for example, those methods are slowly emerging. Where the prospect of working within a previously villainized industry seemed romantic, it has proven to not be easy and requirements for scientific integrity continue to unfold.

Still, routine frustrations and adversities cloud the cannabis testing industry that regularly compromise the validity and utility of analytical test results. For example:

- There is no federal oversight in the United States which defers to individual states for regulatory specification and guidance. At present, the United States has 33 independent operating systems that prohibit the transfer of cannabis with THC concentration greater than 0.3%.
- The regulatory bodies governing the US do not necessarily have experience in laboratory science, or any other science, yet they are responsible for defining the scope and specifications of laboratory testing.
- Unlike most other industries, many cannabis-testing laboratory owners and investors are not scientists and have no previous experience in laboratory work, many of whom are first time business owners.
- Many labs populate their analytical team with young, bright, and very enthusiastic, yet relatively inexperienced scientists. It is not uncommon to find a freshly minted PhD scientist as Chief Scientific Officer, or a bench analyst who rapidly rises to an upper level science position charged with developing and validating test methods.
- As a nascent industry, laboratories cannot rely on established and proven standard test methods and instead are encumbered with developing their own. This requires the young analytical team to quickly master the nuances of method development, method validation, and determination of measurement uncertainty.
- In some cases, regulatory bodies require a cannabis-testing laboratory to be ISO/IEC 17025 accredited before they are issued a business license, prohibiting laboratories access to cannabis matrix necessary to meet ISO/IEC 17025 requirements for method development and method validation. This forces the laboratory to use either neat standards or carefully chosen surrogate to develop test methods, which is less than ideal and depreciates the confidence in developed methods.
- There is an overwhelming absence of matrix-based reference materials.
- The typical cannabis testing laboratory has many risks to impartiality and objectivity, including those introduced by their very customers who threaten to take their business to a different laboratory unless they are provided "acceptable" test results following sampling. This increases the probability of a laboratory producing test results without having analyzed the sample, a practice commonly referred to as "dry labbing".
- The economic value of the matrix prohibits a large sampling base to ensure representative samples are tested, a problem that is only exacerbated by the heterogeneity of the matrix.
- This list of risks along with a host of others, set the stage for compromised test results and questionable Certificates of Analysis.

Legalization

The Drug Abuse Prevention and Control Act of 1970, and the Anti-Drug Abuse Act of 1986 [23] supported decades of cannabis prohibition [24]. The shift to legalization started in 1996 with California being the first state to pass the Medical Marijuana

Law. Over the past several years, cannabis has become a staple in the news and a political platform for many advocates in state and federal legislative bodies.

At the time of this writing, cannabis possession remains federally illegal; however, 11 independent state jurisdictions have legalized cannabis for medicinal (or compassionate) use and recreation for adults over 21 years of age. Also, 33 states have legalized for strictly medicinal (or compassionate) use, although eight states continue to uphold complete prohibition.

In 2018, the USDA differentiated between marijuana and hemp on the basis of THC content. They stated that anything that exceeds 0.3% THC is classified as "marijuana," and those products or materials in which THC is less than or equal to 0.3% THC are defined as "hemp." This distinction inadvertently creates more stress and the opportunity for poor-testing quality in the industry.

Rapidly Changing Regulations

The cannabis industry is centered squarely in a new and expanding universe that changes seemingly every day. The demand for cannabis products is increasing at an enormous rate with every age group represented in the consumer assemblage. Medicinal products are especially in demand since most states still have not legalized recreational marijuana.

With an ever-changing industry, the regulations for that same industry change just as speedily. What is approved as a standard one month can, and will, change the next month. Because of its continued criminalization of all cannabis products, there are no centralized federal regulations for the production, testing, and sale of any cannabis products. Within this regulation miasma is the additional problem of some regulations being developed without solid understanding of the scientific principles underlying all testing. The goal for testing, however, is always the same; to ensure the end consumer has full disclosure of the contents of the product he/she is purchasing.

A Revolutionary Product and Opportunity

All of this fractionalization had led the cannabis industry into very muddy waters in the presence of counterproductive, sometimes subversive, always destabilizing politicization of cannabis. While standardization of testing must be maintained, there remains an absolute necessity for a safe and quality-derived product where consumers have a right to know, unquestionably, what is and is not in their product. How that can be established is the conundrum that the cannabis industry now faces. However, within a very small interior of these unregulated regulations, there is the standard for which consumers cry out. Therefore, the testing laboratory works within the confines of the state within which it resides while expanding those regulations to meet a standard of quality and safety that must be attained for the sake of the consumer.

The cannabis industry must maintain its commitment to the testing mandate of safety concerns for the consumer. Consistency is key to success. The manufacturers of cannabis products are driven by multiple and often differing interests including the need for profit and the requirement to maintain a consumer base among the suppliers of their products in the marketplace. As in any industry, the consumer is king. If the product does not meet their expectations or does not prove consistent in effectiveness, the consumer will discontinue purchasing from one product line and search elsewhere for satisfaction.

Therefore, it is and will always be the consumers who will drive the business plan and not the scientists. In many ways, the cannabis industry is operating in an environment not unlike the illegal and unregulated bathtub gin and homemade whiskey of the Roaring Twenties. Each producer of their (at that time illegal) substance met their own standard, which never intersected with any testing regulation or quality standard. Scientists are equipped to come to the rescue by working almost *sub rosa* in their attempts to maintain quality standards for cannabis products. International standards exist and can be leveraged to achieve quality, consistency, and safety. Communication among similar laboratories can assist in designing a standardization within and in addition to the superfluity and counterproductivity of some regulations.

Chapter Summary

This chapter has provided the reader with a small fraction of the more significant processes and systems that laboratories must plan, implement, evaluate, and adjust as necessary, if they are to provide the community with viable test results. In fact, the single product of a cannabis-testing laboratory is its data, or more specifically their analytical test results. No test result is a self-standing value; its credibility is based on properly characterized performance criteria, process controls, and quality assurance.

ISO/IEC 17025:2017 is the gold standard for laboratory quality by creating a pathway for each laboratory to employ competent people who work impartially using sound methods to yield consistent results. However there are no guarantees that an ISO/IEC 17025:2017 accredited laboratory is poised to perform the 'best' or even 'good' science. The depth of commitment to quality assurance varies from laboratory to laboratory and is subject to keen attention to details and willingness to critically evaluate one's own laboratory to ensure the best science drives activities. This is the best place to start in the industry's pursuit of laboratory standardization where cannabis samples can be tested and yield the same analytical accuracy and precision, and reliability across all laboratories, offering the end user confidence in the product(s) they use.

References

1. https://www.iso.org/home.html. Accessed 15 Jul 2020
2. https://www.ilac.org. Accessed 15 Jul 2020
3. ISO 9001:2015 (2016) Quality management systems—requirements. International Organization for Standardization (ISO), Geneva
4. https://davidson.weizmann.ac.il/en/online/sciencepanorama/straight-crooked-ruler. Accessed 15 Jul 2020
5. https://www.washingtonpost.com/archive/politics/1986/06/10/challenger-disaster-blamed-on-o-rings-pressure-to-launch/6b331ca1-f544-4147-8e4e-941b7a7e47ae/. Accessed 15 Jul 2020
6. https://instrumentationtools.com/o-ring-failure/. Accessed 15 Jul 2020
7. https://infogalactic.com/info/Rogers_Commission_Report. Accessed 15 Jul 2020
8. https://fortune.com/2020/02/07/hyundai-recall-2020-engine-fires/
9. ISO/IEC 17025:2017(E) (2017) General requirements for the competence of testing and calibration laboratories. International Organization for Standardization (ISO), Geneva
10. ISO/IEC 17034:2016 (2016) General requirements for the competence of reference material producers. International Organization for Standardization (ISO), Geneva
11. ISO Guide 35:2017 (2017) Reference materials—guidance for characterization and assessment of homogeneity and stability. International Organization for Standardization (ISO), Geneva
12. ISO/IEC Guide 99:2007 (2007) International vocabulary of metrology—basic and general concepts and associated terms (VIM). International Organization for Standardization (ISO), Geneva
13. Ellison SLR, Barwic VJ, Duguid Farrant TJ (2009) Practical statistics for the analytical chemist: a bench guide. RCS Publishing, Cambridge
14. Eurachem (2014) The fitness for purpose of analytical methods: a laboratory guide to method validation and related topics, 2nd edn
15. Prichard E, Barwick V (2007) Quality assurance and analytical chemistry. Wiley, West Sussex
16. ISO Guide 98:1995 (1995) Guide to the expression of uncertainty in measurement. International Organization for Standardization (ISO), Geneva
17. Bettencourt da Silva R, Williams A (eds) (2015) Eurachem/CITAC guide: setting and using target uncertainty in chemical measurement. Available from https://www.eurachem.org
18. Eurachem/CITAC guide 2000: quantifying uncertainty in analytical measurement
19. Swift JA (1995) Introduction to modern statistical quality control and management. St. Lucie Press, Delray Beach, FL
20. ILAC G8:09/2019 (2019) Guidelines on decision rules and statements of conformity. International Laboratory Accreditation Cooperation, Silverwater
21. Decision rules and statements of conformity, Lab 48, Edition 3 Jun 2020, United Kingdom Accreditation Service (2020)
22. ISO/IEC 19011:2018 (2018) Guidelines for auditing management system. International Organization for Standardization (ISO), Geneva
23. H.R.5484—Anti-Drug Abuse Act of 1986, 99th Congress (1985–1986)
24. The Comprehensive Drug Abuse Prevention and Control Act of 1970, Pub.L. 91-513, 84 Stat. 1236, enacted 27 Oct 1970

Cannabis Laboratory Management: Staffing, Training and Quality

Robert D. Brodnick, William A. English, Maya Leonetti, and Tyler B. Anthony

Abstract Cannabis laboratory proprietors risk overexposure to staffing, training, quality management, and leadership culture failures. As the technical startup burden is eased by an increasing body of standardized testing methodologies, laboratories must establish programs for continuous improvement to remain competitive. The fastest path to sustainable high-reliability cannabis testing operations incorporates the regulatory history, technical strengths, and modernized quality practices from established relevant testing sectors. This chapter will discuss critical components of cannabis laboratory management including hiring competent personnel, ensuring and documenting adequate training, and promoting a culture of quality and its relationship to ISO/IEC 17025 certification.

You should feel confident, having now checked many of the boxes to launch a state of the art cannabis and hemp testing laboratory. You have secured a great building, raised plenty of capital for the best equipment, and hired a PhD to do the work. Depending on your ISO/IEC 17025 assessor and your jurisdiction's regulatory environment, you could be a few paper shuffles from capturing your first licensed testing revenue. To estimate your chances for long term success, you—the aspiring cannabis laboratory manager—ought to ask yourself the following: am I uniquely prepared to solve the problems with cannabis and hemp testing? Knowing the answer depends on your understanding of the problems in the first place.

Challenges to Cannabis Testing

The United States is emerging from cannabis (and hemp) prohibition. But without federal quality and safety standards, state governments are obligated to reinvent approaches to cannabis regulations and enforcement. The aspiring laboratory

R. D. Brodnick (✉) · W. A. English · M. Leonetti · T. B. Anthony
Titan Analytical Inc., Huntington Beach, United States
e-mail: robert.brodnick@titananalytical.com; william.english@titananalytical.com; maya.leonetti@titananalytical.com; tyler.anthony@titananalytical.com

© Springer Nature Switzerland AG 2021
S. R. Opie (ed.), *Cannabis Laboratory Fundamentals*,
https://doi.org/10.1007/978-3-030-62716-4_6

manager may rightfully feel confused and anxious while not knowing exactly how the regulators will treat them. However, if there is any confusion regarding enforcement priorities, look no further than labs against whom the regulators have taken enforcement actions. The shutdown order below was issued by the California Bureau of Cannabis Control (BCC) during their first 3 months of cannabis testing license oversight.

Cannabis testing laboratories that cut corners, falsify results, or otherwise pose a threat to public safety should expect to face swift action by regulators to protect consumers. In the shutdown order displayed in Fig. 1, the regulators did not describe the specific violations in detail, but they clearly point to the company's unwillingness to appropriately contemplate their predicament and cite them generally for failing to prioritize public safety.

If regulatory uncertainty is the major legal and administrative risks laboratories face, then the paucity of standard testing methods is your greatest scientific challenge.[1] Cannabis testing labs are generally required by the states to develop and validate fit for purpose chemical and microbiological testing methods in accordance with ISO/IEC 17025:2017 [1] and the U.S. Food and Drug Administration's (FDA)

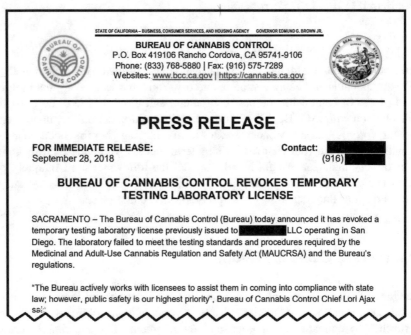

Fig. 1 September 2018 press release from the California Bureau of Cannabis Control announcing the revocation of a prominent testing lab's license to conduct commercial cannabis activity

---

[1] The AOAC Cannabis Analytical Science Program (AOAC CASP) [4] has established several standard method performance requirements in its aim to publish consensus methods for testing cannabis and hemp samples. https://www.aoac.org/scientific-solutions/casp/.

guidelines [2, 3]. Building the team needed to deploy and manage a system of Laboratory Developed Tests (LDT) for cannabis analysis deserves your best possible effort.

The industry's history of pre-regulated (black market) operations makes the quality culture within cannabis and hemp supply chains another important consideration. The public's confidence in cannabis testing laboratories' ability to protect consumers is compromised when unprincipled customer service expectations overrule ethical data generation. For example, unscrupulous laboratories may inflate Δ9-THC potency results because clients will often "shop" for laboratories who will provide results showing the highest potency. What's worse, criminals posing as laboratory leaders are incentivized to sell falsified passing results for tainted products containing Aspergillus, heavy metals, pesticides or other threats to cannabis consumer heath.

Why should the public ever trust human health to a supply chain that promotes such unconscionable violations of data integrity? Thanks to a handful of bad actors, cannabis testing labs generally have a reputation for perpetual dishonesty and incompetence. Your well-intentioned startup will need to deal with these realities.

Finally, your cannabis testing lab will live and die by your resource management decisions regarding mission critical analytical instrumentation, facilities, and staff. Even the best scientists and the most ethical leaders will struggle to accurately estimate their operating budget if the assumptions underlying it are not evidenced-based or not based on adequate cannabis testing experience. Resource management decisions that contemplate corrective actions involving matters of compliance are especially risky. Your challenge is to fully understand and accurately translate the technical risks in cannabis testing into business intelligence.

In our view, successful laboratory operations management is only sustainable with agile stewardship of three fundamental laboratory systems: staffing, training, and quality. While the challenges to cannabis testing operations are plentiful and uniquely complex, they are all still tractable with old school best practices.

Staffing

In this chapter, we refer to you as the manager of the laboratory business itself who has broad responsibility for building the company's systems. You will decide who will do the work, how it's checked, and who calls the shots under various circumstances. You also have full financial authority.

The leadership burden within high-reliability laboratories can be divided into three practical and distinct functional areas: Laboratory Directorship, Operations/ Quality Control, and Quality Assurance. The size of the staff under each group is a function of the size and scope of the laboratory. We often refer to this leadership structure as the "Laboratory Triumvirate." Like the original trio of Roman leaders Caesar, Pompey and Crassus, the Laboratory Triumvirate is a regime ruled by three powerful individuals who together work to increase their control of the operation.

The powers vested in each role are equally impactful, but like Caesar, the Laboratory Director maintains final decision making authority.

The Triumvirate

The Laboratory Director (LD) is your Commander on the ground who bears full responsibility for all scientific operations and related business outcomes. This most important hire should initially report directly to the Chief Executive Officer (CEO), or to whatever business manager has the authority to fully delegate budget execution responsibilities. While you and the corporate head shed set the culture of the business, the LD is responsible for the climate within the lab. The Director wears every hat imaginable from time to time, from the lab's cheerleader and coach, to janitor, to the lead ambassador to customers, vendors, and regulators. This individual must be competent at the bench, but more importantly, they should be the first to volunteer to stay late when duty calls. Your LD must be a values-driven leader that inspires confidence and demonstrates integrity when faced with a challenge. As a key member of the company's executive staff, the LD coordinates with corporate leadership to champion quality policies, and secures all necessary resources to complete the cannabis and hemp testing mission. Consider including the following roles and responsibilities in your LD's job description:

- Maintains an effective culture of laboratory safety; develops and implements environmental, physical, chemical and biological hazard and safety programs.
- Maintains strict confidentiality.
- Recruits, trains and leads staff to support and execute high-reliability testing operations.
- Oversees the laboratory and administrative staff performance management; manages evaluations, incentives, rewards, and performance improvement programs.
- Sustains mission critical laboratory core competencies.
- Coaches and mentors QC/Operations Manager, Quality Assurance Manager, and section leads and supervisors into roles of greater responsibility.
- Maintains client relationships.
- Manages financial, consumable, and capital resources, and proactively reports needs and fiscal health to leadership.
- Ensures strict security protocols are maintained.
- Ensures the laboratory is cleaned and sanitized to standard.
- Develops and implements a laboratory training program.
- Develops and implements a laboratory Quality Assurance (QA) program.
- Oversees the production of monthly and annual quality reports.
- Secures and maintains an ISO/IEC 17025 scope of accreditation for all required tests.
- Develops and executes on operational and quality improvement plans.

- Develops, validates, monitors, and enforces Standard Operating Procedures (SOPs) for all tests and matrices. Updates SOPs with necessary changes and revisions.
- Ensures proper document control.
- Conducts a tertiary data review and certifies data results on a Certificate of Analysis (COA). Delegates COA signature authority and accepts responsibility for all results released from the laboratory.
- Establishes policies and procedures for monitoring laboratory personnel conducting pre-analytical, analytical, and post-analytical phases of testing.
- Utilizes risk assessment and makes evidence-based decisions in the conduct of all work.
- Performs other duties as assigned.

Qualified LDs should have a doctoral degree in chemistry, microbiology, food science, agriculture, biology, environmental, or related sciences from an accredited college or university. However, practical laboratory leadership experience is often more valuable than a degree. Consider Master of Science degreed candidates with at least 2–4 years of laboratory supervisory experience, or Bachelor of Science degreed candidates with prior experience serving as a lab directors. Be sure to confirm the minimum staff qualification requirements of ISO/IEC 17025 and applicable regulations.

The Laboratory QC/Operations (QC/OPs) Manager is one of LD's two deputies. QC/OPs works under the direction of the Laboratory Director and coordinates with other executive staff to oversee and manage the technical operations and Quality Control functions of the laboratory. The QC/OPs staff plans the analytical production batches for the day, maintains and prepares control samples and Certified Reference Material standard lots, conducts secondary review of the data, and provides direct supervision of section leads and bench level staff. QC/OPs coordinates heavily with the Quality Assurance section to fulfill requests for data pertaining to audits and investigations. Consider including the following roles and responsibilities in your QC/OPs Manager's job description:

- Maintains an effective culture of laboratory safety; implements environmental, physical, chemical and biological hazard and safety programs.
- Maintains strict confidentiality.
- Tracks mission critical laboratory core competencies.
- Coaches and mentors supervisors, section leads and bench level staff into roles of greater responsibility.
- Maintains client relationships.
- Proactively communicates financial, consumable, and capital resource needs to the Laboratory Director.
- Ensures strict security protocols are maintained.
- Ensures the laboratory is cleaned and sanitized to standard.
- Ensures compliance with ISO/IEC 17025 and applicable testing regulations.
- Recruits, trains and leads staff to support and execute high-reliability testing operations.

- Oversees laboratory technical staff performance management; manages evaluations, and makes recommendation for incentives, rewards, and performance improvement programs.
- Directly oversees the development, validation, monitoring, and enforcement of Standard Operating Procedures (SOPs) for all tests and matrices. Makes recommendations for SOP updates to the Laboratory Director and the Quality Assurance Manager.
- Ensures proper document control.
- Develops and implements the laboratory training program.
- Manages all technical operations for the laboratory. Directs the analytical production batches for the day.
- Ensures the proper calibration, function and preventive maintenance of instruments and equipment.
- Executes policies and procedures for monitoring laboratory personnel conducting pre-analytical, analytical, and post-analytical phases of testing.
- Conducts a secondary review of bench generated data. Certifies data results on a Certificate of Analysis (COA), and accepts responsibility for all results released from the laboratory.
- Ensures compliance with established Turnaround Times (TAT), and ensures COAs include all accurate, pertinent, and required information for interpretation.
- Oversees the Quality Control functions of the laboratory. Maintains and prepares control samples and Certified Reference Material standard lots.
- Reviews all quality control results, and documents corrective actions in response to control failures.
- Identifies and documents non-conforming events and supports Quality Assurance investigations for Corrective Action and Preventative Action (CAPA) procedures.
- Develops and executes on operational and quality improvement plans.
- Utilizes risk assessment and makes evidence-based decisions in the conduct of all work.
- Under the direction of the CEO, may temporality assume responsibility for all laboratory operations in the absence of the Laboratory Director.
- Performs other duties as assigned.

Laboratory QC/Operations Manager candidates should have at minimum an advanced degree in the sciences plus at least 2 years of full-time supervisory experience, and should otherwise have exceptional leadership abilities and be able to step up and lead the lab in the absence of the Director.

The cannabis testing lab's Quality Assurance (QA) Manager works at the direction of the Laboratory Director, but should maintain parallel lines of communication with top managers of the business itself. While the Ops/QC Manger is your #2 leader of people and processes, your QA Manager is responsible for the most impactful activities you will put into practice under your ISO/IEC 17025 Quality Management System (QMS): continuous risk assessment, auditing, and documenting your response to errors and complaints. The QA Manager has the authority to stop work in the laboratory. Consider the following roles and responsibilities for QA job description:

- Maintains an effective culture of laboratory safety; develops and implements environmental, physical, chemical and biological hazard and safety programs.
- Maintains strict confidentiality.
- Maintains client relationships.
- Manages financial, consumable, and capital resources, and proactively reports needs and fiscal health to leadership.
- Ensures strict security protocols are maintained.
- Ensures the laboratory is cleaned and sanitized to standard.
- Develops and implements a laboratory training program.
- Tracks all employee certifications. Prior to testing, certifies that the staff is properly trained, that they have the appropriate education and experience, and that they can perform all testing operations reliably according to SOP.
- Develops and implements a laboratory Quality Assurance (QA) program. Reviews and interprets laboratory records in all areas, and reports findings to the LD and corporate Head of Quality.
- Builds the ISO/IEC 17025 quality management system and maintains a scope of accreditation for all tests.
- Produces monthly and annual quality reports.
- Develops and tracks operational and quality improvement plans. Reports on quality improvement initiative progress and generates new continuous improvement projects.
- Develops, validates, monitors, and enforces Standard Operating Procedures (SOPs) for all tests and matrices. Updates and publishes SOPs with necessary changes and revisions.
- Maintains proper document control for the laboratory.
- Conducts retrospective data review and investigates non-conformances.
- Establishes a schedule for auditing all laboratory procedures and products related to pre-analytical, analytical, and post-analytical phases of testing.
- Develops, leads, and completes Corrective Action and Preventive Action (CAPA) procedures. Leads investigations of Non-Conforming Events (NCEs). Completes Corrective Action Reports (CARs) and Preventive Action Reports (PARs).
- Conducts a tertiary data review and certifies data results on a Certificate of Analysis (COA) at the direction of the CEO from time to time. Accepts responsibility for results released from the laboratory.
- Utilizes risk assessment and makes evidence-based decisions in the conduct of all work.
- Under the direction of the CEO, may temporality assume responsibility for all laboratory operations in the absence of the Laboratory Director.
- Performs other duties as assigned.

Cannabis testing lab QA Managers are uniquely challenged by the technical diversity of tests, the lack of consensus methods in the industry, and the industry's fledgling culture of quality. The incumbent will help carry the torch for the lab's quality culture and policies, so recruit carefully. The QA Manager has the same education requirements as the QC/OPs position; however, previous experience leading an ISO/IEC 17025 QMS, organizational skills, and an obsessive passion for quality management are imperative.

Technical Sections

The technical sections within a well run cannabis testing laboratory are stratified into discrete, manageable workflows (Fig. 2). Each section lead or supervisor provides daily open- and shift-readiness reports to the QC/OPs Manager that necessarily include updates on workload, supplies, equipment, and personnel. QC/OPs then takes a daily account of all of the day's incoming tests, staff, and resources, and prioritizes the work. Let's follow along with a sample in the testing process:

> *Under strict chain of custody procedures, our sample is collected by a courier who is trained in statistical batch sampling and aseptic collection. The accompanying test orders are entered in the Laboratory Information Management System (LIMS) by the intake section back at the lab.*

The sample intake's section's primary mission is to maintain accountability of the sample, which is of considerable economic value and evolving legal status. Intake ensures the accuracy of all administrative data and places the samples into temporary storage to await testing. The intake section is also the lab's customer service call center and cashier. Cash transactions are still common in cannabis business thanks to restrictive Federal banking laws. Fill these roles with dependable professionals like medical office workers, bank tellers, retired police officers, and Veterans.

> *QC/OPs reviews the tests ordered in LIMS: HPLC-MS trace-cannabinoid potency, and Aspergillus PCR screening. QC/OPs assigns our sample to appropriate test method batches, and delivers the batch control samples to the processing section.*

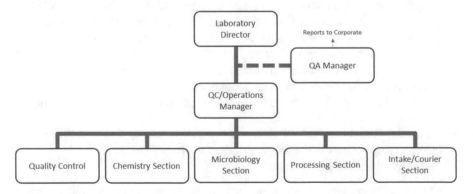

Fig. 2 Organizational chart for cannabis testing laboratory technical operations

QC/OPs is the Quality Control department. Assign your most talented and responsible bench level chemists and microbiologists to the roles of creating and tracking laboratory controls, calibrators, and reference materials. These are the scientists that loved spending their weekends in the lab back in school. They are commonly great chemistry and microbiology section leads, and hopefully handy with a wrench.

> *The processing section assembles and processes the batches, and delivers them with controls inserted to the chemistry and microbiology sections for analysis. The sections perform their analysis, conduct primary data reviews, and report the results back to QC/OPs.*

Processing chemists and microbiologists will perform physical measurements and sample homogenization, but can also manage DNA extractions, dilutions, internal standard addition, etc. The processing section will need a handful of hard-charging technicians, especially those adept at programming robotic liquid handlers. The processing section also provides a venue to train interns, and newly recruited non-technical staff who would benefit from a couple weeks of laboratory experience.

So far, we've described a Triumvirate-led organization with key roles and responsibilities that are a perfectly good starting point for nearly any non-clinical testing laboratory; however, successful management of your cannabis testing lab will require careful planning to recruit a myriad core competencies that are not commonly assembled within other testing industries.

The Problem with Cannabis Testing Laboratories: Competency

No matter the product tested, most routine laboratories have common characteristics: advanced degree scientists, equipment to test products quality and/or safety, and systems to publish results via certificates of analysis (COAs). However, each routine testing industry clearly has its own unique technical requirements, or what we will refer to as 'core competencies'.

As a thought experiment, let's imagine you are a hospital clinical lab that needs to add new Coronavirus testing services. Take an inventory of your readiness. You have multiple compatible test platforms, and strong vendor relationships. You have a staff of highly trained medical laboratory technologists who can handle validation, and a world class quality staff that keeps us compliant with ISO 15189, the Clinical Laboratory Improvement Amendments (CLIA), the College of American Pathologists (CAP), the Joint Commission, and the Food and Drug Administration (FDA). Your customers are still the doctors and patients, and we know exactly how to communicate our results to them. Sample surges would be no sweat given your stable LIMS, automated SOP-driven work environment, and staff with experience managing everything from testing bottlenecks to quality initiatives to mitigate mislabel events. You have the knowledge, skill, and ability to agilely adapt the operation to fit the new need. It should not be a surprise that your clinical lab was well prepared to expand to COVID-19 testing needs.

Now let us assume your experience is in forensic drug testing. You don't have any Medical Laboratory Technicians or Medical Technologists; is that a problem for your COVID lab plan? Your chemists can probably safely and aseptically process biological patient samples. You constantly defend chemistry data and quality systems in court, but would you need a virologist to design your validation plan? Can your current LIMS adequately protect patient identifying data? Do you need CLIA waivers from the Centers for Medicare and Medicaid Services (CMS)? You will need to make heavy investments in people, quality systems and processes to join this fight. A sudden transition to coronavirus testing might be a challenge for forensic toxicologists. Could you instead reposition your assets for food testing? Not without hiring an army of microbiologists, and you better know what H.A.C.C.P. stands for. Those coming to cannabis testing from neighboring laboratory industries face similar learning curves.

There are several ways to recruit the Triumvirate and bench staff for cannabis testing. Strategically recruiting skill sets from adjacent industries can add value to key hiring decisions by checking off multiple competencies. In Fig. 3, we score familiar testing verticals' relative readiness to begin reliable cannabis testing given typically existing core competencies. Individual laboratory operator candidates should be scored similarly.

Quality Management

Your intent to operate a high-reliability cannabis laboratory will be reflected in your key quality management recruiting priorities. Consider laboratory leadership candidates with a scientific degree and ISO/IEC 17025 (or ISO 15189) experience. Clinical laboratories and the pharmaceutical and food supply chains breed quality champions with natural instincts to perform quality functions in the cannabis lab space. *Pro tip: Prioritize candidates who have led laboratories through ISO/IEC 17025 assessments, and who have quality-specific training such as an American Society for Quality (ASQ) certification.*

	Quality Management	Highly Regulated Testing	Legal Data Defensibility	Analytical Chemistry	Food Microbiology	Food Safety Expertise	High Throughput/ Automation	Supply Chain Empathy
Cannabis Labs	✓	✓	✓	✓	✓	✓	✓	✓
Clinical	✓	✓	☐	✓	☐	-	✓	-
Forensic Toxicology	☐	✓	✓	✓	-	-	✓	-
Pharmaceutical	☐	✓	☐	✓	☐	-	☐	☐
Food/Ag	-	-	-	☐	✓	✓	☐	✓
Environmental	-	☐	-	✓	☐	-	☐	☐

Fig. 3 Examples of established lab industries' varying readiness for competency-driven high reliability cannabis testing

Highly Regulated Testing

Regulated cannabis may be the most tested product pound-for-pound available for human consumption. Hundreds of contaminants and quality attributes are tested under varying rules across legal cannabis jurisdictions. Cannabis regulatory bodies are as new as the labs they regulate, and regulators do not typically have the resources to investigate each corner of every laboratory. Regulators therefore settle into a "trust-but-verify" mode with most laboratories, and reserve targeted enforcement efforts for operations that inspire less confidence. Cannabis labs that suffer license revocations generally demonstrate an underappreciation for transparency, and miscalculate their bargaining positions. *Pro tip: Trust lab Directorship to those with a proven ability to communicate with regulators competently, respectfully, and with a sense of urgency. Be sure your leadership on the ground is always ready to host the occasional surprise audit.*

Legal Data Defensibility

When traditional food safety risks and cannabis-borne illnesses lead to cannabis product recalls, your lab ought to be prepared to defend the integrity and support-ability of its data in formal legal proceedings, which can take several forms: perhaps a client who hired you to produce a COA for their cannabis product has had that product recalled because it was shown to contain a substance your laboratory's analysis confirmed was absent, or perhaps a state regulator has found your sample destruction protocols are out of compliance with state requirements.

Whether facing a regulatory shut down order or defending against a lawsuit, the function and performance of your complex testing and quality systems must be thoroughly documented but translated into relatable, non-technical concepts for judges, juries, and administrative hearing officers. Relevant evidence includes Standard Operating Procedures (SOPs), validation reports, control charts, Corrective and Preventive Action Reports (CAPAs), audit records, Good Laboratory Practices (GLPs), and the lab's record of compliance with ISO/IEC 17025, among others. Allegations of incompetence or negligence asserted by an opposing party must be met with an evidence-based strategy and closely coordinated with your company's lawyers. Your lawyer's challenge is to convey the lab's intent in the face of errors or missing documents. Your future may hang in the balance of your expert's testimony during a hostile cross examination, and there is no substitute for having experienced counsel, knowledgeable expert witnesses, and prepared business principals when facing a potential legal judgment. *Pro tip: Hopefully you will never need to defend your laboratory in court; however, having a qualified lawyer and expert witness on speed dial can be helpful, including to provide training for junior quality assurance and supervisory staff.*

Analytical Chemistry

Your chemistry section needs more experience than an advanced degree alone, but a Master of Science or Ph.D. from a reputable analytical chemistry program is a good place to start. In addition to being able to educate the rest of the organization on equipment maintenance, separations, and data analysis, your senior chemistry staff must be able to develop and validate in house chemical methods per the FDA's guidelines. You could instead hire chemistry consultants to develop and validate your scope's initial testing methods, but beware that new validation requirements will continue to surface as the lab is required to comply with ever-evolving standards and matrices. It's better to have a team that learns to "catch its own fish." If you hire a consultant to help with validation, be sure that they are also contracted to help staff and train your bench with permanent qualified support. *Pro tip: You alone are responsible for learning all your consultants have to offer while on site, so be aware that this core competency will evaporate once their contract ends if you aren't considering your lab's technical independence.*

Food Microbiology

Microbiological testing requirements on cannabis vary among jurisdictions with early regulators requiring only product safety indicator tests like total yeast and mold count. More recently, detection of specific pathogens like Aspergillus species, Salmonella, and shiga toxin producing *Escherichia coli* (STEC) have been required. The experience and skill required for cannabis microbiology can be recruited from the upper technical echelons of the food testing industry. The cannabis lab microbiology section must be able to design and execute a validation plan for cultural and molecular LDTs, and similarly evaluate manufacturer test kits if not yet AOAC International or United States Pharmacopeia (USP) approved for detection on the cannabis matrix of interest. Build your microbiology bench to manage your biological safety program as well as your in-house environmental monitoring program. *Pro tip: Recruit and cross train courier/sampling technicians with bench level microbiologists to ensure statistically relevant aseptic sampling is conducted throughout the preanalytical phase while adding flexibility to address intake surges and bottlenecks.*

Food Safety Expertise

Medical and lifestyle cannabis products have different risk profiles, and questions remain concerning the FDA's inevitable approach to cannabis product regulation. Pharmaceutical-style Good Manufacturing Practices (GMP) regulations may govern medical cannabis products one day, but any U.S. based facility producing edibles for consumption may already be required to register with the FDA under the Bioterrorism Act, and could be subject to FDA enforcement under the Food Safety Modernization Act (FSMA). If you can anticipate having edible producers as clients, consider the value of a Preventive Controls Qualified Individual on staff to help clients catch up with existing food safety and quality standards. Sophisticated edible manufacturers with existing food safety programs have begun to expect cannabis labs to conduct nutritional analysis and generate FDA compliant labels. *Pro tip: The cannabis lab of the future will likely contain a full scope of standard food testing methods, so know how your testing results fit into your clients' Hazard Analysis and Critical Control Points (HACCP) plan, and be prepared to offer value added studies like chemical and microbiological endpoint shelf-life studies, process validation studies, and microbial challenge studies.*

High Throughput Testing and Automation

Naturally, local cannabis supply lines have become commoditized in jurisdictions where supply keeps up with demand. If our industry continues to mature toward a supply chain that trends like food, cannabis testing margins will likely be squeezed like they are in highly competitive food testing labs. Automation is not critical for the cannabis lab start-up, but it will quickly become a necessary core competency for labs that wish to achieve respectable testing volumes at a reasonable time and cost. To prepare for victory in a sustained price war, your laboratory must look to automate repeatable processes. A high-quality, high-throughput lab is not only physically automated, but the facility layout and workflow is optimized for efficiency. Moreover, the cannabis lab is required to capture and handle massive amounts of machine generated data for an extensive list of methods and analytes. We discuss the value of LIMS-driven cannabis lab workflows and data handling in detail later in this chapter. While machines are not supposed to call off work or make sloppy mistakes 7 h into their Friday shift, be advised, fancy automation of a garbage process will *precisely* produce garbage results. Like any other standard process in the lab, automated processes must be designed for purpose, validated, and consistently monitored for performance. *Pro tip: Pick the right LIMS solution and recruit a staff that has automated processes in high throughput forensic or clinical toxicology labs, for example.*

Supply Chain Empathy

Cannabis labs are required in many jurisdictions to provide couriers that physically visit with the supply chain to conduct sampling for compliance testing. Even when a lab pickup is not required, offering the service is a very good idea because it puts a personal touch on the front end of your business. Exceptional technical customer service delivered with a soft touch and a sense of urgency strengthens crucial relationships with margin-driven, quickly expiring supply chains like food and cannabis. *Pro tip: Ideally, you'll have a staff that has walked in your clients' shoes and understands their challenges and decisions; however, if cannabis supply chain experience is not available to you, look to the expectations and behaviors in the food industry to calibrate your approach to customer service and sense of urgency.*

Cannabis

What about experience in *our* industry? If all of the core competencies are covered on our roster, is cannabis experience still important? Most definitely. Delivering useful data to the cannabis supply chain is unlike anything experienced in most other industries. As public support for the cannabis and hemp industries increases, highly qualified talent is beginning to take note of the myriad opportunities created in these industries. However, not every one of your clients was managing food distribution, running a testing lab, or scaling pharmaceutical active ingredient synthesis in a previous career. Your lab must be trusted to help interpret scientific results, especially if your client lacks the technical background to assess product liability risks independently. You must be technically competent, solution oriented, and empathetic during a difficult call explaining a surprise product failure, lest you risk losing that client for the wrong reason. *Pro tip: Honor the cultural realities and traditions of the cannabis supply chain by making every effort to recruit high-level, qualified cannabis experience to lend perspective and translation to difficult customer facing conversations.*

Solving the competency problem is not an easy task. Budget limitations may require you to make tough, risk-based hiring decisions early on. *Pro tip: Recruit C-suite and business partners with leadership experience across the competencies who can fill the ranks with trainable junior recruits at a sustainable cost.*

Training Program (Employee Certification)

A proper training program is vital to the success of a cannabis laboratory. Similar to how we run batches outlined in a method validation plan to validate a testing method; your training program is where you are going to validate that your personnel have the knowledge, skills, and abilities (KSA) to perform those test methods.

There is no professional organization, certification program, or educational degree that automatically qualifies an employee as proficient in testing cannabis. This burden falls squarely on the shoulders of the laboratory leadership. Assume this burden willfully and cheerfully. Training is usually the first extended contact you will have with a new employee. It is a time you can impart your standards and expectations. Starting the training with well written SOP's, work instructions, and checklists is vital to establishing an environment of professionalism. Now is not the time to "wing it," this cheats both the employee and the laboratory.

Have a Training SOP

A Training SOP is similar to a Validation SOP. It should answer the generic questions one would ask about your training plan. Such as:

- Who can train a new employee?
- What is in a practice/validation batch?
- How many batches need to be prepared?
- What is the passing criteria for a practice batch?
- Where will results be recorded?
- Does certification expire?

Specific to each test method your Training SOP should contain training checklists. The checklist should contain a bullet point training item, the trainer's initials, the trainee's initials, and training date for each item. For example:

Item	Trainer	Trainee	Date
Location of safety equipment	WAE	RDB	16 July 2020
Location of dry chemicals	WAE	RDB	16 July 2020
Location of solvents	WAE	RDB	16 July 2020

The simplest way to start a training checklist is to go through the method SOP and make a bullet point for each major step. Include safety, chemical/biological hygiene, preventive maintenance, and Quality Checks in addition to all the batch preparation and testing steps.

Emphasize Quality Checks throughout training. The bench level employee is your first level quality manager. The majority of non-conforming events should be caught by this person. Therefore, they must be proficient with the established quality norms. They will also start the majority of Corrective Action Reports. List bench-level quality assurance actions as checklist items and designate QA as the trainer for these tasks.

The trainer should train from the method SOP. In well-established labs it is common to find out while training a new employee that the method has "drifted" from the SOP. Use this time to double check practice versus SOP. If the SOP needs to be changed, then change it. If lab practice needs to be corrected, then correct it. Documented either in accordance with your QMS rules.

Have a Written or Oral Test

Q: When should I test the newly trained employee?

A: When you trust them to perform the test without supervision. All reported results should come from data generated by fully certified employees. If your lab allows a trainee to work under the supervision of a certified employee on customer samples, then the supervisor is responsible for ALL of the trainee's actions and will be the employee of record in your LIMS.

Q: How hard should the test be?

A: The lab is certifying that the employee has the KSA's to generate reportable data. Someday an inspector is going to look over the completed test. The test should be difficult enough to assure an inspector that the employee possesses the requisite KSAs.

Q: What if the test is given orally?

A: The tester is then responsible for generating a Memorandum for Record (MFR) that details the topics covered and the employees strength/correctness of response.

Q: Should the test be graded?

A: Yes. The standard for success should be in the Training SOP.

Q: Is giving a test required by law?

A: No. It is optional but having a complete record of your employees' training records will come in handy if the results they help produce ever come into question.

Q: Why are you recommending this to us?

A: We have decades of leadership experience that unfortunately required the management of high-risk crises like: miss-reported results, catastrophic lab failure, malfeasance, and employee death. In the Root Cause Analysis (RCA) that follows, we have seen the following play out over and over:

Leadership: "We didn't train you that way!"
Employee: "Yes you did! We've always done it this way!"

Think of your training program as the first step in your risk mitigation plan. A well laid out plan not only mitigates risk through proper training, it puts the employee on notice that we have documented that we trained you properly and we are going to hold you to the standard we trained you to. It also puts the trainer on notice that their reputation is on the line as well. The trainer is putting their name on the line that they trust the trainee to generate supportable data.

There is another reason to train to a level higher than required. Pride. Is your lab the best? How good are your employees? How would you know? It's not just bragging rights for you, the employee also gets something out of it. The harder certification is to obtain, the more pride the employee will take in it. Pride is directly proportional to earned accomplishment.

Demonstrate Competence

In addition to a knowledge test, certification should include a demonstration of competence with actual samples run in a quality/testing batch. The samples should have a known amount of analyte. They can be spiked control material, customer samples that have been "destroyed" in the inventory, or Certified Reference Material.

The batches should be set up in accordance with SOP just as a batch of customer samples would be.

While the trainee is preparing, running, and analyzing the batch the trainer should be silently observing the trainee's adherence to the SOP.

All batch data and results should be saved for inclusion in the Employee's Training Folder.

The standard for pass/failure should be directly stated in the Training SOP.

Record Competence

Competence is recorded in an Employee's Training File. For each method the file should contain:

- The original training checklist
- The original written test or oral test MFR
- The certification batches with a worksheet or MFR compared to expected results.
- MFR from senior laboratory leadership granting competence. Include:

  - Test/Matrix competence is granted for
  - Date granted
  - Date or condition for expiration

- Other Supporting documentation

Competence cannot be assumed without documentation. An employee with an advanced educational degree or prior work experience may not meet all of the requirements that your Training Manual outlines. If management accepts that an employee's education and experience meets some or all of the requirements then management must document, in the employee's training file, which tasks on which methods were assumed.

A common practice is to award "competence" to the bench level employee that validates a test method. This first employee then becomes the trainer for subsequent employees for that test method. This must still be documented in the employee's training file.

Reward the Employee

At the end of the certification process all of the following should be met.

- The Employee has confidence in their KSA's
- Management has confidence in the employee's KSA's
- Through documentation, any future inspector has confidence in the Employee's KSA's

At most labs, this is the end of the process and the employee goes on to generate supportable data.

Our challenge to you is to be better than that. Be a leader, not a manager. Your employee has accomplished something, reward them. Options include:

- Removal of probationary status
- Monetary raise
- Praise at the next all-hands meeting
- A special patch for their lab coat
- Take the team to lunch
- Buy them a cake

If you want to establish the highest-level employee satisfaction present the employee with a Certificate of Accomplishment. This should look like a diploma and be signed by a PhD and/or a company officer. Accompany with an MFR that details the scope of training. Many object to this method of reward by, correctly, pointing out that you have given your employee something they can take to a competing lab for immediate employment. This is true, you have made it easy for an employee to leave. They will be back. The best teams attract the best players, the elite colleges attract the smartest students, and the best companies attract the best employees. Trust us, if your company has refined operations to the point where you are acting as the industries unofficial credential granting agency, you have already won.

Quality

Dying in the hospital because a laboratory technician carelessly swapped your specimen with someone else's is clearly unacceptable. The Joint Commission, which is the driver for quality improvement and patient safety in healthcare, would characterize such a catastrophic clinical failure as a "Sentinel Event" [5]. As such, highly redundant and heavily audited systems of pre-analytical checks are standardized and trained into a state of high-reliability to keep innocent sounding missteps from harming patients. However, clinical lab operations did not transform into a high-reliability industry overnight. Starting the late 1980s, clinical laboratories underwent an industry-wide standardization effort under the Clinical Laboratory Improvement Amendments (CLIA). Administered by the Center for Medicare and

Medicaid Services (CMS), CLIA established quality standards for laboratories to ensure the accuracy, reliability, and timeliness of patient test results regardless of where the test is performed [6].

The pharmaceutical industry saw major legislative changes to its quality standards decades earlier. The thalidomide tragedy in the late 1950s led to the Kefauver-Harris Amendments which established the scientific safeguards used today by the FDA to protect consumers from unsafe and ineffective medications [7]. The FDA began requiring Good Manufacturing Practices (GMP) that assures proper design, monitoring and control of manufacturing processes and facilities, and Good Laboratory Practices (GLP) for sustainable, reliable testing laboratories [8].

More recently, food testing laboratories were invited, again by new federal legislation, to embark on a quality-focused transformational journey. According to the Centers for Disease Control estimates, each year roughly 1 in 6 Americans (or 48 million people) gets sick, 128,000 are hospitalized, and 3000 die of foodborne diseases [9]. To counteract alarming figures such as these, the Food Safety Modernization Act (FSMA) was signed into law in 2011. Under FSMA, the FDA regulated supply chain began an industry wide transition to a more proactive, risk-based approach to food quality and safety. In addition to requiring human and animal food manufacturers to assess risk and put preventive controls in place, FDA inspection efforts increased significantly, and a rule was made requiring food testing laboratories to maintain ISO/IEC 17025 accreditation. Although the FSMA food lab accreditation rule is still pending, most food testing laboratories voluntarily embrace the spirit of ISO/IEC 17025, with many using it to enable continuous improvement as intended. It probably doesn't hurt that food producers do a lot more testing to comply with FSMA, which motivates food testing quality improvement with a competitive carrot instead of a heavy handed regulatory stick.

By mid 2018, state regulators in newly established legal cannabis markets were focused on preventing black market diversion, and ensuring licensees complied with established administrative processes. Before some of the California labs could pronounce "calibration", test packages were selling at $800, even $1200 apiece to perform all mandated tests to release 10–50 lb of cannabis flower to retail. The regulators may have assumed that an ISO/IEC 17025 accreditation was an adequate technical guarantee that their licensed laboratories' methods were fit for purpose and fully validated for every test matrix. They would be 100% correct. However, to this day the industry lacks standard testing protocols, making cannabis labs responsible for deploying LDTs for pesticides, mycotoxins, heavy metals, residual solvents, salmonella, and more. A long list of labs were awarded licenses to test, but only a small fraction of those labs have timestamped validation reports predating their entire cannabis testing scope of accreditation. Similar chaos and controversies mark the regional histories of cannabis lab development across the country.

Thanks to a minimal ISO/IEC 17025 effort and lucrative margins (fueled by a dizzying list of required tests), cannabis testing labs are not externally incentivized to improve beyond prescribed minimum procedural standards. This presents the quality-minded laboratory manager with a unique opportunity for differentiation. Invest now in procedures for data defense and continuous improvement rather than waiting on competitors or lawmakers to raise the cannabis lab quality bar for you, or worse, lead you with the stick.

> *Quality Management Principles (QMPs) are the backbone of your cannabis testing laboratory QMS. The systems you build and the actions you take should be consistent with the fundamentals of quality.*

Vignette of a High-Reliability Cannabis Testing Operation

Your world class cannabis laboratory's quality policies are set by the CEO and evangelised from the top-down. Staff performance expectations are tied directly to the organization's quality objectives from the C-suite to the bench. The ISO 9001 Quality Management Principles (QMPs) are baked into laboratory policies, operational plans, and the culture. Test method validation reports are meticulously generated for the sake of defending the data rather than for myopic purposes like obtaining a testing license. The quality culture is reinforced through all processes including training, and is synchronized with your mission, vision, and values. Risk is carefully assessed with every error and decision, and your lab has the distinct and rare reputation for being highly reliable at testing cannabis. Your team gushes with pride with each passed audit and ethically closed corrective action, and you celebrate completed process improvements with Quality Champions. Experienced staff are fully engaged within a quality system designed to improve, which translates to a self-learning, innovative, and efficient high-reliability cannabis testing operation with exceptional client retention.

> *Cannabis Lab QMP Highlight - Customer Focus. Supply chain empathy is critical for sales driven cannabis labs. However, the best way to support your customer is with accurate and supportable results. Prioritize laboratory quality improvements along with bedside manner for the most favorable customer service and client retention outcomes.*

Cannabis is both a lifestyle product and one provided to medical patients, many of whom are the most vulnerable among us. Behind the facade of patient-focused themes, few of the cannabis industry's laboratories truly incorporate the tough lessons learned from healthcare and pharmaceutical disasters, and many lack the benefits of modern food safety culture. Sentinel events aren't really considered in cannabis testing operations today, but what if your lab is accused, perhaps wrongly, of botching a test with mortal downstream consequences? You need to be sure your records show the true story.

Policies, Procedures, and Programs

Of the many laboratory policies you will promulgate, few will enhance your chances for success more than your Quality Policy. Your people will know that you are serious about quality because you took the time to publish it on page one of the employee handbook, and you verify that it has been communicated and understood at quarterly all-hands training events. While the Quality Policy adds context to your well crafted Mission, Vision, and Values statements, it is the procedures within your QMS that can make or break your opportunities for success and improvement. Having a quality manual is no longer required with the 2017 republication of ISO/IEC 17025; however, it is in your best interest to design and detail programs to identify, respond to, and prevent errors, or otherwise mitigate avoidable operational risk.

As discussed, LDT method design and defense is the toughest challenge that *de novo* lab managers face, so it's incumbent upon you to verify the organization's understanding of FDA validation standards, ISO/IEC 17025 fit-for-purpose requirements, and some basic best practices from the start. Publish an FDA-style chemical and microbiological method validation SOP and include guidance on method modifications, matrix extensions, and periodic performance reverification. Decide how often methods will be recalibrated including updates to quantitation and detection limits, and establish rulemaking for upper linear limits, dilutions, and repreparations. While we're on the topic of methods, be sure to validate an onsite sampling SOP and have it assessed for accredited scope inclusion.

Obviously, the instruments need to keep running, so consider risk management policies on capability redundancy requirements, preventive maintenance schedules, vendor service contracts, and new versus refurbished equipment acquisitions.

Cannabis Lab QMP Highlight—Relationship Management. As with the food supply chain, the cannabis industry operates in a high-context culture, and a failure to maintain good working relationships can translate to mission failure. Take your vendors to dinner sometimes. Give your customers access to technical scientific staff when issues arise. Don't give your landlord a reason to sue you. Be a professional.

As with your staff training and certification program, we prefer to *certify* methods on each individual instrument for service once the QA team has double checked validation, reverification, calibration, limits, and maintenance logs for QC/Ops.

The Triumvirate should review and endorse all newly published technical SOPs and work instructions prior to promulgation and staff training.

There are a number of non-testing and administrative workflows that should be codified to keep the operation compliant, such as Chain of Custody procedures, and controlled substance and sample retention inventories. One critical workflow that helps ensure the defensibility of the data itself the laboratory's tertiary data review process. Your tertiary review flow chart should start with the primary reviewer, often the instrument analyst, who reviews chromatographic acceptability (raw data), method acquisition parameters, timestamps, sequence tables, control performance, and otherwise checks for nonconformances. The secondary data review is performed by the QC department, who rechecks the data and makes decisions on rework, dilutions and standard corrective actions, and passes-on the data packet and draft COA for tertiary quality assurance review. The tertiary data reviewer certifies that all systems, methods, and personnel involved with testing the sample are valid and defensible, and that all administrative data are correct and client requirements are met prior to COA release.

A number of laboratory processes can be automated to enhance reliability, like liquid handling for validated extractions and dilutions. Automated liquid handling processes can make your lab more efficient by reducing errors without breaks, and may improve your method measurements of uncertainty thanks to the added precision. While your physical automation procedures are important, your LIMS can be used to automatically detect test results that are out of specification, and drive quality processes in response, too.

LIMS

The cannabis lab's Laboratory Information Management System (LIMS) should enable high-reliability quality practices. Due to the sheer volume of data your lab will collect and the high likelihood of investigations, we recommend that you either develop your own LIMS, or generate a wish list for a commercial platform to drive high-reliability workflows and audit-friendly data management. For example, your LIMS requirements document might include:

- *Electronically automates and drives testing workflows*

  - Customizable dashboards for lab broadcast
  - Bench sheets and other workflow forms provided
  - SOPs embedded with intuitive step by step instructions
  - Tablet-compatible electronic notebooks with barcode scanning capability
  - Facilitates sampling and courier intake processes
  - Tertiary data review and release processes are driven and tracked
  - Collates standardized and custom reports and COAs

- *QC/Ops Dashboard*

  - Tracks all control material, media, consumable and reagent lots that have any impact on method performance
  - Sample barcode verification is married with QC data and work orders to generate analytical instrument sequence tables
  - Customizable control charts automatically produced
  - Audit trails established for every SOP-driven process
  - Dilution and rework orders available to management

- *QA Dashboard*

  - Enables Corrective Actions and root cause analyses
  - Generates memoranda for record and other custom reports
  - Links related data sets like reagent lots, customer results, instrument-specific records to facilitate audits and NCE root causes analysis (RCA)

Audits

Mistakes are opportunities to improve, but they can only be found if you are open to looking for them, and the best way to do that is by rolling up your sleeves and grabbing a clipboard. The job of an auditor is to verify that the laboratory work processes are carried out in accordance with written instructions, and that the work products meet the documentary evidence standards required by the laboratory's stakeholders. More frequent audits can reduce the scope of errors identified; by virtue of less time having passed and fewer samples being run since everything checked out last time, less damage may have occurred since the departure from conformity. However, if everyone is auditing, nobody is testing. Balance your auditing efforts with the risks, and post the audit schedules next to the control charts where section supervisors should be spending at least part of their day. Deputize all staff members as quality auditors from time to time, and grade them on their performance.

Work Product quality audits verify that products (in our case, paper and electronic files) meet designed specifications and other quality metrics related to production of defensible data. These include any generated reports and records collected in the normal course of laboratory operations. Be sure to verify analytical traceability by establishing the work product audit linkages between laboratory results, calibrators, controls, and reference material records. Consider regular Work Product audits of these records:

- Standard operating procedures
- Work instructions
- Training files
- Maintenance logs
- Data packets
- Badge access records
- Customer communications

Work Process audits verify that the actions carried out in the normal course of laboratory data generation meet documented quality standards. In other words, Work Process audits reveal whether or not written instructions are being followed.

Synchronize section Work Process audit schedules to follow recent method validations and other events requiring SOP revisions to ensure the newest processes to the organization are on track. The resulting audit reports can be used to bolster new method validation files with the new processes, and training program demonstrations of competency.

> **Cannabis Lab QMP Highlight—Process Approach.** *Mitigate avoidable risks by bringing order to chaos. Stratify discrete data generating systems and steps, and monitor performance against established metrics.*

One of the many formal roles entrusted to QA is that of audit manager, both as the facilitator of internal QMS reviews, and as the chief liaison to external auditors, investigators, and assessors. Have QA establish audit schedules, like the annual

Work Process audit plan in Fig. 4. Note that the audit can be completed by any staff member QA deems to have the appropriate knowledge, skill, and ability to determine whether the steps within the process match what's in the SOP.

Comprehensive internal audits are like practice drills for the big game. When an external visitor enters your laboratory to determine its reliability, your staff's preparedness will be reflected in their performance. Just like with football's best athletes and teams, the Super Bowl is just another day to execute on the playbook, but there is no denying that the stakes can be highest at the kick off of an external visit. Transparency, respect, and trust are prerequisites to success during an unscheduled visit for cause, so encourage the Triumvirate to make special efforts to maintain strong professional relationships with assessors and regulators. Invite clients to visit and audit the lab as well, and buy them lunch. Give your customers an open invite and full access to all of your audit reports and corrective actions, method validation and client specific data, and let them eat with your staff in the breakroom. They should walk away from the lab knowing you are their high-reliability testing partner, in it for the long haul.

External auditors commonly establish data quality by working backwards from the data packet for a reported sample. By starting with the certificate of analysis, an auditor should be able to connect all the documented administrative data, tests, control statistics, employee training records, and method validation data to the sample. Thankfully, QA has handy all of the necessary internal audit findings, corrective

2020 Work Process Audit Schedule	J	F	M	A	M	J	J	A	S	O	N	D
Sample collection												
Homogenization												
Foreign Material Inspection												
Moisture Analysis												
Cannabinoid Testing Batch Prep	ML - 21JAN2020											
Cannabinoid Testing Results Analysis		RDB - 06FEB2020										
Microbiological Testing Plating												
Microbiological Testing PCR											AR - 30NOV2020	
Microbiological Testing Results Analysis												
Metals Testing Batch Prep												
Metals Testing Results Analysis												
Pesticide/Mycotoxin Testing Batch prep												
Pesticide/Mycotoxin Testing Result,												
Residual Solvent Testing Batch Prep												
Residual Solvent Testing Results												
Terpenes Testing Batch Prep												
Terpenes Testing Results Analysis	ML - 13JAN2020											
Quality Control Standards Preparation		WAE - 26FEB2020										
Quality Control /Ops 2nd Level Review												
Results Final Review and Release												
Sample Destruction												

Fig. 4 Cannabis laboratory work process audit schedule. The auditor places their initials and the date upon completion of regularly scheduled audits (shaded boxes), or after a QA-directed or off-cycle process audit (white boxes)

action reports, employee training files, and periodic quality summary reports to quickly provide the exact evidence needed to support your results.

Prepare for external audits by directing QA to conduct tertiary review of around 10% of the results released by the laboratory. QA should be able to authenticate the availability of supporting records for all samples chosen by corporate or the Laboratory Director, and report to the CEO or Board of Directors.

Nonconforming Events and Corrective Action Reports

A nonconformance is a failure to conform to accepted standards of behavior, and a non-conforming event (NCE) is an instance of failing to meet a requirement, or otherwise having things not go according to plan. The measures you take in response to NCEs are referred to as Corrective Actions, and Preventive Actions are those efforts you make to avoid nonconformities in the first place. Corrective and Preventive Actions (CAPA) separate the good labs from the bad, and the devil is in the details. The Corrective Action Report (CAR) is one of your laboratory's most powerful tools that memorializes the circumstances surrounding an NCE, and documents your response and rationale for key evidence-based decisions.

> **Cannabis Lab QMP Highlight—Evidence-based decision making.** *You are a scientist now. Your opinion as the laboratory manager is important but will be ignored if you fail to consider the data when making decisions for the business. Labs that care about evidence-based decisions will adopt a quality system that measures its own efficacy, such as audit programs geared to mitigate intelligently assessed risks, and corrective actions that consider the costs of poor quality.*

The CAPA procedures we put into practice are not necessarily required by any regulation, or by ISO/IEC 17025.

Many cannabis laboratory CARs point to ISO/IEC 17025 assessment deficiencies and customer complaints. Unfortunately, NCEs identified by external parties can mean that you might have already let one of your stakeholders down. Your opportunities for continuous improvement are best identified internally, and hopefully prior to any negative impact on testing readiness. Authorize and empower your entire staff, as they are all truly members of your Quality Assurance team, to identify and document NCEs. Monitor QA's histograms and Pareto charts carefully to get a feel for the frequency and significance of errors that are occurring. If you can find the courage to celebrate and incentivize the self reporting of mistakes, you will be promoting the culture needed for high reliability lab testing.

When a nonconformance is observed, the staff must know what to do. The first action taken is always communication. Inform management (section lead, QA, etc.) of the situation, and record the 4W's, the who, what, where, and when, pertaining to the NCE as soon as the conditions permit. No matter how obvious the cause may appear, save the 'why' for your root cause analysis. For example, if a processing technician drops a set of controls on the ground, their 4W report may be as simple as:

On 23 May 2020 @ approx 1545, I (RDB) tripped over a cord in the processing laboratory and dropped QCs for batch XYZ with lot numbers 123, 789, and 456 on the ground, spilling their contents.

Initial correction:

Notified section lead that batch XYZ controls were destroyed by laboratory accident, and placed batch XYZ on hold in LIMS. Obtained a new set of batch XYZ controls (lot numbers 123, 789, 456) from QC/Ops who directed batch rework in accordance with SOP…".

Notice that if the work can continue, QC/Ops has the authority to decide on the best course of action to proceed with defensible procedures. However, with each unique or recurring NCE, the Triumvirate and other laboratory leaders must excel at their #1, highest priority, mission-critical duty: laboratory risk assessment.

We define laboratory readiness as the state of being fully prepared to deliver accurate and defensible results within customer specifications. Your risk matrix does not have to look like ours, but let us agree that any threat to laboratory testing readiness is directly proportional to the threat to the business itself.

In the laboratory accident case, we might agree that there appears to be little risk to the collection of supportable data, but for a minor delay to distribute replacement controls for the same lots. If the technician had not tripped but instead only dropped the controls, this level of NCE would fall into the Minor risk category. Even if the same mistake reoccurs, if it does not represent any appreciable loss or impact to laboratory testing readiness or employee wellbeing, then the entire event can be documented by detailed MFR and electronically attached to batch and lot records. Careful review of minor risk NCEs is worth some analysis in monthly or quarterly quality reports to consider quality improvement tweaks that unlock efficiencies or improve safety; such as, transporting the QCs to the processing lab on a cart or in a cooler. Carefully consider the allowable minor risks, and determine how you are going to efficiently monitor those NCEs without creating a 'death by CAPA' scenario.

The presence of an unmarked tripping hazard clearly violates the laboratory safety SOP. Let's suppose that a nearly identical NCE took place last week, only with a different member of the processing section who sustained a minor scrape. Today's NCE risk level was just elevated to Severe by the Laboratory Director. QA has been directed to conduct initial fact finding, and to probe for commonalities among the events. The QA Manager, who is currently inspecting the microbiology freezer, recruits a member of the intake team to investigate the space and talk to the staff. Thankfully, you had the sense to make quality program support at least 10% of every employee's job description, which keeps the operations manager from getting bent out of shape about people's time.

It is often helpful to keep those involved with the NCE engaged throughout the investigation, and corrective action implementation. The sense of ownership and pride for improving the system will surpass any lingering grief around a team member's correctable technical misstep.

The level of effort and resources dedicated to any investigation should be based on the assessed risk and the probable costs associated with an unaddressed nonconformance. Based on your guidance, the laboratory will know when a five-why (5Y) Root Cause Analysis (RCA) is appropriate, or if it's time to order pizza and pull up your lab's Ishikawa fishbone template in the conference room.

The QA manager analyses the results of the investigation, and considers classifying the NCE root cause as a failures related to:

1. System
2. Training
3. Behavior

Specifically, in that order. Bench level laboratory staff unfortunately suffer the harshest consequences for a lab's mistake sometimes. To avoid creating a punitive quality culture, always look at the system first, and determine if broken SOPs are actually setting the staff up for failure. If a mistake was actually a violation of SOP, then look next to your training program's adequacy and compliance, and consider retraining as an appropriate corrective measure.

Rarely, an individual's under-performance contributes to an NCE. Classify the NCE as 'Behavioral' if there is an instance of code of conduct violations or insubordination, and deal with the individual separately through the performance management system. The CAR should not include any performance management data, and should always prove whether or not systems and/or training failures were involved.

> *Cannabis Lab QMP Highlight—Engagement of People. The work is done at the bench, so is your bench properly supported? No training program or quality check will overcome the detriment of a poor climate in the trenches if you act in a manner that lets the staff believe that revenues are more important than their job satisfaction and safety. Keep everyone's head engaged in the quality program to foster employee growth and satisfaction.*

Before considering any remedy, measure the costs of poor quality, and write these figures down. Direct costs of errors include rework supplies, labor, losing the customer, service call costs, and equipment replacement. But that's not all; make an honest assessment of the indirect or potential costs associated with the error, such as reputational damage, sales challenges, increased regulatory pressure, lawsuits, and lost licenses. Take a moment to re-verify your risk assessment with these costs in mind.

Next, consider the various courses of action available to the laboratory. Understand the costs, timeline, and manpower required to take effective action for each option. When a necessary course of action comes with a price tag beyond the LD's authority, have faith that the information collected thus far will enable evidence-based decision making at the corporate capital resource level.

> *Cannabis Lab QMP Highlight—Leadership. Don't look around the table now because leadership starts with you. Own your actions, inspire integrity in your staff, and give them the proper equipment, systems and training to accomplish the mission.*

Periodically follow up and reassess the efficacy of your corrective actions, and document your findings in your monthly quality reports.

Preventive Actions and Quality Improvement Initiatives

High quality Preventive Action Reports (PAR) are a hallmark of self-learning, innovative, and continuously improving cannabis and hemp testing laboratories. While corrective actions deal with a nonconforming event that has occurred, a preventive action addresses the potential for the NCE in the first place. With a little audit and CAR practice, your lab will enjoy a level of quality that deals with fewer instances of rework than in the early months, and employee performance objectives for quality can now be achieved by contributing to PARs. Using the cannabis laboratory risk matrix (Fig. 5), determine the impact and probability of the error(s) you wish to prevent, and assess the risk of an actual event. Empirical estimates for the costs of poor quality are essential for prioritizing continuous improvement efforts.

The most impactful PARs are inexpensive pilot projects that generate laboratory Quality Improvement Initiatives. For us, corporately sponsored improvement initiatives are derived from convincing PARs that require corporate management because of the unplanned expenses associated with the prescribed preventive measures, or a requirement for additional manpower. Although everyone in the laboratory is responsible for quality, certain employees will discover their passion and knack for quality improvement. Bestow the title of Quality Champion on these hard charging, future leaders. Show off your now reputable, high-reliability cannabis testing culture by rewarding your Quality Champions with tradeshow and conference travel

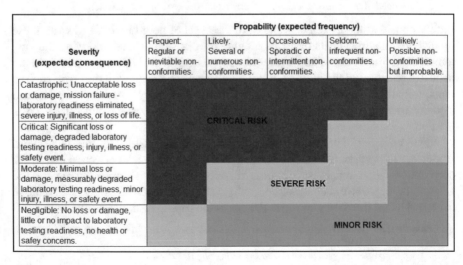

Fig. 5 Example cannabis laboratory risk matrix. (Adapted from the Department of the Army Pamphlet 385-30 on Risk Management)

where they can present their results, and invite these scientific warriors to give short presentations at key sales and corporate meetings.

Cannabis Lab QMP Highlight—Continuous Improvement. *Establish a culture that never stops learning by maintaining fiscal support for quality and training, and enforce executive credence for purpose built audits, CAPA, and customer complaint response systems.*

We have described relatively straight-forward systems for comprehensive laboratory auditing and objective CAPA processes. However, the key ingredient for synchronizing resource management with risk is your leadership. Count on all laboratories making mistakes. Embrace minor errors as essential data points in your program for continuous improvement, and simultaneously enrich your laboratory's quality culture and competitive advantages.

Cannabis Testing 2.0

Mainstream access to cannabis has sparked incredible interest in related agricultural, food and analytical sciences. Garage cannabis testers tinkering with thin layer chromatography and 30-year old gas chromatographs have been replaced by $3–5M start-up efforts with HPLCs, mass spectrometers, and molecular microbiology. The stakes have changed, the technology has improved, and the culture of quality in cannabis analytics is in your hands.

Bright spots in cannabis testing modernization are encouraging. Testing technology providers are developing much needed solutions for microbiological pathogen detection, and pushing state of the art mass spectrometers to their performance limits to combat cannabis matrix interferences. Many of the industry's laboratories answered the call to develop tocopheryl (vitamin E) acetate detection methods to mitigate the health threat and public policy crisis caused by deadly adulterated black market vaping products. ISO/IEC 17025:2017 has been adopted as the official cannabis lab QMS standard in most jurisdictions. Finally, the most impactful effort in cannabis testing today might be the work of the AOAC International Cannabis Analytical Science Program (CASP), which aims to provide consensus in cannabis and hemp testing methodologies, proficiency programs, and training.

Although you must go to war with what you have, the *ready-fire-aim* approach to cannabis testing has an expiring shelf-life. Instead of copying the work of your competitors, seek to learn from the struggles and success of modern laboratory quality systems within other routine testing sectors. Like soldiers and first responders, train your team to fall back on the hours you've spent training them, and to trust (but re-verify) the performance of equipment they expertly maintain.

Give me six hours to chop down a tree and I will spend the first four sharpening the axe.—Abraham Lincoln.

Like "Web 2.0", "Cannabis Testing 2.0" refers to a better organized, interactive, quality-driven means for sharing and using cannabis laboratory data (Fig. 6). By

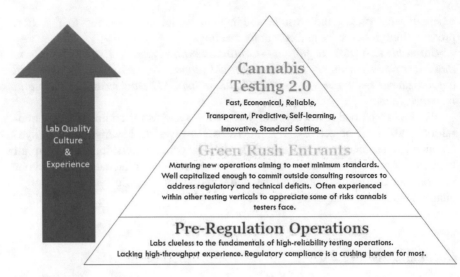

Fig. 6 Cannabis testing 2.0 represents a new high-reliability expectation for cannabis testing operations

encouraging the entire staff to generate knowledge regarding opportunities to improve, your cannabis testing laboratory will teach itself to be more predictable, transparent, innovative, and efficient.

References

1. ISO 1025:2017: General requirements for the competence of testing and calibration laboratories. Available from: https://www.iso.org/ISO-IEC-17025-testing-and-calibration-laboratories.html
2. Guidelines for the validation of chemical methods in food, feed, cosmetics, and veterinary products, 3rd edn. U.S. Food and Drug Administration Foods Program, Oct 2019. Available from: https://www.fda.gov/media/81810/download
3. Guidelines for the validation of microbiological methods for the FDA foods program, 3rd edn. U.S. Food and Drug Administration Foods Program, Oct 2019. Available from: https://www.fda.gov/media/83812/download
4. The AOAC Cannabis Analytical Science Program (AOAC CASP). Available from: https://www.aoac.org/scientific-solutions/casp/
5. Available from: https://www.jointcommission.org/en/resources/patient-safety-topics/sentinel-event/
6. Available from: https://www.cms.gov/Regulations-and-Guidance/Legislation/CLIA
7. Available from: https://www.fda.gov/consumers/consumer-updates/kefauver-harris-amendments-revolutionized-drug-development
8. Available from: https://www.ncbi.nlm.nih.gov/pmc/articles/PMC3853691/
9. Available from: https://www.cdc.gov/foodborneburden/2011-foodborne-estimates.html

Laboratory Information Management Systems (LIMS)

Kyle Boyar, Andrew Pham, Shannon Swantek, Gary Ward, and Gary Herman

Abstract Even in the year 2020, the majority of well-established analytical laboratories are utilizing paper documents and excel spreadsheets to generate and compile data. While it is possible to run a lab with these basic tools, it makes it subsequently very difficult to scale up operations when operational efficiency is tied to pen and paper. Additionally, with the nationwide rise in legal and regulated cannabis markets, a significant amount of effort and attention to detail must be placed on laboratory personnel in order to ensure regulatory compliance. In this chapter, the functionalities of basic LIMS are discussed and how these functionalities can improve productivity, enhance traceability and testing quality, as well as reduce errors. Improvements to these systems are discussed and how the implementation of these enhancements can better inform laboratory personnel, encourage good laboratory practices, and ultimately, increase efficiency and reduce regulatory paperwork.

Perspective

Even in the year 2020, the majority of well-established analytical laboratories are utilizing paper documents and excel spreadsheets to generate and compile data. While it is possible to run a lab with these basic tools, it makes it subsequently impossible to scale up operations when operational efficiency is tied to pen and paper. At some point, laboratory executives have to ask—is there a better way? When considering your capital investment, the typical things that make a pitch deck are instrumentation, personnel, permits, consumables, training, method

K. Boyar · G. Herman (✉)
Tagleaf, Inc., Santa Cruz, CA, USA
e-mail: kyle@tagleaf.com; gary@tagleaf.com

A. Pham
ILP Scientific LLC, Westminster, CA, USA

S. Swantek
Enlightened Quality Analytics LLC, Boulder Creek, CA, USA
e-mail: shannon@cnlightenedquality.com

G. Ward
GK Ward Associates LLC, Vancouver, WA, USA

© Springer Nature Switzerland AG 2021
S. R. Opie (ed.), *Cannabis Laboratory Fundamentals*,
https://doi.org/10.1007/978-3-030-62716-4_7

development and sales. Often left off is the choice of a LIMS system; the attractive "cheaper" option on paper is to use paper and spreadsheets.

However, the true value of a LIMS goes far beyond the initial one time investment. Perhaps you've used many LIMS in the past and had the unfortunate experience of being exposed to overly complicated, archaically designed or underpowered tools that did nothing more than get in your way. But good LIMS will protect your laboratory from shutdowns, increase your testing output, and allow your organization to focus on what's important—the science.

So with all this being said, what does a LIMS do and why can it be such a difference maker in the overall success of a lab? Well that's exactly what we're about to answer in the following chapter, but bear in mind that not all LIMS are cut from the same cloth.

Is a LIMS By Any Other Name Still a LIMS?

Over many years, the word LIMS has grown to encompass a vast number of different software applications that contain a varying number of features that range from the simplest data reporting tools to the most robust application suites. In fact, the wide variation in feature sets, complexity, and technical prowess can be quite confusing. So the first thing to understand is that with all this complexity there are a few distinctions that can help you focus on the right product for your organization (Fig. 1).

As with any industry, technology solutions in laboratories have facilitated compliance with strict record-keeping requirements and created solutions for tracking

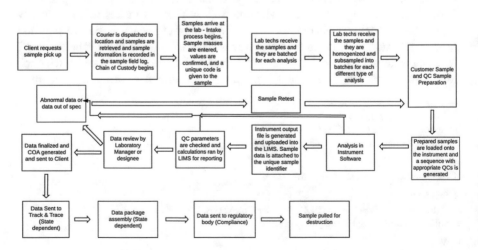

Fig. 1 A typical cannabis testing laboratory's workflow that is tracked through a Laboratory Information Management System

data quality indicators more efficiently and consistently than with paper or spreadsheets. These advances have made it possible to effectively address gaps in data.

Paper records can increase risk of errors and non-compliance and makes it difficult to assess large amounts of data for trends. Some regulatory standards require secondary review on any data entered from hand-written records due to the high probability of transcription errors. In response, a comprehensive software system that can track raw data from sample intake to the production of a certificate of analysis has become the standard in all types of laboratories.

Types of Laboratory Information Software: Differences in Functionality

There are currently many different types of laboratory software being used, and each has their own depth of functionality. It is important to distinguish the differences between them in order to define expectations for software and assess their fit-for-purpose.

Sample Inventory Tracking and Reporting Software

This software tracks the physical sample from collection to sample receipt within the laboratory. Sampling conditions such as temperature and location are recorded in conjunction with field records. This software has the functionality of assigning unique identifiers to the samples in order to maintain proper chain of custody. These systems often have limited functionality in the sample lifecycle through preparation, storage, and analysis. Final results must be entered or uploaded through other platforms for reporting, and often lack proper review points to ensure that non-conforming results are not reported.

Quality Management System (QMS) Software

A quality management system (QMS) is defined as a formalized system that documents processes, procedures, and responsibilities for achieving quality policies and objectives. In a laboratory, these objectives include meeting the requirements in the ISO/IEC 17025:2017 standard which includes provided documented information, often called "tracking records" to demonstrate compliance [1].

QMS software helps to address quality assurance aspects of laboratory activities such as control of documented information, control of externally provided products and services, tracking of non-conformance corrective action activities as well as audit findings and tracking of personnel training. QMS software does not address technical activities within the laboratory scope of accreditation. These systems can vary and are often customized based on the accreditation and regulatory standards the laboratory is held to. Some provide a simple repository for documents and records that is well-organized and easily accessible to regulatory and accreditation authorities. More complex systems include workflows and automation that trigger alerts when the elements being tracked require monitoring on timelines or by status. It is important to note that these systems will not always update automatically when regulations or requirements are updated. For example, when the ISO/IEC 17025:2005 updated to ISO/IEC 17025:2017 the standard significantly reorganized their requirements, causing some compliance software built for the 2005 standard to become obsolete. When considering QMS or "compliance" software, one must consider the level of automation desired to avoid missing annual reviews or closing Corrective Actions, as well as the need for costly updates should the requirements in regulations and standards change or the laboratory expand into areas of work covered by different standard requirements.

Laboratory Instrumentation and Informatics Software

Laboratory Instrumentation Software is specific to a manufacturer's platform and can vary widely in functionality. If data from this software is not imported into a system that can assess quality control and other data quality indicators, the laboratory must ensure that these results are reviewed manually, inevitably introducing potential error and requiring significant redundancy checks.

Laboratory Information Management Systems (LIMS)

The LIMS is the core backbone of a laboratory's operations. An ideal LIMS can drive laboratory workflows and allows for maximum throughout the testing process. The LIMS manages all data generated during the lab testing process pertaining to client samples. This data is ultimately used in the generation of a laboratory's final product for its clients, the certificate of analysis (COA).

Laboratory analysis of cannabis products incorporates multiple fields of scientific expertise ranging from analytical chemistry to microbiology. This requires trained staff that all have to operate within a complex framework of rules and regulations. One of the primary challenges in cannabis testing that laboratories face is the illegality of cannabis on the federal level means there are no consistent or unifying rules governing cannabis testing. Instead, any state determining cannabis to be legal has to create its own framework of regulations. When testing regulations are drafted, often they must refer to rules from the environmental, food, and pharmaceutical industries, though none are particularly suitable for cannabis. Ultimately, all of these labs must operate within the appropriate regulatory framework. A LIMS can help a lab ensure they are within compliance if it is tailored properly to the appropriate standard.

It is important to note that a LIMS is more than just a software; it is a process of integration that encompasses laboratory workflow combined with user input, data collection, instrument integration, data analysis, and end user reporting. A LIMS can also help inform those running the business by providing insights about sample workflow, consumable inventory, and staffing needs.

Common LIMS Features

Many LIMS features are common to a wide variety of laboratories. Requirements of a typical LIMS systems in an analytical testing laboratory might include the following:

- Sample receipt and sample login—Sample information is logged into the system and assigned a unique identifier. Typically matrix type, sample mass, analyses requested, and other items are entered at this time. Additional parameters may be tracked as well.

  - Barcoding may be employed for optimal tracking of the sample during its life cycle in the laboratory.
  - Sample Storage and Sample Integrity Indicators are evaluated against requirements.

- Chain of custody—This is a record documenting the possession of the samples from the time of collection to receipt in the laboratory. This record generally includes the number and type of container, the mode of collection, the collector, time of collection, preservation, and requested analyses. See also Legal Chain of Custody Protocols.
- Instrument integration—This ensures that the analytical tools being utilized in the lab can easily interface with the software. Integration can be achieved in a variety of ways either by physical connections to the computer or by generating files from the instrument software that can be read by the LIMS. For the latter, data output files are generated from the software in a format that is easily read or parsed by the LIMS which is then used to calculate the final result.
- Result entry—During sample processing a record is kept of those handling the sample and the data collected during the testing process.
- Quality control samples—Ensures that all quality control sample data generated meets the acceptance criteria.
- Result reporting—Performing calculations from sample prep masses and raw data for use in the generation of a Certificate of Analysis.
- Archiving and data warehousing—Most requirements (i.e., EPA) state to maintain records for 5 or more years.

Functionality to Reduce Error

Many LIMS will automate or integrate aspects of workflow that are potentially error prone. For example, barcode scanning ensures that sample identifiers such as lab codes and track and trace identifiers are always being properly transcribed and sample mismatches do not occur. This is especially important in the in track and trace systems where errors in reporting can invite regulatory enforcement.

Licensing is another realm within which a LIMS can reduce error. A valid state or city license is needed in order to do any kind of cannabis business, and access to a state database can help ensure the validity of a license before a sample pick up or will remove the possibility of accepting samples from an expired or fraudulent licensee.

In states where there are numerous assays required for compliance, a LIMS can also prevent incomplete testing being sent to regulatory bodies. For example in California there are numerous pesticide analytes that typically necessitate the use of both LC and GC methodologies. If an analyst uploads all of the LC analyte data but forgets to upload the GC analytes a LIMS can help flag that the data is incomplete for compliance purposes and can prevent the sample data from being transmitted to track and trace or the state's regulators. In particular, many track and trace softwares do not allow the submission of data multiple times so a LIMS that properly integrates with these systems can ensure that a data set is both complete and only being sent once. Furthermore, various quality checks can be employed by the LIMS to ensure data is of known and documented quality.

Another area that is critical to ensuring accurate data is points at which samples are sub-sampled for preparation or analysis. In particular, it's very easy for transcription errors to occur when entering weights into a system which is why LIMS scale integration is another valuable tool to reduce technician error.

Build Your Own or Buy Off the Shelf?

Given the varying nature of cannabis testing regulations in each state, an organization may consider developing their LIMS in-house to suit their needs rather than repurpose an out of box system designed for a different industry. Before this decision is made, the laboratory needs to consider a variety of different factors, including size, personnel, complexity, client needs, and of course, cost.

The primary advantage of developing an in-house LIMS system is that it can be customized to meet the specific needs of the laboratory. For instance, a LIMS system developed for a Research and Development laboratory may look very different from one developed for a third party lab, as their end goals are very different. The R&D system may need statistical tools and allow flexibility in its data collection processes. In contrast, a highly regulated analytical laboratory may desire a LIMS system that is designed to conform to regulations throughout the testing process and therefore is designed with specific traceability tools, control limits and data checks to ensure that non-conforming data is not approved for publication to clients or regulatory bodies.

It is important to note that if the laboratory data is reported to a governmental agency as well as the client, the laboratory's client is technically both parties. In these cases, the ability of the LIMS system to report data properly to the appropriate end user needs to be the primary focus of system development.

Despite the benefits of a customized in-house LIMS system, the laboratory must consider that an in-house LIMS system will always require IT personnel resources to support, maintain, and perform on-going development as the laboratory's operational needs change. The initial development and implementation is a very costly and time intensive process, and thus only recommended if off-the-shelf systems cannot meet the needs of the laboratory.

When qualifying a vendor of a commercial LIMS system, consider the level of long-term support being offered. It is imperative that the vendor plans for support beyond the initial purchase of the software, even if the vendor themselves are no longer operating.

The decision to purchase or develop a LIMS system is just one of the many decisions that must be carefully evaluated by laboratory management. The process should include establishing criteria and specifications that support the goals the laboratory hopes to achieve with the system. This purchase will affect every activity the laboratory undertakes, from compliance with regulatory requirements to the clients' business needs.

Development of LIMS Specifications and Planning

Planning, using a risks and opportunities approach, is a key element of a good Quality Management System (QMS). Assessing the needs the LIMS is meant to address in the laboratory and creating specifications before beginning the procurement process is important to ensure that the product can be easily integrated into laboratory operations in a ways that meets the objectives of ensuring compliance, traceability, control of records, and reporting data of known and appropriate quality to the client. The process by which a laboratory develops specifications for a LIMS system should not vary from the processes contained in the laboratory's purchasing procedures regarding control of externally provided products and services. The laboratory should be careful to follow procedures that meet ISO or other related requirements for both products and services, as installation and on-going support and development of the LIMS system are often part of the contract.

In the development of specifications, the laboratory should consider the following:

1. The type of data that the laboratory produces and the identification of its clients.
2. How the data as the laboratory's products are used by its clients.

3. Meeting laboratory QMS, Regulatory, and Accreditation Requirements.
4. The size of the laboratory—number of locations and personnel that will be using the system.
5. The complexity of the analyses and associated quality control.
6. Unique analyses or activities performed by the laboratory, such as statistical sampling design.
7. The ancillary functions desired by the laboratory.

Lab Culture and LIMS Implementation: Why Laboratory Staff Engagement is Paramount

While a LIMS can provide a framework for the organization, it requires that all laboratory personnel are vested in that framework and follow the correct protocols. In many cases, laboratory staff may be hesitant to adopt new protocols because they are most comfortable with what they are familiar with and simply don't want to conform to a new way because that new way requires them to learn something new and it may seem cumbersome. Change and transitions are not always easy but in business, just as in Darwinism, only those that adapt to change can survive.

It is essential that all staff follow the procedures being set forth when integrating a new LIMS system. For example, say that one technician decides they prefer manually entering their weights rather than use the automatically transcribed ones that the system was built with. Suddenly there are tons of transcription errors that can confound all subsequent calculations, and your data ultimately becomes invalid. Your organization has lost its integrity due to personal preference of one rogue team member. Management personnel play a particularly important role in keeping the system together by acting as crucial points of communication in your laboratory. Miscommunication between departments can lead to disaster. Say a Quality Manager traditionally keeps paper only records, but the LIMS system only functions with digital uploads. If this miscommunication is uncovered during an audit, it could lead to a deficiency and potentially a lack of accreditation if not properly remedied. This just goes to show that your LIMS can only go as far as your staff will take it.

Implementation

Account Setup

Client notification—As with any major operational change, the laboratory should notify its clients of the value they can expect to see when the LIMS system is implemented as well as any delays or changes they can expect during the

implementation process itself. Clients should be prepared with realistic expectations for any future changes in receiving deliverables.

Planning within the laboratory QMS: implementation timelines—Establishing a
properly phased implementation timeline should be done to ensure there are no
operational interruptions. It is important to properly follow lab procedures relating to personnel training and evaluation of products for suitability that are common in ISO-based laboratory environments. Standard Operating Procedures on
the use of the software must be written, reviewed, and approved before using the
system for client-reported data to ensure consistency of use and allow for identification of non-conformance to these procedures. The requirements in this area
must be taken into account when implementation timelines are developed. The
timeline for implementation can vary heavily depending on the system itself;
simple systems may be implemented in a matter of weeks, while others may
require multiple phases of installation and development over the course of
months or even years before the system is fully installed. Depending on the
amount of processes and instruments within the operation that the LIMS must
integrate with and how deeply will largely dictate the amount of time it takes to
complete an implementation.

Personnel training—Laboratories in regulatory and ISO-based accreditation environments have, as required, detailed procedures on personnel training on any
activities that could affect the quality of the data. At a minimum, this includes
that staff provide documents indicating that they have read and understood the
SOPs associated with the relevant activity. Management, quality assurance, and
technical staff often have additional requirements associated with their training
as they are assigned specific responsibilities and authorities over those activities.
While the standard lists the requirements, it is up to the laboratory to establish
how their system will satisfy those requirements.

Audit trail—Audit trails are one of the strongest tools for ensuring compliance
through a LIMS system by tracking all changes made within it. Audit trails
should be configurable such that any actions taken by users that could potentially
impact results are logged with enough detail to trace back to any source of error.
An audit trail is meant to also protect against unauthorized and untracked changes
made by an individual, whether intentionally or unintentionally, and therefore
will track each change within the system and the user making that change so that
the original result or value can be restored and the identity of anyone manually
making changes is known.

User setup and permissions—When planning for this aspect of the implementation,
be sure to consider the personal and shared responsibilities of each team member
for the purposes of setting permission levels. It helps when a laboratory has
already outlined specific responsibilities and authorities of lab management that
necessitate the restriction of access to certain modules and prevent unapproved
changes. Ask yourself where in the organization's hierarchy does each user sit,
and what permissions should be granted to them based upon their role? The
range of responsibilities can vary greatly depending on the size of the company,

so it is important to ensure that all of the different scenarios are considered for each user. Figure 2 is an example of security permissions from Tagleaf LIMS.

Reporting setup—When considering this aspect of the implementation one should consider what their governing federal or state regulations say about reporting requirements. Are there track and trace identifiers that should be incorporated? Should license numbers be listed? Which assays need to be run to achieve regulatory compliance and how are the values reported? Compare the listed data on the certificate to your regulatory requirements to make sure they are all satisfied; even one missing data point can cause serious headaches for the laboratory and its clients.

Analysis setup—What types of tests are going to be conducted by the laboratory where the LIMS is being implemented? The typical suite of tests offered by full service cannabis labs include potency, residual solvents, terpenes, microbial contaminants, heavy metals, mycotoxins, water activity, and pesticides. Here are some helpful questions you can ask when considering how to configure your LIMS for each analysis:

1. What type of instrumentation is utilized for this analysis?
2. What analytes are measured by the analysis and what are the reporting requirements for each? For example when reporting isomers compounds should these be reported individually or summed?
3. What types of quality control measures should be recorded for this type of analysis? Do they require continuing calibration verifications, check standards, field duplicates, or method blanks? How many of each type must be run with every batch?

Security Roles — Disabled — Changed

Roles	Admin	Chemist	Courier	Default	IT Section	Lab Assistant	Lab Director	Lab Manger	Lab tech	Sales	Sr. Chemist	+ Add Role
General settings												
All abilities	✓			✓	✓		✓	✓				
COA settings												
Full access to coas	✓			✓	✓		✓	✓				
List coas	✓	✓	✓	✓	✓	✓	✓	✓	✓	✓	✓	
View coas	✓	✓	✓	✓	✓	✓	✓	✓	✓	✓	✓	
Create coas	✓	✓		✓	✓	✓	✓	✓	✓	✓	✓	
Edit coas	✓	✓		✓	✓	✓	✓	✓	✓	✓	✓	
Delete coas	✓			✓	✓		✓	✓				
Send for Review	✓	✓		✓	✓	✓	✓	✓	✓	✓	✓	
Finalize coas	✓			✓	✓		✓	✓		✓	✓	
Send to Track & Trace	✓			✓	✓		✓	✓		✓	✓	
Send for Compliance	✓			✓	✓		✓	✓		✓	✓	
Send to Client	✓			✓	✓		✓	✓		✓	✓	
Publish coas	✓			✓	✓		✓	✓				

Fig. 2 Security roles in a cannabis laboratory are essential for ensuring that lab personnel can access areas of the LIMS pertinent to their role while preventing access to areas that are not relevant to their function within the organization

4. Should you establish control charts to track possible changes in quality control results over time?
5. What Data Quality Indicators indicate "gateways" that need checks against established criteria to ensure the results conformed to the laboratory's data quality objectives for the method?
6. What types of calculations should be performed to achieve the result in the required reporting units?
7. What are the reporting requirements for the end user? What needs to be identified with the results on the final report? If clients require reporting of raw data or additional records, different report formats must be established in LIMS and validated for compliance before use.

Client project set-up—ISO/IEC 17025:2017 requires that the laboratory assess requests by the client to meet their specific data needs. These requests can include additional quality control, tighter control limits, or additional analytes that are not usually reported with a normal suite. These requests can be made for single sample events or on entire projects, and are part of the mutually agreed upon contract between lab and client. If there are on-going client projects during LIMS implementation, it is crucial that the laboratory ensures they are captured in the new system before transitioning.

Laboratory internal testing and validation—It is considered good practice to identify a core group of beta users within the laboratory that test the system while existing prior compliance and reporting systems are still in place to ensure that there are not workflows within the laboratory that do not fit into the new system, resulting in loss of essential records or reporting of non-conforming data that is not flagged or identified. Examples of possible "holes" in a new system are the transfer of sample collection data, data and records review tracking, and traceability of factors that contribute to the uncertainty of the final result. It is also productive to establish a sandbox staging area, where developers and core users can collaborate to provide necessary changes uncovered during the internal testing process. This allows for future changes to be made to the system to be tested without overriding the previous version until fully developed.

Implementation Case Studies

Without a well-organized process, implementing a newly purchased LIMS system can be painful and costly. Some labs have a paper system or some combination of paper, Excel spreadsheets, and Word. Reliance on such spreadsheets without the proper safeguards in place can put the laboratory at risk of liability. Upon auditing a number of labs, some spreadsheets were not locked or monitored, and often the calculation cells became corrupted without anyone being aware for months. In one lab, a spreadsheet meant to output "pass" or "fail" was incorrectly mapped such that any result would yield a "Pass" result, resulting in a serious nonconformance in QC procedures. In any LIMS calculations must always be validated or checked prior to implementation and locked to prevent future errors.

Planning for implementation should include the huge effort to load in static data such as methods, analytes, matrices, reporting limits, and all other information requiring records. Keep in mind that without consistent spelling of client names, searches and queries will not work. Some labs have had to go back and meticulously clean up their database because of spelling errors, which while seemingly minor can launch outsized regulatory headaches. This is another example of where barcoding and other methods used to limit manual transcription can greatly reduce errors to the benefit of the laboratory.

Some processes within a new LIMS system may seem to add an increased burden on personnel, causing some to feel that the system makes their job more difficult. However, the overall process is more efficient and less prone to error, resulting in significantly less time spent in reviews and corrective actions. When the staff is properly engaged in the process, the benefits of the different processes are easily perceived by the laboratory and the process changes are more easily accepted.

Implementation requires running the LIMS side by side to current processes. This is a lot of work but usually can be completed in 2–3 weeks with appropriate effort. This process unveils errors or logic issues and gets the lab used to the LIMS, while validating that the system works correctly. One implementation strategy is to assemble a small team composed of a representative sampling of every department, for example the IT person and at least one lab person from each lab department (i.e., organics, inorganics, micro, etc.). For labs with multiple sites, a team can be assembled at an initial sitehen travel to subsequent sites for implementation and training.

Sometimes, even when training is complete, personnel may be reluctant to use the new system. In some cases, when the system has been tested and online, management had to simply turn off the older system and force the issue. Implementing a new LIMS is almost always painful unless properly planned for. One lab, for example, complained bitterly about the LIMS replacing their systems. Three months later, they said they didn't know how they lived without it. Persistence, patience, and problem solving are key to a successful implementation. Labs should have a single point of contact with errors or problems, then list and prioritize the fixes based on impact.

Data Storage and Backup

In this section we discuss the reasons why good data storage and backup plans are a necessity in the form of various case studies from previously audited labs by the one of the authors of this text. Most requirements (i.e. EPA) are to maintain records for 5 years [2]. Some contracts require 7 years. Labs usually archive records after 2 years and maintain them off-site because it does not use expensive lab space and is safe in case of a catastrophic event at the lab site.

Data back-up must be stored off-site and in today's world this is often done in the Cloud. One lab had all of their backups maintained in the Lab Director's office. This particular lab had an accident which resulted in an explosion that burned up the lab. They never recuperated and all their data was lost. This highlights the importance of redundancy and storing laboratory data in multiple locations to prevent loss.

As experienced lab owners know, redundancy in human capital is also important to ensure uninterrupted sample throughput. Similarly, it is highly important to ensure that knowledge of the current system is kept by more than one person. For example in one lab, the IT manager was the person who programmed the system and maintained it. Unfortunately, while riding his bike, he got in an accident, hit his head and died. That lab was plunged into confusion and ultimately never recovered. They couldn't even access their data since major permissions were set to the IT manager. It was eventually sold to another lab company who implemented their own LIMS.

LIMS and ISO/IEC 17025:2017

ISO/IEC 17025:2017 is an international standard of accreditation that many labs are held to in order to be deemed technically competent. Having this accreditation is mandated in most states with lab testing regulations. The numerous requirements contained in the ISO/IEC 17025:2017 and ISO/IEC 17025 based standards are discussed below. The laboratory's Quality Management System (QMS) must meet these requirements in addition to technical requirements. The ability of a LIMS system to meet these requirements in an efficient and standardized process can save the laboratory time and money. More importantly, it eliminates the need for multiple software systems or paper systems which can minimize risk of regulatory repercussions.

Management Responsibility

ISO standards pertaining to assigning responsibilities and authorities for each activity ensure that Laboratory Management, however designated, are engaged in maintaining data integrity. This includes a requirement for assigning stop work authority and responsibilities pertaining to the prevention of unauthorized access to confidential client data. Good practice dictates that access to certain levels and statuses of laboratory documents and data are limited to only the personnel necessary to reduce risk of unintentional and unauthorized changes. Each person working in the laboratory should be provided with a certain level of access in order to perform their functions within the organization. Access should also be restricted to prevent tampering with data involved in areas outside the specific person's role. For example, a lab technician or analyst should not have authorization to finalize a certificate of analysis which is typically reserved for a Laboratory Director, Quality Officer, or Laboratory Manager. However, analysts need access to the areas of LIMS relevant to their assigned activities such as entering sample collection data, peer review of analysis data, and performing calculations within the system to generate results. Ensuring proper access can greatly minimize risk of inadvertent alteration of crucial data.

Document Control and Records Management

Additionally, a LIMS may assist with document control to ensure that only current revisions of an SOP are being utilized in the lab. It also can help with retaining important documents such as training records for lab personnel. This aspect also ties into things like data packages where a clear trace of who handled a given sample and their credentials to perform such a task should be included. Data packages will be discussed in more detail in a later section of this chapter.

Metrological Traceability

Metrological Traceability is defined as a "property of a measurement result whereby the result can be related to a reference through a documented unbroken chain of calibrations, each contributing to the measurement uncertainty." Achieving metrological traceability includes identifying all possible contributing factors to the uncertainty of the result and ensuring that those components have their own traceability. Using Certified Reference Materials from companies accredited to ISO/IEC 17034, ensuring that support equipment is calibrated by an ISO/IEC 17025:2017 calibration company, and verifying calibrations in between intervals are all part of maintaining this documented unbroken chain. These components all contain their own uncertainty measurements that are used to calculate the final measurement uncertainty of the result.

Any good LIMS should always be focused on providing maximum transparency and traceability for laboratory processes to ensure procedures are correctly followed and results are valid. This is typically embodied through the use of audit trails which allow for the determination of which personnel handled a sample at a given time, which instruments were utilized, when they were last calibrated, and what reagents or solutions were used to prepare the sample.

When a measurement is captured from any of the devices being used in the laboratory it's also important to ensure that such measurements are validated through proper quality control measures. A LIMS should be able to house and potentially analyze these measurements to spot any patterns that could cause a laboratory to produce erroneous data. This provides the laboratory with a mechanism to support their data in a manner that is legally defensible.

Let's use an example to underscore how a properly configured LIMS system can assist in maintaining metrological traceability. For instance, if it was found that a particular batch of continuing calibration verification samples was created using an expired certified reference material as one of the components, a LIMS can prevent that standard being entered into the system for the analysis. If this expired standard was not caught by LIMS checks, the standard would be logged with an unique identifier as part of the analysis and would enable a Quality Manager to determine which samples were tested with that particular CCV. ISO requires client notification if the results of a nonconforming analysis such as this and the impact on data quality is evaluated to determine the need for re-analysis in each instance of broken traceability.

Corrective and Preventive Actions (CAPA)

Any time that major nonconformance to procedures, regulations, and standards occur that possibly affect data quality, formal Corrective Action is required. LIMS platforms are instrumental in identifying non-conformances as it evaluates Quality Control (QC) windows and other analytical requirements that must be met such as acceptability of calibration criteria and method requirements. If these are caught by the LIMS platform and the sample is marked for re-sampling, re-preparation, or re-analysis, a quality control or method failure can be handled within the laboratory's procedures before becoming a non-conformance requiring formal Correction Action.

Formal corrective action required by the ISO/IEC 17025 standards consists of investigation to determine impact on data quality and the need for stop work orders. Root cause analysis is performed to determine the root cause of the nonconformance, and corrective action is applied to ensure that the nonconformance does not recur. One piece of the ISO requirements that is often missed is the documented follow-up to verify the corrective action applied was effective in preventing recurrence.

A LIMS should assist the lab in conducting such root cause analyses by making the records involved in the nonconformance easily traceable from sampling to reporting. Often, laboratories do not have the ability to trace possible root causes to sampling nor the ability to communicate with the sample collection personnel. In cannabis laboratories that perform their own sampling, this weakness observed in other industries can be mitigated. A LIMS system that includes compliance components and automation may also help track any corrective or preventive actions that are implemented. ISO standards require that a running log of CAPAs be continuously updated and available for review to auditors; having a digital log can help immensely in this process.

Data Control

The control of laboratory data and information is a critical component for laboratories in order for them to perform activities. Section 7.11 of ISO/IEC 17025 titled "Control of Data & Information Management" outlines the specific requirements that a LIMS must meet in order to be used at the laboratory. Labs need to verify that they have the necessary access to data and information needed to perform all of its activities. The lab needs to ensure that the information management system used for collection, processing, recording, reporting, storing and/or retrieving data is validated for functionality. If there are changes made to software configuration or modifications to commercial software, they need to be authorized and validated before use.

When a laboratory information management system is managed and maintained off-site or through external providers, the lab needs to ensure that they comply with all applicable requirements. The laboratory must ensure that the instructions, manuals, and reference data relevant to the management system are made available to personnel. Furthermore, calculations and data transfers need to be checked in an appropriate and systematic manner.

Statistical Analysis and Calculations

A LIMS should be capable of performing and automating any calculations that are used in the testing process to provide quantitative results. The final numbers depend on many previous inputs, such as dilution factor calculations, percentage conversions, mass densities, and unit dosage weights. Additional outputs may need to be processed further, such as conversion of total Potential THC and CBD based on a theoretical decarboxylation factor of 0.877 to account for the loss of carbon dioxide. Often the types of calculations to be performed are dependent on the requirements of the client, so it is important to maintain the appropriate lines of communication between staff.

Laboratory Data and Litigation

Laboratory data has historically been the tipping point in both civil and regulatory litigation. Many times, the basis of litigation is in claims that are groundless if the data cannot be used in court. The cannabis industry has seen early litigation regarding label claims and stolen product Intellectual Property. For these cases, to prove or defend against a false label claim, the data that demonstrates the "truth" must be deemed "admissible" in court by both parties and the presiding judge.

Experienced attorneys will hire a data validation expert to try to have the data deemed admissible or have the data evidence thrown out as appropriate to the case. Data validation activities include the review and evaluation of the following elements, as stated in both ICH and AOAC International guidelines:

- Specificity
- Accuracy
- Precision
- Repeatability
- Linearity
- Limit of detection (LOD)
- Limit of quantification (LOQ)
- Range
- Ruggedness/Robustness

Aside from protecting a laboratory against audits a LIMS should be able to house the data generated in such a way that all equipment, personnel, and reagents used that leads to the final result produced on the COA is fully tracked and traceable. This ensures that should a dispute arise between a client and a testing laboratory that all steps in the testing process have been properly documented should the case need to be settled in a court of law.

In the context of things like pesticides this would be embodied by demonstrating that all laboratory QC samples such as method blanks, sample field duplicates, and continuing calibration verifications (CCVs) all passed the acceptance criteria to ensure that the instrument was functioning properly at the time of analysis. Another question that would likely be covered in a court case is did the method definitively prove that the pesticide in question was present? How would you prove this? This would require that the lab can source the sample's instrument file and demonstrate that both the parent ion for the compound in question was present along with its qualifying ions or mass transitions to confirm the compound in question.

If the question appeared in the context of possible mishandling by the lab leading to contamination of the sample, this is where chain of custody and audit trails would largely benefit the defendant. Knowing when the sample exchanged hands and who handled the sample for each portion of the analyses provides accurate timelines as to when contamination may have occurred.

Audit Trails

It is paramount that a LIMS makes the data of a sample's lifecycle through the lab easily accessible to those who require access such as laboratory management, quality assurance staff for client data inquiries, a regulator, or an ISO/IEC 17025:2017 assessor. This should not only apply to the actions occurring with the sample by laboratory technicians but also the hardware and reagents employed to perform the testing. For example, in a laboratory batch it is always good practice to make sure that all components used to create quality control samples are tracked, especially the lot numbers and expiration dates of certified reference materials (CRMs). Audit trails ensure that any action taken within the LIMS is always being recorded and hence.

Certificate of Analysis (COA)

The final product produced by a LIMS for a customer is the Certificate of Analysis (COA). Different states have different reporting requirements for different product types to be listed on the COA, although ISO/IEC 17025:2017 has standardized requirements that apply to any ISO/IEC 17025 accredited laboratory. To meet state requirements, information about the license holders for which the testing is being performed is required. Some states require that each of the respective LOD and LOQ for the instrument method for each analyte be listed along with amount in a particular metric such as percent, parts per billion (ppb), µg/g, or µg/mL depending on the sample matrix type. When it comes to testing data, manufacturers and distributors often like to provide their customers with easy access to testing data so QR codes are often utilized to achieve this function.

Ultimately, a LIMS should be able to customize the COA template to meet all the requirements of each different state ideally with a QR code for client marketing purposes as well as templates for upload into regulatory or client databases. This ability to customize is especially important for laboratories that are operating in multiple states and have to adhere to different requirements.

LIMS 2.0 and Smart Labs

A basic LIMS system should achieve the goal of integrating with lab instrumentation to house analytical data and drive forward testing processes. However, technology can greatly assist with operations beyond the testing process that are critical to the success of a laboratory. Currently many labs are using numerous laboratory information softwares to achieve compliance and many laboratories are beginning to recognize that many of the processes performed by these additional systems can be streamlined and vastly improved when they are all housed in a single solution. This concept of a Smart Lab with expanded LIMS capability or LIMS 2.0 is likely where laboratory software is heading.

Data Package Automation

In some state regulations the generation of a data package is mandated. These data packages contain all of the data pertaining to a given samples analysis including but not limited to: personnel sign offs for each test, training records for each employee performing a certain test, the chromatograms for those tests, sample masses, dilutions, calculations, instrument calibrations, quality control samples, and more.

The assembly of these data packages can be extremely labor intensive and time consuming which reduces the overall output of a laboratory. A forward thinking cannabis LIMS should have the capability to automatically assemble and generate these data packages. This will greatly free up laboratory resources from regulatory paperwork and allow them to focus on the testing process itself.

How a LIMS Can Help Maintain Customer Satisfaction

For any of us that have worked in a cannabis testing lab, a common question that is asked by clients is, "When can I expect my test results?" A LIMS should be able to assist account managers in tracking a sample's status throughout the testing cycle.

A LIMS system can also assign hold times or turn-around times dictated by methods or client contracts and can move expiring or fast-turnaround samples to the top of an analysis worklist. Some LIMS systems embody this through different test statuses or queues to depict the progression of a sample through the testing and review process and this status can be viewed through the use of a client portal that is connected to the LIMS where a client can see the statuses of each test being performed.

Some LIMS companies have created ways to reduce call volumes to the lab by offering a client portal. A client portal is a separate client facing version of the LIMS that is specific to a particular customer. Some client portals will provide the status of any pending tests but most serve as a repository for a client's COAs. In some instances customers can also place orders and fill out a Chain of Custody for samples that are arranged to be picked up.

Encouraging Strong Data Integrity Practices

The laboratory's commitment to training and practices that promote data integrity and the laboratory's ethics policy are paramount to any strong laboratory QMS. Many poor data practices are often the result of poor training or the desire to please management or the client. Often, poor data practices are simply seen as cutting a corner to save time.

While there are plenty of good actors in the cannabis business there are unique pressures on cannabis laboratories that other industries do not face due to the "point-of-sale" testing scheme and the risk of loss if that product fails action limits. This drives the phenomenon of "client lab shopping" where clients decide to test with the lab least likely to fail their product, and puts undue financial pressure that incentivises results that yield "non-detect" in contaminant testing or higher cannabinoid results.

Let's explore some examples of data tampering.

- *Back-dating (time traveling)*: Back-dating occurs when the analyst changes the date and time of a recorded action in retrospect in hopes of giving the appearance that an action was performed earlier or later than the actual time it was performed. If there are holding times or regulatory limits on times for analysis, this can mean the difference between reporting to the client on schedule or needing to re-sample the batch.
- *Peak shaving*: Manipulating the data within instrument software happens often in chromatographic methods where an analyst will override the software's automatic integration for manual integration to force a quality control sample into passing or a client sample under the action limit. Sometimes peaks are deleted as unconfirmed target analytes even though they had clear confirmation indicators and often, peak windows in the method file can be changed to be too narrow to find any peaks which is common in contaminant testing where "Non-Detected Results" or NDs are desirable.

- *Support analyses and calculation manipulation*: In analysis of matrices that require a support analysis to calculate a final result, such as percent moisture, the support analysis is seen as less visible to oversight. In this example, the temperature can be raised to burn off semi-volatile terpenes, causing the percent moisture result to be higher and the final corrected "dry weight" result of other analyses to be lower.

There are multiple ways a laboratory analyst can work outside laboratory procedures to save time or obtain a more desirable, yet biased result. These methods invariably cost more time than they save and can result in serious consequences for the laboratory that have included millions of dollars in fines and prison sentences for laboratory management in the past. A LIMS should have measures to safeguard against as many negative consequences as possible. The LIMS should be used to notify employees of data that does not indicate common practice, but lab management's commitment to data integrity procedures, as well as the lab personnel's personal and professional integrity can be the ultimate factor in whether laboratory staff are aware of the various types of transgressions and consequences of such.

Integration with Track and Trace

In most states with legal cannabis, part of the legal framework is ensuring that all cannabis and cannabis products are tracked in a seed to sale (S2S) fashion. These softwares allow for end-to-end tracking and tracing of cannabis-containing products through the entire chain of production to sale. Some states have mandated that a specific track and trace program must be utilized throughout the enterprise chain including testing laboratories. One example of this is METRC in the state of California. Below are examples of the different track and trace systems that are currently being utilized in the United States at the time of this publication.

- *BioTrackTHC*: Both a track-and-trace system and an enterprise resource planning (ERP) solution, BioTrackTHC streamlines data management and workflows from cultivation and processing to laboratory testing and dispensation. Compliance features include customized reporting to meet government-specific needs, tracking of destruction and waste activities, transport manifests, recall tracking, regulation labels, workflow management, and more. The software has also been adopted by state governments such as Illinois, Hawaii, New Mexico, and New York [3].
- *Leaf data systems*: Similar to BioTrackTHC, Leaf Data Systems is used by both industry operators and government agencies to regulate the cannabis industry. The system can manage data at all points along the cannabis life cycle, from cultivation and processing to distribution, testing, and sale. Leaf can handle customized reporting depending on state or municipality, as well as customizable alerting to ensure enforcement activities are effective. The software has been adopted by the governments of Pennsylvania and Washington [4].

- *METRC*: Developed by Franwell, METRC represents another solution used by not only business in the cannabis supply chain but also state and local governments. Special features include trend analysis, employee activity tracking, credentialing, and process metrics. States using it include California, Colorado, Massachusetts, Montana, and the District of Columbia, among others[5].

In this process, ID tags with a unique track and trace identifier are assigned to the cannabis product as it travels through the supply chain and a LIMS must be able to maintain a record of this unique identifier with a particular sample. Once the product undergoes the testing process, all of the testing data and results must be conveyed to the track and trace provider at key time points such as when samples are received by the lab and once results have been released. Therefore, it is extremely beneficial for a LIMS to integrate and communicate with these systems to ensure that samples are being tracked in a compliant manner.

Concluding Statements

After reading this chapter, we hope you can now see the true value of a LIMS and what it can do for your organization. By digitizing your workflow and processes your lab can reduce errors, improve testing quality, and ensure regulatory compliance.

References

1. Quality Glossary—Q. Accessed 25 May 2020
2. Fuels registration, reporting, and compliance help. Accessed 25 May 2020
3. Nelson S (2017) A seed-to-sale shakeup. Cannabis Business Times, 30 May 2017. Accessed 25 May 2020
4. Wood S (2018) MJ Freeway raises $10M to improve marijuana tracking software, expand operations. The Inquirer. Philadelphia Media Network (Digital), LLC, 20 Sept 2018. Accessed 25 May 2020
5. McVey E (2018) Chart: dominant player emerging for state cannabis seed-to-sale tracking contracts. Marijuana Business Daily, 10 Sept 2018. Accessed 25 May 2020

Pesticide and Mycotoxin Detection and Quantitation

Anthony Macherone

Abstract In Canada and most states in the U.S., medicinal and adult recreational use of cannabis and cannabinoid products have been legalized. In these regions, safety and quality testing prior to commercial sale is regulated. Testing includes determinations of potency and terpene content, heavy metals testing, analysis for residual solvents in products extracted from cannabis, microbial screening to ensure potentially dangerous bacteria and fungi are not present in the products, and testing for residual pesticides and mycotoxins. Of these required tests, pesticides and mycotoxins testing is the most challenging. It requires sample preparation procedures that remove chemical interferences and improves the sensitivity and robustness of the analysis. Testing in some regions further requires a multi-platform approach that leverages the power of analytical systems in both the liquid and gas phases. Even though cannabis products are subjected to a great deal of testing, there are still opportunities for the use of off-target list pesticides, adulteration with synthetic chemicals, the occurrence natural oxidation and degradation chemicals in the growing and production industries, or while the products are in storage that will not be identified using targeted analysis. As the cannabis industry matures, more comprehensive approaches to testing will be required. This chapter will provide a background of pesticide usage, discuss sample preparation procedures, provide examples of rigorously vetted testing methodologies, and discuss approaches to comprehensive testing.

Introduction

At the Federal level, the United States regulates marijuana (cannabis) as a Schedule I controlled substance [1]. However, many states have elected to legalize medicinal and adult recreational use programs. In these states, testing regulations have been defined to ensure safety and quality of the various products. The testing includes potency to determine cannabinoid content, metals analysis, microbial screening,

A. Macherone (✉)
Agilent Technologies, Inc., Wilmington, DE, USA

The Johns Hopkins University School of Medicine, Baltimore, MD, USA
e-mail: anthony_macherone@agilent.com; amacher1@jhmi.edu

© Springer Nature Switzerland AG 2021
S. R. Opie (ed.), *Cannabis Laboratory Fundamentals*,
https://doi.org/10.1007/978-3-030-62716-4_8

153

residual solvents analysis in refined products, terpenes profiling, and residual pesticides and mycotoxins.

The analysis of pesticides in air, water, soil, food stuffs, and agricultural products has been performed for decades in countries around the globe. In the U.S., E.P.A., F.D.A., and U.S.D.A. regulate pesticides in the air and water, drugs and foodstuffs, and agricultural activities, respectively. Figure 1 is a map of the 48 contiguous states and the District of Columbia illustrating pesticide usage and U.S. population statistics from 2019 [2, 3]. There is a 0.57 positive correlation of pesticide usage and population revealing that some of the most densely populated states use the most pesticides.

The use of some pesticides has been banned in many countries and it is imperative that imported foods are tested to ensure the safety of products entering a nation and that they do not contain banned pesticides. Often referred to as persistent organic pollutants (POP), 29 compounds have been banned or restricted in 183 nation states and the European Union since the original ratification at the Stockholm Convention in 2001. Even so, compounds like dichlorodiphenyltrichloroethane (DDT) have continued to be in use in several countries.

POPs enter the body through inhalation, ingestion, and transdermal routes. Exposures to various pesticides have been associated with many chronic diseases including cancers, Parkinson's, Alzheimer's diseases, and reproductive disorders [4]. POPs like DTT, and their metabolites and degradation products, have been shown to be statistically associated with Diabetes mellitus Type II, hepatic fat, and insulin insensitivity in certain populations [5, 6]. Myclobutanil is an especially dangerous pesticide when applied to cannabis because upon pyrolysis, it generates toxic hydrogen cyanide. In fact, product recalls were performed in California and Colorado due to the presence of myclobutanil in vaporizer cartridges and "cannabis flower, trim, concentrates, and infused-products" [7].

In addition to pesticides, other additives or contaminants have entered the product stream resulting in health-related issues and even death. For example, as of December 2019 there were more than 2400 hospitalizations in the U.S. for electronic-cigarette, or vaping, product use–associated lung injury (EVALI). In a small case-control study, 77% of the EVALI cases were associated with vitamin E

Fig. 1 Left: Map of the 48 contiguous states and the District of Columbia and total pesticide usage in kg (Source: USGS high estimate 2016) and right: map of the US population in 2019. Darker shading equals higher values (Source: U.S. Census Bureau). (Powered by Bing. © DSAT for MSFT, GeoNames, Navteq)

acetate in vaporizing Δ^9-tetrahydrocannabinol (THC) products. Ninety-four percent of the EVALI cases were positive for vitamin E acetate in bronchoalveolar-lavage (BAL) fluid compared to 100% of "healthy comparator" controls [8]. In another case, a post-marrow transplant male was exposed to *Aspergillus fumigatus* which was determined to be a contaminant in the marijuana product he was smoking [9]. A 2017 letter to the editor in the journal *Clinical Microbiology and Infection*, reported a study conducted after observation of 2 "medicinal marijuana users" undergoing chemotherapy developed a "fungal disease". The researchers obtained 20 cannabis samples from Northern California and found them to be contaminated with *Aspergillus* spp. as well as other "gram-negative bacilli and fungal pathogens" [10, 11]. Furthermore, a class of mycotoxins known as aflatoxins are produced by fungi like *A. fumigatus* and its cousins *A. flavus* and *A. parasiticus*. There are typically four aflatoxins regulated in cannabis: aflatoxin B1, aflatoxin B2, aflatoxin G1, and aflatoxin G2. Another mycotoxin, Ochratoxin A, is generally regulated as well. Harmful effects of mycotoxins in humans include carcinogenicity, hepatotoxicity, and nephrotoxicity [12, 13].

Given the above facts about the potentially harmful effects of pesticides, mycotoxins, additives like vitamin E acetate, and bacterial contamination, it becomes evident that rigorous testing must be performed to ensure the safety and quality of cannabis and cannabinoid products derived from cannabis or hemp. However, as of June 2, 2020, U.S. Department of Agriculture who is charged with regulating industrial hemp, only stipulates potency testing in hemp to ensure it does not contain more that 0.3% (wt./wt.) of psychoactive THC [14]. As of yet, there is no mandate for cannabinoid products derived from hemp with an intended use of human consumption to undergo pesticide testing.

Analytical Chemistry

Pesticide and mycotoxin testing in cannabis and cannabinoid products is typically performed with a liquid chromatography-tandem mass spectrometer (LC-MS/MS) or a gas chromatography-tandem mass spectrometer (GC-MS/MS). In most states, only LC-MS/MS is needed to adequately measure the pesticides defined in a state target list. However, California, Nevada, Florida, and Canada have pesticides in their lists not well-suite for typical analyses using LC-MS/MS.

Sample Preparation

Proper sample preparation for residual pesticide and mycotoxin analysis in cannabis and cannabinoid products is critical. The analysis itself is arguably the most difficult of the regulated tests primarily because of the intense matrix effects presented by percent by weight cannabinoids and terpenes, and a host of other endogenous chemicals. This chemical noise interferes with the analysis and therefore, proper sample preparation is critical for reliable and repeatable results at the required part-per-billion (ppb) levels. Further complicating this analysis is the myriad products that must be tested other than cannabis inflorescence. However, in this text we will focus only on inflorescence for brevity.

No QuEChERS for Dry Cannabis Inflorescence

The U.S. Department of Agriculture developed an extraction procedure with the acronym QuEChERS (quick, easy, cheap, effective, rugged, and safe) and it's commonly used for residual pesticides analysis in fruits and vegetables [15]. It has been suggested that QuEChERS is appropriate for residual pesticide analysis in cannabis [16] but there are several caveats to this approach which we will discuss here.

The primary approaches to QuEChERS are AOAC International official method 2007.01 [17] and European Union official method EN 15662 [18]. There are pH labile pesticides in the various target lists. The pH of the AOAC method is 4.8 [19] and the pH of the EN method is 5–5.5 [20], which effects the recovery of pesticides like imazalil. Another concern is the generation of an exotherm when the sample and the QuEChERS salts are mixed and the sample is hydrated with water. This temperature spike effects thermally labile pesticides like fenthion [21]. Lastly, chemicals used for dispersive solid-phase extraction (dSPE) in the QuEChERS process (1) do not have enough extraction capacity for this application, (2) scavenge acidic pesticides with the application of primary-secondary amines (PSA), and (3) scavenge planer pesticides with the application of graphitized carbon black (GCB). The combination of these caveats demonstrate that QuEChERS is not a recommended procedure for residual pesticides analysis in cannabis inflorescence.

Dilute and Shoot and Winterization

"Dilute and shoot" [22] and winterization [23] are two other common sample preparation approaches for residual pesticide and mycotoxin analysis in cannabis. A 5 g sample of homogeneous inflorescence is required for the dilute and shoot method to which 10 mL acidified acetonitrile is added. The suspension is then vortexed and centrifuged. A mix of 30 isotopically labeled compounds is added to an aliquot of the supernatant as internal standards. It should be noted that adding internal standards post-extraction does not correct for extraction bias and the purchase of 30 labelled internal standards is very expensive. In the winterization process, 0.2 g homogeneous sample is sonicated in acetonitrile to extract the pesticides. The samples are then "winterized" in a freezer overnight. The samples are then thawed, decanted, and diluted in 75:25 methanol:water (% v/v) prior to analysis.

Winterization is time consuming and dilute and shoot is expensive yet there is little to no return on the expense since the internal standards do not normalize for extraction and other bias. Neither procedure includes a filtration step which leads to clogging of capillaries, significantly decreases analytical robustness, and increases the need for maintenance through fouling of the electrospray ionization (ESI) or atmospheric pressure chemical ionization (APCI) source if used. The reduction in throughput and increased maintenance costs equates to lost revenue that cannot be recovered.

Single-Stream SPE Cleanup with Dilution

Sample preparation for the analysis of pesticides and mycotoxins in cannabis inflorescence needs to leverage extraction methodologies that efficiently remove chemical interferences from the matrix and employ high dilution factors to further mitigate chemical noise. This latter fact may seem counterintuitive but consider how a tandem mass spectrometer works. Once a chemical species is ionized in the source region of either a LC-MS/MS or a GC-MS/MS system, the ion beam enters the first quadrupole which is operated in selective ion monitoring (SIM) mode isolating specific precursor ions. The precursor ions then enter a collision cell—essentially an ion guide containing an inert gas with various energies applied. In the collision cell, each chemical ion precisely fragments into product ions specific to each precursor. As the product ions exit the collision cell, a second quadrupole, also operated in SIM, isolates each product ion resulting from each precursor. In this way, chemical noise is reduced, and selectivity is increased through the precursor—product ion pairs that are highly specific for each chemical under analysis. This process is referred to as selective reaction monitoring (SRM) or multiple reaction monitoring (MRM) in the case of multiple precursor—product ion pairs. For each analyte, MRM significantly improves the signal-to-noise (S/N) ratio which, is at its essence, equates to sensitivity.

Returning to sample preparation and the removal of chemical noise in the matrix, it becomes evident that removal of chemical noise through clean up and dilution further improves S/N. To this end, a single-stream sample preparation method has been developed that employs solid-phase extraction (SPE) and dilution that supports analysis by both LC-MS/MS and GC-MS/MS. The procedure begins with homogenizing a 1.0 g sample of cannabis inflorescence in a 50 mL polypropylene tube with 15 mL of high-purity LC-MS-grade acetonitrile. The suspension is centrifuged and transferred to an unconditioned SampliQ C18 EC cartridge (Agilent Technologies, Santa Clara, CA) and allowed to elute into a clean collection tube vial via gravity. The original suspension is extracted twice more with high-purity LC-MS-grade acetonitrile into the new collection tube. The collected fractions are then brought to a final volume of 25 mL with high-purity LC-MS-grade acetonitrile. An aliquot of the extracted is then filtered prior to further dilution. For LC-MS/MS, 450 µL of 25/75% water/methanol (v/v) + 0.1% (v/v) formic acid is added to 50 µL of the filtered extract in a fresh auto-sampler vial (250-fold dilution). For GC-MS/MS, 200 µL of the filtered extract is mixed with 800 µL high-purity acetonitrile in a fresh auto-sampler vial (125-fold dilution). This procedure takes about 45 min for a batch of 20 samples.

As of May 2020, the Canadian pesticide list contains 96 target pesticides in 3 product types: fresh cannabis and plants, dried cannabis, and cannabis oil. A quantitative limit of quantitation (not to exceed concentration) is provided for all but about 1/3 of the cannabis oil type. Many pesticides in the Canadian list have a LOQ of 10 ppb which is an order of magnitude lower than that of the California Bureau of Cannabis Control (BCC) for the 66 target pesticides regulated in their adult recreational use cannabis program [24, 25]. California has defined two product types: inhalable cannabis and cannabis products, and other cannabis and cannabis products. They have further stratified the pesticide list in these products into two categories. Detection of any Category I pesticide above the empirically determined limits of detection (LOD) will fail the product in the Certificate of Analysis. Category II pesticides are actionable if detected above the LOQ determined by BCC and provided in the regulations.

LC-MS/MS

The most challenging of the regulated tests is most certainly residual pesticides and mycotoxin analysis. This panel is performed primarily with LC-MS/MS platforms that provide the required selectivity and sensitivity in the complex cannabis matrix which can contain percent by weight concentrations of cannabinoids and terpenes, along with hundreds of other endogenous chemicals. Tandem mass spectrometry also enables limits of detection (LOD) and limits of quantitation (LOQ) in the low ng/g (ppb) range. It must be noted however, that even with the analytical power of these systems, LC-MS/MS can still struggle to identify and quantify some pesticides in this matrix. In these cases, a multi-platform approach that employs GC-MS/MS is required.

LC-MS/MS Method for the Analysis of Pesticides and Mycotoxins

Recently, a method for the analysis of residual pesticides and mycotoxins in cannabis inflorescence was published [26]. The method used a single-stream sample preparation as described above. The method detailed a LC-MS/MS workflow to meet the requirements of U.S. States that have legalized recreational marijuana and published regulations at the time of the method development. The LC system was the Agilent 1260 Infinity II binary pump, 1260 Infinity II multisampler, thermostatted, with 100 µL loop and multiwash options, the Agilent 1260 Infinity II multicolumn thermostat with 6-port/2-position valve. The mass spectrometer was the Agilent Ultivo tandem quadrupole mass spectrometer equipped with Agilent Jet Stream (AJS) ESI Source. The method targeted 67 LC-MS/MS amenable pesticides defined in the U.S. recreationally legal states and five mycotoxins: aflatoxin G1, G2, B1, and B2, and ochratoxin A. The LC-MS/MS system parameters are given in Tables 1 and 2.

Figure 2 is a typical chromatogram for the target pesticides. Analyte recoveries were all between 70% to 130%, with most recoveries close to 100%. The LOQ for all compounds was at least 50% below the defined regulatory levels in each respective state. The method as defined was also applicable to the Agilent 6470 LC-MS/MS system and results were comparable across platforms. The 67 pesticides defined in this method were those determined to be amenable to analysis with LC-MS/MS systems using ESI. However, there are at least 27 pesticides in the combined U.S. and Canadian lists that are better suited for analysis by GC-MS/MS. Therefore, the defined LC-MS/MS method was designed to partner with complementary GC-MS/MS techniques. This is necessary for pesticides like captan, chlordane, pentachloronitrobenzene (PCNB) and methyl parathion) amongst others [27].

Canada and U.S. Method Commonalties

LC-MS/MS methods must be sensitive, selective, accurate, precise, and robust. They must further cover the appropriate concentration range for all the analytes and exceed the defined LOQ requirements. The LC-MS/MS method defined above shares these parameters with published methodology that addresses the regulatory

Table 1 Agilent 1260 LC conditions

Parameter	Value	
Column	Agilent Poroshell 120 phenylhexyl, 3 × 100 mm, 2.7 μm (p/n 695975-312)	
Guard column	Agilent Poroshell 120 phenylhexyl, 3 × 5 mm, 2.7 μm (p/n 821725-914)	
Column temperature	55 °C	
Injection volume	10 μL	
Autosampler temperature	4 °C	
Needle wash	Flushport (100% methanol), 10 s	
Mobile phase	(A) 5 mM ammonium formate/0.1% formic acid in water (B) 0.1% formic acid in methanol	
Flow rate	0.5 mL/min	
Gradient	Time (min)	%B
	0.00	30
	1.00	30
	2.00	75
	8.00	96
	9.00	100
	9.50	100
	9.51	30
Analysis and re-equilibration time	13 min	
Total runtime (sample to sample)	14 min	

Table 2 Agilent Ultivo tandem mass spectrometer parameters

Parameter	Value
Ion mode	AJS ESI, positive and negative polarities
Capillary voltage	5000 V (Positive ion mode) 3500 V (Negative ion mode)
Drying gas (nitrogen)	13 L/min
Drying gas temperature	200 °C
Nebulizer gas (nitrogen)	55 psi
Sheath gas temperature	200 °C
Sheath gas flow	10 L/min
Nozzle voltage	500 V
Q1 and Q2 resolution	0.7 amu [Unit, autotune]
Delta EMV	0 V (Not applicable to Ultivo)

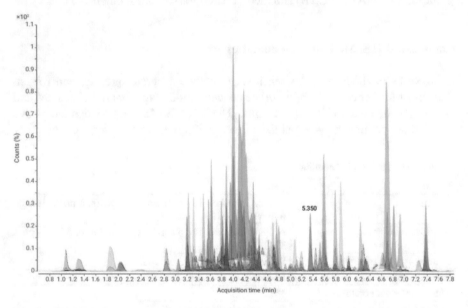

Fig. 2 MRM TIC chromatogram for 62/66 pesticides and 5 mycotoxins in the California target list

requirements in Canada. These similarities include a common sample preparation procedure, identical column phase and dimensions, injection volume, mobile phases, and the LC-MS/MS hardware [28]. These methods, in conjunction with GC-MS/MS, have demonstrated proven success across North America.

Analytical Challenges to Residual Pesticide Testing Using Only LC-MS/MS Systems

For pesticides that either do not respond or respond poorly in LC-MS/MS methods with ESI, APCI in negative ionization mode has been suggested as an alternative. Like ESI, APCI is a soft ionization technique that typically creates $(M+H)^+$ quasi-molecular ions in the positive ionization mode. In the negative ionization modem APCI can form $(M−H)^−$ proton abstraction species, $M^{\bullet-}$ radical electron capture species, M^- dissociative electron capture species, or MX^- anion adduction species [29]. As noted previously, the dual use of ESI and APCI methods significantly decreases analytical throughput resulting in diminished revenue generation per day. However, a primary concern with APCI in negative ionization mode for analysis of certain pesticides is selectivity and pentachloronitrobenzene (PCNB) is a model compound for this argument. The APCI negative ionization mechanism for PCNB was proposed as the loss of HCl followed by the formation of an ammonium ion adduct [30]. An investigation into this claim was made using liquid chromatography-quadrupole time of flight mass spectrometry (LC-QTOF), GC-QTOF, and GC-MS/MS. In that study, the correct ionization mechanism was determined to be a complex *in situ* formation of two new chemical species: 2,3,4,5-tetrachloro-6-nitrophenolate and 2,3,5,6-tetrachloro-4-nitrophenolate in a 3.5:1 ion ratio [31]. Figure 3 illustrates the high-resolution accurate mass (HRAM) spectrum for PCNB obtained on the LC-QTOF system in APCI negative ionization mode and the *in situ* chemical species that are observed. Figure 4 is the HRAM product ion spectrum from the 273.8 m/z precursor ion isolated in the quadrupole region of the LC-QTOF system.

Curtis (2019) further compared the HRAM spectrum for PCNB using GC-QTOF in electron ionization (EI) mode and determined that GC-MS technologies were extremely selective, sensitive, robust, and linear over a large range compared to LC-MS/MS in APCI negative mode. Selectively has also been compared (unpublished data, A. Macherone, Agilent Technologies, Inc.) to the use of the 275.8 → 35.10 (2,3,4,5-tetrachloro-6-nitrophenolate → Cl⁻) SRM transition given in another article [32] and determined it to be extremely non-selective compared to GC-MS/MS in EI mode.

GC-MS/MS

LC-MS/MS technologies are adequate for most pesticides in the various U.S. and Canadian pesticide lists but a thorough evaluation of each compound, revealed many pesticides respond poorly in ESI or APCI. This determination was made through literature review, empirical evidence, and a proprietary a priori algorithm that evaluates physicochemical properties and parses each analyte into an analytical bin: LC-MS/MS with ESI or GC-MS/MS with EI.

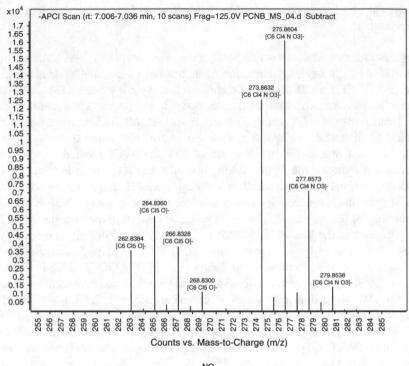

Fig. 3 HRAM spectrum of the new chemical species form in situ from PCNB

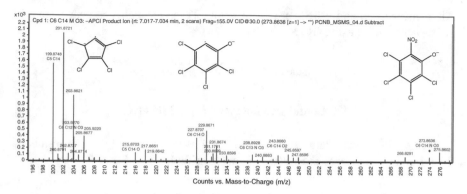

Fig. 4 HRAM product ion spectrum of the 273.8 m/z precursor ion and the collision induced dissociation (CID) product ion species

Recently, a GC-MS/MS method was developed to partner with the LC-MS/MS methods described above (Stone et al. 2020; Roy et al. 2018). The method focused on the five most problematic compounds for analysis by LC-MS/MS: captan, PCNB, chlordane, methyl parathion, and chlorfenapyr [33]. The method used the single-stream sample preparation procedure used in the LC-MS/MS applications and matched all samples, calibrators, etc. in matrix. The method development defined a rigorous vetting paradigm to explicitly determine method statistics such as range, linearity, accuracy, precision, LOD, and LOQ. The vetting paradigm used a multi-day, multi-replicate model of five replicate injections of all calibrators, QCs, and samples over three independent days. The data from the multi-dimensional dataset was evaluated statistically using these primary equations:

$$\text{Average} = \Sigma x_i / n \tag{1}$$

$$\text{Sample standard deviation} (s) = \left[\frac{\Sigma(x - \bar{x})^2}{n-1} \right]^{1/2} \tag{2}$$

$$\text{MDL (LOD)} = (s) * (\text{Student } t-\text{value, } n-1 \text{ df, 99\% confidence}) \tag{3}$$

$$\text{MDL (LOD) Test} = \text{Calculated MDL} < \text{Spike Level} < 10 * \text{Calculated MDL} \tag{4}$$

$$\text{LOQ} = 10 * (s) \tag{5}$$

$$\text{Percent Accuracy} =$$
$$\left[\left(\frac{\text{spiked concentration} - \text{calculated average concentration}}{\text{spiked concentration}} \right) \right] * 100 \tag{6}$$

$$\text{Precision} (\%\text{RSD}) = \left[(s) / \text{Average} \right] * 100 \tag{7}$$

Alternatives to (s) and %RSD are given below. These have also been evaluated in subsequent studies using GC-MS technologies [34]

$$\text{Standard Error}, SE = \frac{(s)}{\sqrt{n}} \qquad (8)$$

$$95\% \text{ Confidence Interval}, CI_{95\%} = 1.96 * (SE) \qquad (9)$$

Figure 5 is the extracted MRM chromatograms for each analyte. The results of the method development and vetting are given in Tables 3, 4, 5 and 6.

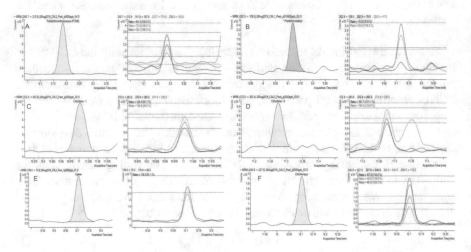

Fig. 5 Extracted ion MRM chromatograms. (**a**) Pentachloronitrobenzene, (**b**) methyl-parathion, (**c**) chlordane 1, (**d**) chlordane 2, (**e**) captan, and (**f**) chlorfenapyr. The left panel is the targeted, quantitative transition, and the right panel illustrates the qualitative ions with horizontal lines defining the acceptable ion ratio range (typically ±20%)

Table 3 Empirically determined in-vial and in-matrix LOD and LOQ

Compound	CA category	CA LOQ (ppb)	Empirical LOD in-vial (ppb)	Empirical LOD in-matrix (ppb)	Empirical LOQ in-vial (ppb)	Empirical LOQ in-matrix (ppb)
PCNB	II	100.00	0.061	7.59	0.16	20.25
Methyl Parathion	I	>LOD	0.031	3.88	0.084	10.50
Captan	II	700.00	1.64	204.75	4.37	546.38
Chlordane 1	I	>LOD	0.23	29.00	0.62	77.38
Chlordane 2	I	>LOD	0.26	32.75	0.70	87.38
Chlorfenapyr	I	>LOD	0.19	23.88	0.51	63.63

Table 4 Average quantitative accuracy (%)

Compound	Day 1	Day 2	Day 3	3-Day average
PCNB	74.18	92.07	105.95	90.73
Methyl Parathion	109.08	113.18	86.72	102.99
Captan	118.19	117.01	105.3	113.50
Chlordane 1	93.74	92.21	72.15	86.03
Chlordane 2	96.86	96.6	104.68	99.38
Chlorfenapyr	99.99	96.96	112.39	103.11

Daily N = 5; average N = 15

Table 5 Inter-day accuracy and precision (n = 15)

Compound	Target calibration level (ppb)	Empirical average and 99% confidence interval	Percent accuracy	Precision (%RSD)
PCNB	0.25	0.23 ± 0.019	91	6.96
Methyl Parathion	0.25	0.28 ± 0.013	111	4.71
Captan	4.00	4.31 ± 0.50	108	10.00
Chlordane 1	1.00	0.86 ± 0.071	86	7.61
Chlordane 2	1.00	0.99 ± 0.081	99	7.02
Total Chlordanes	1.00	0.93 ± 0.076	93	7.32
Chlorfenapyr	1.00	1.02 ± 0.059	102	5.08

Table 6 Range, curve type, and weighting

Compound	In-vial range (ppb)	Number of calibrator levels	Curve	Weighting
PCNB	0.016–64.00	8	Linear	1/x
Methyl Parathion	0.016–64.00	8	Linear	1/x
Captan	1.00–64.00	5	Linear	1/x
Chlordane 1	0.016–64.00	8	Linear	1/x
Chlordane 2	0.016–64.00	8	Linear	1/x
Chlorfenapyr	0.016–64.00	8	Linear	1/x

All linear coefficients of determination (r^2) were >0.998

Comprehensive Pesticide Screening

A 2018 publication described a compressive approach to pesticide screening in cannabis [35]. The method used a single-stream sample preparation technique to analyze 215 pesticides in cannabis matrix using LC-MS/MS and GC-MS/MS systems. The target pesticide list fully encompassed the Oregon list of 59 pesticides plus others from states that had published regulatory requirements at the time, and many more commonly used in agriculture. This work demonstrated the need to test for more pesticides than regulated by a state or country. The reason for this is the fact that tandem mass spectrometry (MS/MS) is a targeted methodology—meaning that only the pesticides defined in the analysis will be identified and quantified if present in the sample. MS/MS data with defined MRM transitions for each target analyte is static and cannot be interrogated to identify compounds not in the target list. For example, Oregon regulates 59 pesticides. If a laboratory defines a MS/MS method for only those 59 pesticides, any off-target pesticides will be missed. This opens the door for growers to use pesticides not defined in their jurisdiction because they know they probably won't be found. This poses a potentially serious safety risk. One way to address this concern is to test for as many pesticides as possible within a reasonable and profitable model as demonstrated in the comprehensive screening publication.

An alternative option would be to employ time of flight (TOF) mass spectrometry in either the liquid phase or gas phase or both [36]. TOF mass spectrometers acquire high-resolution-accurate mass (HRAM) data over a defined mass range. These systems can be used for targeted, quantitative analysis and the same data can be interrogated to identify other pesticides in the sample [37]. More information about TOF systems can be found in chapter "Cannabinoid Detection and Quantitation" of this text. TOF systems are not common in the day-to-day cannabis testing lab but as general screening tool, they may be quite powerful in providing more quality and safety information about a sample. As an example, as cannabis plants and products age, chemical changes occur in the sample. The most common example of this is spontaneous decarboxylation of acid phytocannabinoids like Δ^9-tetrahydrocannabinolic acid (THCA) and cannabidiolic acid (CBDA) to their corresponding neutral compounds THC and cannabidiol (CBD), respectively. Other processes like oxidation occurs as in the case of CBD conversion into cannabidiolquinone (CBQ) [38]. CBQ was originally synthesized from CBD as an anti-cancer drug but was found to be cytotoxic and its development was abandoned [39, 40]. Screening with TOF systems could also be employed to identify illegal drugs or adulterants that may be present in a sample including synthetic cannabinoids, cathinones (bath salts), designer drugs and chemicals that may leach into a product from packaging materials. Other uses for TOF systems in cannabis laboratories include characterization of endogenous chemicals like flavonoids and fatty acids, and comparison of chemical profiles in different cannabis strains.

Summary

Pesticides and mycotoxin testing in cannabis and cannabinoid products is essential to ensure safe products are available where medicinal or adult recreations use programs have been legalized. In the U.S., each state has a unique target list and definitions of action limits and LOQ. In Canada, the Federal Government controls the regulations for the entire country. In many regions, LC-MS/MS systems provide the required selectivity and sensitivity to analyze the target pesticides in complex cannabis matrices. However, in some states and Canada, pesticides in the target lists are not amenable to LC-MS/MS using ESI. In these cases, GC-MS/MS provides the appropriate platform to work in conjunction with LC-MS/MS. Using a single-stream sample preparation procedure that is quick and efficient for both LC-MS/MS and GC-MS/MS analyses provides reliable, repeatable, and robust workflows to support the demands of high-throughput laboratories. It is further important to properly vet a methodology before implementation in the day-to-day laboratory operations. This vetting procedure supports the validation protocols cannabis testing laboratories must perform prior to licensure.

Even with defined regulations there still exists multiple vectors for the use of off-target pesticides, adulteration with dangerous drugs, *in situ* formation of degradation or oxidation compounds, or contamination from microbes and chemicals they synthesize. The need for more comprehensive testing that quantifies target compounds but can also test for as many possible pesticides and other chemicals as feasibly possible is becoming more evident. Laboratories must test for pesticides, more cannabinoids, more metals, and more solvents than defined in their regional lists. A comprehensive approach to cannabis testing is the only real way to provide safe products to participants in medicinal or recreational use programs.

Acknowledgements The author would like to thank Anastasia Andrianova, Matthew Curtis, Sue D'Antonio, Eric Fausett, Wendi A. Hale, Jeffery S. Hollis, Bruce Quimby, Peter J.W. Stone, and Jessica Westland of Agilent Technologies.

References

1. Drug Scheduling. Available via DEA.gov. https://www.dea.gov/drug-scheduling. Accessed 14 Apr 2020
2. USGS high end estimates for 2016. Available via ZeroHedge.com. https://www.zerohedge.com/s3/files/inline-images/StatesMostLeastPesticides.png?itok=UiTEGnRg. Accessed 15 Apr 2020
3. U.S. Census. State population totals and components of change: 2010–2019. Available via Census.gov. https://www.census.gov/data/tables/time-series/demo/popest/2010s-state-total.html. Accessed 10 Apr 2020
4. Mostafalou S, Abdollahi M (2013) Pesticides and human chronic diseases: evidences, mechanisms, and perspectives. Toxicol Appl Pharmacol 268(2):157–177
5. Daniels SI et al (2018) Elevated levels of organochlorine pesticides in South Asian immigrants are associated with an increased risk of diabetes. J Endocrine Soc 2(8):832–841
6. La Merrill MA et al (2019) Exposure to persistent organic pollutants (POPs) and their relationship to hepatic fat and insulin insensitivity among Asian Indian immigrants in the United States. Environ Sci Technol 53(23):13906–13918
7. Biros AG (2018) Two recalls hit California cannabis market. Cannabis Industry Journal. https://cannabisindustryjournal.com/tag/myclobutanil/. Accessed 2 Jun 2020

8. Blount BC et al (2020) Vitamin E acetate in bronchoalveolar-lavage fluid associated with EVALI. N Engl J Med 382(8):697–705
9. Hamadeh R, Ardehali A, Locksley RM, York MK (1988) Fatal aspergillosis associated with smoking contaminated marijuana, in a marrow transplant recipient. Chest 94(2):432–433
10. Nunley K (2017) Study finds medical marijuana samples to be contaminated with mold and bacteria. Available via Medical Marijuana, Inc. News. https://www.medicalmarijuanainc.com/news/study-finds-medical-marijuana-samples-contaminated-mold-bacteria/. Accessed 2 Jun 2020
11. Thompson GR III et al (2017) A microbiome assessment of medical marijuana. Clin Microbiol Infect 23(4):269–270
12. Ellis WO, Smith JP, Simpson BK, Oldham JH (1991) Aflatoxins in food: occurrence, biosynthesis, effects on organisms, detection, and methods of control. Crit Rev Food Sci Nutr 30(4):403–439
13. Pleadin J, Frece J, Markov K (2019) Mycotoxins in food and feed. Adv Food Nutr Res 89:297–345
14. Federal Register Rules and Regulations (2019) Establishment of a domestic hemp production program, vol 84(211), pp 58522–58564
15. Anastassiades M, Lehotay SJ, Stajnbaher D, Schenck FJ (2003) Fast and easy multiresidue method employing acetonitrile extraction/partitioning and "Dispersive Solid-Phase Extraction" for the determination of pesticide residues in produce. J AOAC Int 86:412–431
16. Daniel D, Silva Lopes F, Lucio do Lagoa C (2019) A sensitive multiresidue method for the determination of pesticides in marijuana by liquid chromatography–tandem mass spectrometry. J Chromatogr A 1603:231–239
17. Lehotay SJ (2007) Determination of pesticide residues in foods by acetonitrile extraction and partitioning with magnesium sulfate: collaborative study. J AOAC Int 90(2):485–520
18. CEN-EN 15662 (2018) Foods of plant origin—Multimethod for the determination of pesticide residues using GC- and LC-based analysis following acetonitrile extraction/partitioning and clean-up by dispersive SPE—Modular QuEChERS-method. Available via CEN.eu. https://www.cen.eu/Pages/default.aspx. Accessed 2 Apr 2020
19. Lehotay SJ, Mastovská K, Lightfield AR (2005) Use of buffering and other means to improve results of problematic pesticides in a fast and easy method for residue analysis of fruits and vegetables. J AOAC Int 88(2):615–629
20. Anastassiades M, Scherbaum E, Tasdelen B, Stajnbaher D (2007) Crop protection, public health, environmental safety. In: Ohkawa H, Miyagawa H, Lee PW (eds) Pesticide chemistry: crop protection, public health, environmental safety. Wiley-VCH, Weinheim, p 439
21. Stevens J, Jones D (2010) QuEChERS 101: the basics and beyond. Available via Agilent.com. https://www.agilent.com/cs/library/eseminars/Public/QuEChERS_101_10_11_01.pdf. Accessed 30 Mar 2020
22. Dalmia A, Hariri S, Jalali J, Cudjoe E, Astill T, Schmidt C, Qin F (2020) Novel ESI and APCI LC/MS/MS analytical method for testing cannabis and hemp concentrate sample types, 78758 (69954A). PerkinElmer, Inc., Waltham, MA
23. Butt CM, Di Lorenzo R, Tran D, Romanelli A, Borton C (2019) Analysis of the Massachusetts cannabis pesticides list using the SCIEX QTRAP® 6500+ System, Document number: RUO-MKT-02-8937-A, AB Sciex
24. Publication 190301 (2019) Mandatory cannabis testing for pesticide active ingredients. Government of Canada
25. Bureau of Cannabis Control Text of Regulations. California code of regulations title 16 division 42 (2019). Available via BCC.ca.gov. https://www.bcc.ca.gov/law_regs/cannabis_order_of_adoption.pdf. Accessed 19 Sept 2019
26. Stone PJW et al (2020) Determination of pesticides and mycotoxins in cannabis flower as defined by legalized U.S. state recreational cannabis regulations. Application note 5994-1734. Agilent Technologies, Inc.

27. Andrianova AA et al (2020) Analysis of twenty-seven GC-amenable pesticides regulated in the cannabis industry in North America with the Agilent 8890/7010B triple quadrupole GC/MS system. Application note 5994-1786. Agilent Technologies, Inc.
28. Roy J-F et al (2018) A sensitive and robust workflow to measure residual pesticides and mycotoxins from the Canadian target list in dry cannabis flower. Application note 5994-0429. Agilent Technologies, Inc.
29. McEwena CN, Larsen BSJ (2009) Am Soc Mass Spectrom 20:1518–1521
30. Tran D et al. (2018) DuoSpray™ ionization: a novel approach to analyzing the California mandated list of pesticides in cannabis, Document number: RUO-MKT-02-7607-A, AB Sciex
31. Curtis M et al (2019) Up in smoke: the naked truth for LC–MS/MS and GC–MS/MS technologies for the analysis of certain pesticides in cannabis flower. Cannabis Sci Technol 2(5):6–11
32. Dalmia A et al (2019) A novel ESI and APCI LC/MS/MS analytical method for meeting the Canadian cannabis pesticide residues regulatory requirements, 21563. PerkinElmer, Inc., Waltham, MA
33. Hollis JS, Fausett E, Westland J, Macherone A (2019) Analysis of challenging pesticides regulated in the cannabis and hemp industry with the Agilent Intuvo 9000-7010 GC/MS/MS system: the Fast-5. Application note 5994-1604. Agilent Technologies, Inc.
34. Hollis JS, Harper T, Macherone A (2020) Terpenes analysis in cannabis products by liquid injection using the Agilent Intuvo 9000/5977B GC/MS system. Agilent Application note 5994-2032EN. Agilent Technologies, Inc., in press
35. Jordan R, Asanuma L, Miller D, Macherone A (2018) A comprehensive approach to pesticide residue analysis in cannabis. Cannabis Sci Technol 1(2):26–31
36. Macherone A. Cannabinoid quantitation using HPLC-UV & multi-platform, comprehensive quality testing. In: PittCon conference & expo. Orlando, 26 Feb–1 Mar 2018
37. Wylie PL et al (2020) Screening for more than 1,000 pesticides and environmental contaminants in cannabis by GC/Q-TOF. Med Cannabis Cannabinoids. https://doi.org/10.1159/000504391
38. Caprioglio D et al (2020) The oxidation of phytocannabinoids to cannabinoquinoids. J Nat Prod. https://doi.org/10.1021/acs.jnatprod.9b01284
39. Natalya M et al (2007) HU-331, a novel cannabinoid-based anticancer topoisomerase II inhibitor. Mol Cancer Ther 6(1):173–183
40. del Rio C et al (2016) The cannabinoid quinol VCE-004.8 alleviates bleomycin-induced scleroderma and exerts potent antifibrotic effects through peroxisome proliferator-activated receptor-γ and CB2 pathways. Sci Rep 18(6):21703

Cannabinoid Detection and Quantitation

Anthony Macherone

Abstract In countries were medicinal or adult recreational use cannabis programs have been legalized, analytical testing including cannabinoid quantitation, pesticides and mycotoxins analysis, heavy metals analysis, residual solvents analysis, terpene profiling, and microbial screening has been regulated. Other tests like moisture content, water activity, and "filth" may also be regulated. Performance of these tests requires a suite of analytical testing platforms. This chapter will focus on cannabinoid quantitation using high performance liquid chromatography, gas chromatography, liquid chromatography mass spectrometry, and gas chromatography mass spectrometry. Other topics of discussion will include sample preparation, caveats to some of the applications, and non-targeted comprehensive screening methodologies using time-of-flight mass spectrometry. Although focused on cannabis and cannabinoid product testing, these methods are easily translated to hemp and hemp products. At its essence hemp is the same species as cannabis but differentiated only by its psychoactive potential as defined by total Δ^9-tetrahydrocannabinol content.

Biosynthesis of Cannabinoids

Cannabis spp. do not directly synthesize Δ^9-tetrahydrocannabinol (THC), cannabidiol (CBD) or any of the non-acid phytocannabinoids in the living plant. The in vivo genesis of cannabinoids begins through two polyketide fatty acid pathways that synthesize divarinic acid and olivetolic acid. These acids react with the substrate geranyl pyrophosphate and geranyl-diphosphate:olivetolate geranyltransferase to synthesize cannabigerovarinic acid (CBGVA) and cannabigerolic acid (CBGA), respectively. Through the actions THCA synthase, CBDA synthase, and cannabichromenic acid synthase (CBCA) the former synthesizes Δ^9-tetrahydrocannabivarinic acid (THCVA), cannabidivarinic acid (CBDVA), and cannabichromevarinic acid (CBCVA), respectively. The latter synthesizes THCA, CBDA, and cannabichromenic acid (CBCA), respectively. Post-harvest exposure

A. Macherone (✉)
Agilent Technologies, Inc., Wilmington, DE, USA

The Johns Hopkins University School of Medicine, Baltimore, MD, USA
e-mail: anthony_macherone@agilent.com; amacher1@jhmi.edu

© Springer Nature Switzerland AG 2021
S. R. Opie (ed.), *Cannabis Laboratory Fundamentals*,
https://doi.org/10.1007/978-3-030-62716-4_9

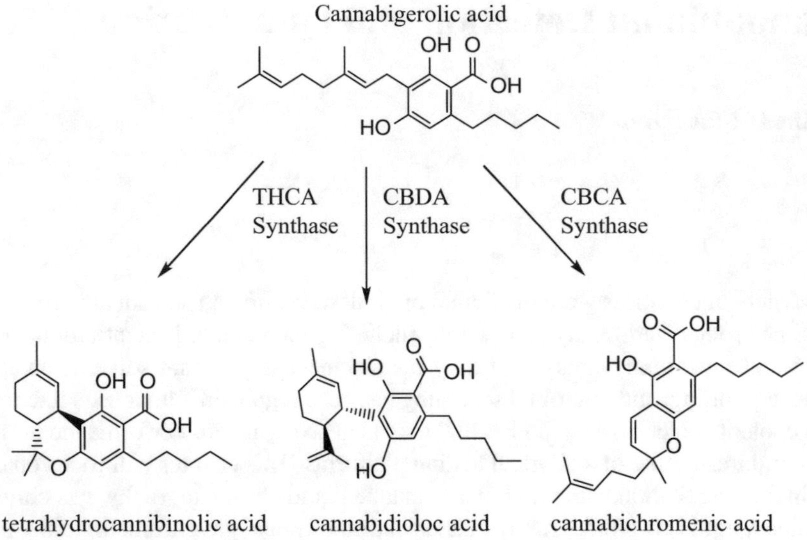

Cannabigerolic acid

Fig. 1 Phytocannabinoid acid biosynthesis from CBGA

to drying, heat and light, cause spontaneous decarboxylation of THCA, CBDA, and other acid phytocannabinoids to yield THC, CBD, cannabichromene (CBC), cannabigerol (CBG), etc. Other cannabinoids such as cannabinol (CBN) and cannabicyclol (CBL) are oxidation or photo-irradiation products [1, 2]. Figure 1 illustrates THCA, CBDA, and CBCA biosynthesis from CBGA.

Cannabinoid (Potency) Testing

In the past, cannabis has been treated as a U.S. Drug Enforcement Agency (U.S. DEA). Schedule I narcotic that is highly illegal to possess. Schedule I drugs have no "accepted medical use" and include marijuana (cannabis), heroin, lysergic acid diethylamide (LSD), methaqualone, peyote, and 3,4-methylenedioxymethamp hetamine (ecstasy) [3]. From this legal perspective, most laboratory testing was for post-consumption toxicology or criminalistic identification of bulk drug seizures. However, over the past several years, in addition to medicinal cannabis programs, Canada and 11 states and the District of Columbia in the U.S. have passed legislation legalizing recreational use of cannabis or cannabinoid products for adults over 21 years of age.

Regardless of region or country where medicinal or recreational cannabis use has been legalized, potency testing is always required. Potency specifically refers to the total Δ^9-tetrahydrocannabinol (TCH) content of the plant or product. Total potential THC content is determined through Eq. 1.

$$0.877 * [THCA] + [THC] \tag{1}$$

where [THCA] is the concentration of tetrahydrocannabinolic acid, and [THC] is the concentration of THC in the product. The factor 0.877 is derived from the molecular weight ratios of the decarboxylated chemical species and the phytocannabinoid acids (314.2 m/z/358.2 m/z = 0.877).

In the U.S. and Canada, a total THC content > 0.3% by weight of plant material is considered marijuana or a potentially intoxicating cannabinoid product. In the European Union the cut-off is 0.2% (wt/wt), and in Switzerland, it's 1.0% by weight. Many hemp products tend to be higher in CBD concentration and the total CBD content is determined through Eq. 2

$$0.877 * [CBDA] + [CBD] \tag{2}$$

where [CBDA] is the concentration of cannabidiolic acid, and [CBD] is the concentration of CBD in the product. Since the molecular weights of THCA and CBDA are the same, as are the molecular weights of THC and CBD, the same factor of 0.877 applies to Eq. 2.

The results of cannabinoid analysis are generally determined in µg/mL as calculated from regression statistics of a linear calibration curve. However, many regulatory agencies require the cannabinoid testing results to be reported in percent weight of the product. To convert from µg/mL to % weight Eq. 3 is used.

$$\% \, wt./_{wt.} = \left[\frac{Concentration * V * DF}{M * 1 \times 10^6} \right] * 100 \tag{3}$$

where, Concentration = concentration of analyte from linear regression analysis (µg/mL), V = volume of extraction solvent for solid samples or dissolution solvent for oils, tinctures, resin, etc., in mL, DF = dilution factor (unitless), M = mass of sample (g), and 1×10^6 = conversion from µg/g to g/g, and 100 determines % wt./wt.

In December 2018, the U.S. Agriculture Improvement Act was signed into law. This legislation legalized hemp as an industrial crop. Unlike medicinal and adult recreational use cannabis programs, which are regulated at the state level in the U.S., U.S.D.A. regulates industrial hemp. The Federal Register (FR 53365) defined hemp as any part of *Cannabis sativa L.* with a dry weight concentration of THC not greater than 0.3% (wt./wt.) [4]. U.S.D.A. has further defined the "Establishment of a Domestic Hemp Production Program" in the Federal Register [5]. Therein, they positioned both high performance liquid chromatography (HPLC), and gas phase chromatography (GC) as the preferred testing equipment and stipulated all testing must occur in a U.S. DEA registered laboratory. However, a moratorium was placed on this stipulation until October 2020.

Analytical Chemistry

All regions where some form of cannabis or cannabinoid products have been legalized for medicinal or adult recreational use programs, require identification and quantification of THCA, CBDA, THC, and CBD. The State of California further regulates CBN and CBG [6]. However, most laboratories identify and quantify the regulated cannabinoids and as many as 5–17 more that are commercially available as certified reference materials.

Sample Preparation for Inflorescence and Oils, Tinctures, and Concentrates [7]

Inflorescence: Accurately weigh and record approximately 0.20 g homogenized inflorescence into a 50-mL centrifuge tube. Add 20 mL methanol and shake for 10 min to extract. Centrifuge at 5000 rpm for 5 min. Filter 2.0 mL of the supernatant into a fresh vial with a 0.45 μm regenerated cellulose syringe filter. Transfer 50 μL of the filtered supernatant into a 2 mL auto-sampler vial. Add 950 μL methanol and vortex to mix. In this example, M is the actual mass of the sample, the initial volume V, of the extract is 20 mL, the dilution factor DF, is 20, and the total dilution is 2000. An example of this and the conversion to % wt./wt. as given in Eq. 3 is given in Table 1 for ten virtual inflorescence samples (Vs).

Oils, tinctures, resins, and concentrates: Transfer 100 μL of sample into a tared 10 mL volumetric flask. Accurately record the weight in the flask. Add approximately 8 mL ethanol and cap to mix well. Bring to volume with ethanol. Centrifuge and filter 2.0 mL of the supernatant into a clean glass vessel using a 0.45 μm regenerated cellulose syringe filter. Transfer 100 μL of the filtered solution into a 2 mL auto-sampler vial and add 900 μL high purity ethanol. Cap and vortex briefly to mix. In this example, M is the actual mass of the sample weighed into the tarred vial, the initial volume V, of dissolution solvent is 10 mL, the dilution factor DF, is 100, and the total dilution is 1000.

Table 1 Percent weight determinations for 10 virtual samples

Sample ID	Target mass (g)	Actual mass (g)	V (mL)	DF	Concentration (ug/mL)	% wt./wt.
Vs-1	0.20	0.19	20	20	1.15	0.25
Vs-2	0.20	0.22	20	20	4.20	0.76
Vs-3	0.20	0.18	20	20	5.40	1.20
Vs-4	0.20	0.20	20	20	2.53	0.52
Vs-5	0.20	0.22	20	20	16.49	3.07
Vs-6	0.20	0.19	20	20	0.98	0.21
Vs-7	0.20	0.19	20	20	0.82	0.17
Vs-8	0.20	0.18	20	20	0.99	0.22
Vs-9	0.20	0.19	20	20	1.05	0.22
Vs-10	0.20	0.20	20	20	15.80	3.13

Caveats to sample preparation: Solvent selection, pH, and temperature should be considered when extracting cannabinoids from plant material and solubilizing oils, tinctures, resins, and concentrates. In the presence of sulfuric acid, muriatic acid, and acetic acid, CBD transforms into THC through an acid catalyzed stable carbocation intermediate [8]. A similar phenomenon has been observed with trifluoracetic anhydride (TFAA) and 1,1,1,3,3,3-hexafluoroisopropanol (HFIP) derivatization of THC and CBD [9]. Under these acidic conditions, THC and CBD yielded trifluoracyl derivatives of both Δ^9-THC and Δ^8-THC isomers. This was not observed with trimethylsilyl (TMS) derivatives using N-Methyl-N-(trimethylsilyl) trifluoroacetamide (MSTFA) which yielded the expected THC-TMS and CBD-TMS derivatives.

CBD to THC conversion has also been observed when using halogenated solvents like dichloromethane (DCM), which becomes acidic over time through hydrolysis processes as given in Eq. 4. Even the slightest amount of HCl can initiate and propagate the conversion since the acid catalyst is regenerated as the reaction proceeds. It is therefore suggested to avoid halogenated or other organic solvents that may become acidic when extracting/solubilizing cannabis and cannabinoid products, especially those with high CBD content.

$$CH_2Cl_2 + H_2O \rightarrow CH_2O + 2HCl \tag{4}$$

The temperature of the extraction/solubilization solvent must also be considered as should heating that may occur through friction during the homogenization process using mechanical grinders. As noted in section "Biosynthesis of Cannabinoids," decarboxylation of a phytocannabinoid acid occurs spontaneously over time and with exposure to heat and light. Using HPLC analysis, a temperature dependent experiment was performed to determine the extent of THCA conversion to THC and other cannabinoids [10]. In that study, the authors intentionally decarboxylated THCA in an oven at 120 °C, 140 °C, 160 °C, and 180 °C for 15 min. They determined that up to 140 °C, THCA to THC conversion was about 70%. At 160 °C and greater, THCA fully decarboxylated to THC and two degradation products, CBN and dihydroxycannabinol. Temperature dependent acid phytocannabinoid decarboxylation is also observed in the hot inlet of a GC system and in the heated zones of an electrospray source.

Given the information above, a laboratorian must control for three potential transformations that may occur during sample preparation that will affect quantitative outcomes. These are:

1. CBDA acid catalyzed conversion to THCA followed by decarboxylation with heat to THC
2. CBDA decarboxylation with heat to CBD followed by acid catalyzed conversion to THC
3. CBD acid catalyzed conversion to THC

Proper use of extraction /solubilization solvents, temperatures, and using cold solvents and cryo-freezing when grinding where possible may mitigate these unwanted conversions.

High Performance Liquid Chromatography (HPLC)

In regulatory testing laboratories, analytical chemistry to identify and quantify cannabinoids in each plant strain or product is nearly always performed with HPLC systems with UV detectors. The analysis typically yields concentration results in µg/mL. Most reporting agencies require the results to be reported in percent per dry weight of the product. For plant material, the dry weight is generally determined gravimetrically as moisture content pre- and post-drying in a vacuum oven. Dry material is considered to have a moisture content <= 13%. To convert from µg/mL to % wt./wt., Eq. 3 is used.

To perform cannabinoid testing using HPLC-UV, the system must include a binary or quaternary pump, an auto-sampler, an analytical column and column compartment, and an UV detector. The mobile phases typically consist of an aqueous channel and an organic channel which are buffered with a low percentage (v/v) of formic acid or millimolar concentrations of ammonium formate. An "iso-gradient" method with a long hold time of the initial conditions followed by a fast ramp to high organic phase is most useful. Superficially porous particle (SPP) columns with sub 3-micron C_{18} particles further increase chromatographic resolution compared to traditional large bore, large particle column packings. Diode array detectors (DAD) are the best choice for cannabinoid analysis. A DAD can monitor multiple wavelengths simultaneously and collect a UV spectrum over a defined wavelength range for each analyte and offers some structural information to distinguish acids from neutrals as an example. When plotted on an x-y axis, the concentration of each analyte is determined through linear regression from Eq. 5.

$$Concentration\left(mass\,/\,volume\right) = \left(\frac{Response - Intercept}{Slope}\right) \qquad (5)$$

Storm et al. (2019) used a self-contained HPLC system with a "gradient pump (maximum pressure 400 bar) with integrated degassing unit, autosampler, column oven, and variable wavelength detector (VWD) with standard flow cell." The VWD monitored 230 nm. A superficially porous column was used (Agilent InfinityLab Poroshell 120 EC-C18 3.0 × 50 mm, 2.7 µm) at 50 °C. A binary mobile phase comprised of A) 0.1% (v/v) formic acid aqueous phase, and B) 0.05% (v/v) formic acid organic phase was used at 1.0 mL/min. The mobile phase gradient was 60% B at time zero, hold 1.0 min. Then, linearly increase B to 77% at time 7.0 min, no hold time. Then, linearly increase B to 95% at 8.2 min, hold 1.3 min. Eleven cannabinoids were analyzed and included Cannabidivarin (CBDV), Tetrahydrocannabivarin (THCV), CBD, CBG, CBDA, CBGA, CBN, THC, Δ^8-THC, CBC, and THCA. Method statistics such as accuracy, precision, linearity, and range were

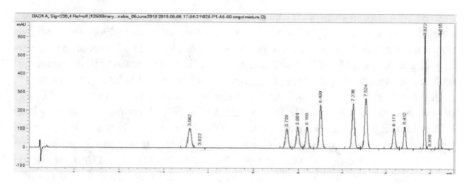

Fig. 2 Typical HPLC chromatogram for 11 cannabinoids monitored at 230 nm

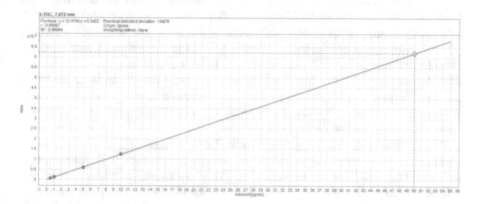

Fig. 3 Calibration curve for THC

determined. Figures 2 and 3 illustrate a common chromatogram obtained from the method and calibration curve for THC, respectively.

Liquid Chromatography-Mass Spectrometry (LC-MS)

The addition of MS to the analysis of cannabinoids facilitates increased selectivity for more accurate identification of cannabinoids in each product. MS also provides differentiation of co-eluting chemicals that may be encountered and unduly influence analytical results when using non-selective detectors like UV. Adding MS to the analysis generally requires no changes to the HPLC method parameters, and the methods can simultaneously collect both UV and MS data. Like a DAD, collecting spectral information, MS information is 3-dimensional data but in this latter case, the third dimension is a unique mass/charge ratio of ions formed in the electrospray ionization (ESI) source of a LC-MS system. Quasi-molecular ions formed in ESI include $(M + H)^+$ or $(M-H)^-$ in ESI positive (ESI+) or ESI negative (ESI-) modes, respectively where M is the molecular weight of the molecule and H is a proton.

Ionized adducts can form with metal ions such as Na^+ or K^+ in ESI+ and in these cases the ion is in the form $(M + X)^+$ where X is the metal ion adduct. Anion adduction can occur in ESI- as in the case of $(M + Cl)^-$ or $(M + Br)^-$. The charge is generally ± 1 but dimers $(M + 2H)^{2+}$ and trimers $(M + 3H)^{3+}$ can form with certain small molecules. Macromolecules like proteins are typically multiply charged $(M + nH)^{n+}$.

A recently published LC-MS method used both DAD and MS [11]. The LC-MS system was the Agilent 1260 Prime Infinity II HPLC coupled to the Agilent InfinityLab single quadrupole mass selective detector (iQ MSD). The HPLC system was equipped with a vial sampler with a 20 µL metering device and 20 µL loop, an integrated column compartment, sample cooler, and DAD. The DAD monitored 228 nm and 270 nm and scanned over the range of 220 nm through 450 nm at a 40 Hz sampling rate. Agilent OpenLab CDS Software (version 2.4) was used for data acquisition and data processing. A superficially porous column was used (Agilent InfinityLab Poroshell 120 EC-C18, 3.0 mm × 100 mm, 1.9 µm) at 50 °C. A quaternary mobile phase system comprised of A) 100% LC-MS grade water, B) 100% LC-MS grade acetonitrile, C) 100% LC-MS grade methanol, and D) 0.1% (v/v) formic acid in LC-MS grade water was used. The mobile phase gradient was 25% A, 70% B, 0% C, and 5% D at time zero and these conditions were held until 3.2 min. Then, A was decreased to 5%, B was decreased to 0%, C was increased to 90%, and D remained constant at 5% until 8.2 min. These conditions were held constant until the end of the analysis at 12 min.

The iQ MSD parameters were as given in Tables 2 and 3.

Figure 4 shows the DAD and SIM TIC chromatograms for the 11 cannabinoids defined for the HPLC method in section "High Performance Liquid Chromatography (HPLC)." Many laboratories however, desire to identify and quantify as many cannabinoids as possible. The method of D'Antonio, et al. (2020) was redefined to analyze 16 commercially available cannabinoids by adding the proper SIM ions and SIM windows. Table 4 lists the target cannabinoids and their chromatographic resolution, and Fig. 5 is the DAD and SIM TIC chromatograms.

Caveats to LC-MS analyses: Phytocannabinoid acids like THCA, CBDA, THCVA, CBDVA, etc., spontaneously decarboxylate after harvest and upon exposure the heat and light. In recent work, the question was posed: do acid phytocannabinoids decarboxylate in the heated zones of an ESI source? [12]. In that study, the stability of phytocannabinoid acids analyzed with positive and negative electrospray LC–MS was evaluated as a function of drying gas temperature. Both flow injection analysis (FIA) and chromatographic conditions were used to measure insource acid decarboxylation over a drying gas temperature range of 75 °C through

Table 2 iQ MSD parameters

Source parameter	Value
Gas temperature	325 °C
Gas flow	13 L/min
Nebulizer	55 psig
Capillary voltage	3500

Table 3 iQ SIM/scan parameters

Segment	Name	$(M + H)^+$ (m/z)	Fragmentor	Polarity
Scan	Scan	200–700	100	ESI+
SIM	THCA and CBDA	359.2	100	ESI+
SIM	CBN	311.2	100	ESI+
SIM	CBD, THC, CBC	315.2	100	ESI+
SIM	CBG	317.2	100	ESI+
SIM	CBDV and THCV	359.2	110	ESI+
SIM	CBGA	361.2	110	ESI+

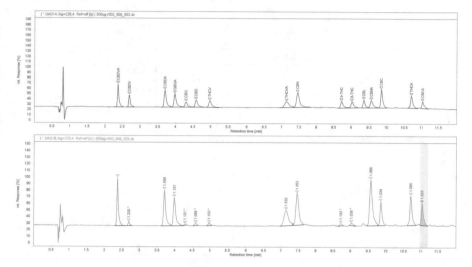

Fig. 4 Top: LC-MS SIM TIC chromatograms for 11 targeted cannabinoids. Bottom DAD chromatogram with compound labels and retention times

300 °C. The mass spectrometers used in this study included single quadrupole LC-MSD, time-of-flight (LC-TOF), and quadrupole time-of-flight (LC-QTOF) systems. In ESI+ mode, no decarboxylation was observed under FIA or chromatographic conditions using the LC-MSD and the LC-QTOF systems. However, as a function of drying gas temperature, ESI- mode did reveal a decrease in the relative abundance of THCA and CBDA with an associated increase in the relative abundance THC and CBD on the LC-TOF system. "These data highlight a potential concern for the quantitation of acid phytocannabinoids using ESI- mode LC-MS technologies" (Stone et al. 2020).

Table 4 Sixteen target cannabinoids and chromatographic resolution using the method of D'Antonio et al. (2020)

Compound in retention order	Chromatographic resolution
CBDVA	
CBDV	1.220
CBDA	1.558
CBGA	1.101
CBG	1.101
CBD	1.089
THCV	1.103
THCVA	1.532
CBN	1.052
Δ^9-THC	1.193
Δ^8-THC	1.038
CBL	1.100
CBNA	1.066
CBC	1.034
THCA	1.095
CBCA	1.033

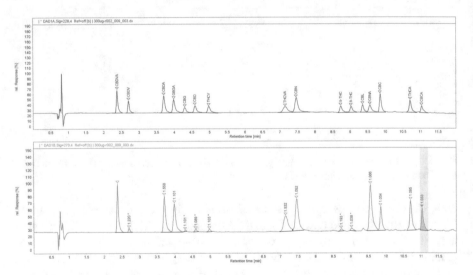

Fig. 5 Top: DAD chromatogram with compound labels and retention times. Bottom: LC-MS SIM TIC chromatograms for 16 target cannabinoids

Gas Chromatography (GC) and GC-Mass Spectrometry (GC-MS)

From a forensic point of view, marijuana (cannabis) testing has been conducted for decades. Common post-consumption toxicology testing includes identification and quantification of psychoactive THC and common biological phase I metabolites like 8-OH-THC, 11-OH-THC and 11-COOH-THC. Phase II metabolic glucuronide conjugates are common in urine samples, and phase I metabolites are found in blood, hair and oral fluid. Figure 6 illustrates in vivo metabolism of THC.

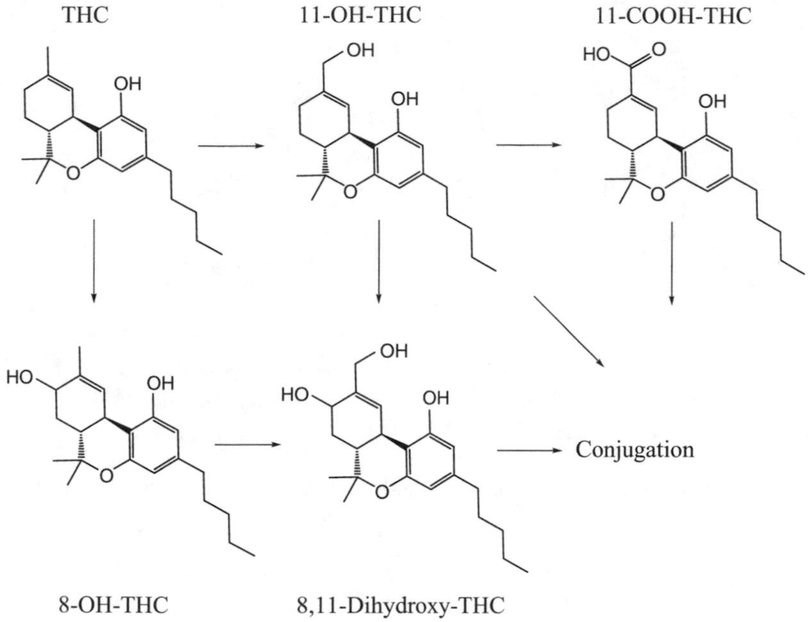

Fig. 6 Phase I and II biological metabolism of THC [13]

When presented with bulk drug seizures, criminalistic labs traditionally have asked the question: "is it marijuana"? A semi-quantitative GC method has been employed to answer this binary question. However, since the December 2018 signing of the Agriculture Improvement Act (the 2018 Farm Bill), hemp has been legalized as an industrial crop. Industrial hemp is defined as any part or derivative (including seeds) of the plant *Cannabis sativa L.* with a dry weight concentration of tetrahydrocannabinols not greater than 0.3% wt./wt. (Macherone 2019). This change in the law at the federal level now requires criminalistic and state testing labs to quantitatively determine the total potency (Eq. 1) of any presumptive cannabis material to either classify it as hemp or a hemp product or as marijuana or a cannabis product.

Decarboxylation of acid phytocannabinoids is readily observed when using gas phase systems for cannabinoid analyses. This reaction has been used in semi-quantitative GC analysis with mass spectrometry (GC-MS) for the binary determination of THC content for forensic purposes. In this methodology, THCA to THC conversion is presumed to be 100% and the THC chromatographic peak area is compared to that of isotopically labeled THC added to the sample at a known concentration.

Trimethylsilyl derivatization with MSTFA inhibits decarboxylation in the hot GC inlet and allows for the determination of total potential THC and total CBD as given in Eqs. 1 and 2. Figure 7 shows TMS derivatized THCA and THC.

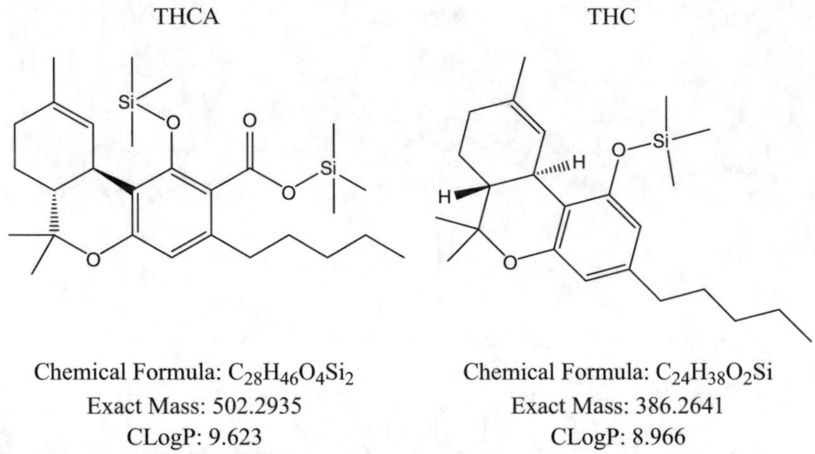

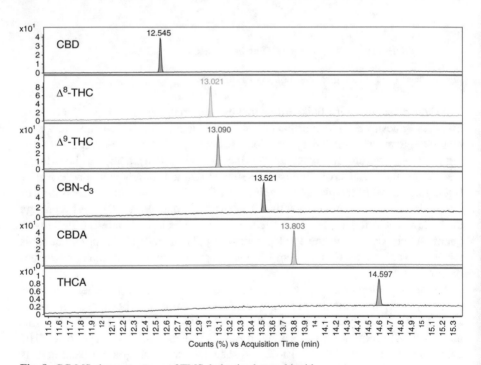

THCA

Chemical Formula: $C_{28}H_{46}O_4Si_2$
Exact Mass: 502.2935
CLogP: 9.623

THC

Chemical Formula: $C_{24}H_{38}O_2Si$
Exact Mass: 386.2641
CLogP: 8.966

Fig. 7 Left: TMS-THCA derivative. Right: TMS-THC derivative

CBD — 12.545

Δ^8-THC — 13.021

Δ^9-THC — 13.090

CBN-d_3 — 13.521

CBDA — 13.803

THCA — 14.597

Counts (%) vs Acquisition Time (min)

Fig. 8 GC-MS chromatograms of TMS derivatized cannabinoids

Recent unpublished work has demonstrated a quantitative methodology for cannabinoid analysis and the determination of total THC and total CBD. Figure 8 illustrates GC-MS chromatograms for CBD, Δ^8-THC, Δ^9-THC, deuterated CBN-d$_3$, CBD, and THCA derivatized with MSTFA. The method used an Agilent Intuvo 9000 GC coupled to a 5977B MSD GC-MS system. The GC was equipped with a Multi-Mode Inlet (MMI) operated at 280 °C in split mode with a 100:1 split ratio. The Intuvo system also included the Mid-Column Backflush Flow Chip and Guard Chip. Other Intuvo GC setpoints were Guard Chip Track Oven, Bus Temperature 280 °C, MSD Connector 300 °C, MSD Transfer Line 300 °C, and PSD Purge 5 mL/min, Supplies Column 2. Column #1 was a 15 m × 250 µm × 0.25 µm Agilent HP-5MS UI INT operated at 1.0 mL/min constant flow. Column #2 was also a 15 m × 250 µm × 0.25 µm Agilent HP-5MS UI INT operated at 1.2 mL/min constant flow. The post-run backflush conditions were −5.6833 mL/min on column #1, and 6.000 mL/min on column #2 at 300 °C. The GC oven parameters were 70 °C, hold 1 min. Then, the oven was ramped at 20 °C/min to 300 °C, hold 4 min. The MSD was operated in electron ionization (EI) mode in scan from 35 m/z to 600 m/z with a 3-min solvent delay.

Although this method is well-suited for the analysis of cannabis and cannabinoid products, most cannabis testing labs performing the state or country regulated tests, use HPLC with UV detection. HPLC methods are simple, robust, do not require derivatization chemistries, and they are relatively cheap with respect to cost per sample.

Time-of flight mass spectrometry: Time-of-flight (TOF) mass spectrometry is not typically found in the day-to-day regulatory testing lab. Single quadrupole and tandem quadrupole (MS/MS) mass spectrometers generally provide the sensitivity and selectivity required to successfully perform the required testing at the LOD and LOQ defined by the various state and country regulatory entities. TOF systems generate high-resolution accurate mass (HRAM) information that is very useful for discovery-based experiments like metabolic identification, metabolomics, proteomics, and large-scale exposomic studies.

There are two independent properties to consider when discussing resolution: resolving power, and mass resolution [14]. Resolving power is given by Eq. 6.

$$Resolving\ Power = \frac{M_1}{M_1 - M_2} \tag{6}$$

where M_1 and M_2 are the m/z masses for two adjacent ions of equal height with an overlap of 10%.

Mass resolution is defined by Eq. 7.

$$Mass\ Resolution = \frac{M}{\Delta M} \tag{7}$$

where M is a singly charged m/z ion, and ΔM is the ion width at full-width half height.

These two properties of a HRAM time-of-flight mass spectrometer define the power of the system to differentiate unique chemical species in sample matrices from neighboring ions with similar molecular weights. As an example, the typical mass resolution of a quadrupole mass spectrometer approximately 0.6 atomic mass units (amu) while on a TOF system it is 0.01 amu or smaller. This means that for an ion with a mass of 500 m/z, the quadrupole system mass resolution is about 833, while the TOF mass resolution is 50,000.

Another unique property of HRAM mass spectrometers is mass accuracy. Since the kinetic energy and distance an ion travels in the flight tube are fixed, the time in microseconds it takes for the ion to travel the distance of the flight tube can be accurately measured and the mass m, of the ion is determined by Eq. 8.

$$m = \left(\frac{2E}{d^2} \right) \left(t_m - t_0 \right)^2 \tag{8}$$

where E is the applied kinetic energy, d is the distance of the flight tube, t_m is the measured time, and t_0 is the difference between the start and stop delay times of the ion as it travels through the flight tube. A true TOF instrument can measure mass defects within ±0.00005 amu and the difference between the measured mass and the true mass calculated from an empirical formula is measured in parts per million (ppm) as given in Eq. 9.

$$\Delta ppm = \left(\frac{measured\ mass - true\ mass}{true\ mass} \right) * 1 \times 10^6 \tag{9}$$

As an example, a difference in the measured mass of 292.0404 m/z and the true mass of 292.0403 m/z is −0.3424 ppm. The power of high mass accuracy is the ability to generate putative empirical formula from the measured masses for identification purposes. Even with this level of mass accuracy, at least 5 empirical formulae are generated for a measured mass of 292.0403 m/z using common organic chemical elements C, H, N, O, S, and Cl. This is far better than the dozens or hundreds that could be generated from a quadrupole instrument with unit mass resolution (±0.5 amu). Table 5 lists empirical formulae, m/z, and Δppm generated from a mass of 292.0403 m/z from common organic elements.

GC-TOF analysis of CBD oil pet supplements: Many pet owners are supplementing their animals with CBD products. The number available products for pets

Table 5 Elemental analysis for a measure mass of 292.0403 m/z

Empirical formula	m/z	Δppm
$C_{17}H_9ClN_2O$	292.0403	−0.14
$C_6H_{16}Cl_2N_5O_2S$	292.0402	0.42
$C_{13}H_6N_7S$	292.0405	−0.82
$C_{14}H_{12}O_5S$	292.0405	−0.84
$C_{10}H_{19}Cl_3O_3$	292.0400	1.10

is growing every day but quality control may be questionable in some cases. It is therefore important for the same screening protocol that is regulated in medicinal and adult use recreational programs be applied to products manufactured for pets.

Data-driven investigations are common in omics (big data) studies of plants, microbes, insects, animals, and humans for hypothesis generation. For example, a recent investigation used LC-QTOF, GC-QTOF, and qPCR to apply a discovery-based exposome paradigm to the study of honeybee health and disease [15]. The results of that work identified pesticides that were significantly associated with *Nosema ceranae* prevalence, dysregulated biological pathways, and postulated a causal-reactive feedback loop that increases disease susceptibility.

Another study used GC-TOF mass spectrometry in a non-targeted, data-driven survey of 6 commercially available CBD oil pet supplements to statistically examine and compare the products based on their intrinsic chemical profiles [16]. In that work, 5 representations of 6 commercially available pet supplements (n = 30) were purchased to account for lot variations in each product where possible. The samples were prepared with fresh dichloromethane to avoid the acid issues noted in section "Sample Preparation for Inflorescence and Oils, Tinctures, and Concentrates [7]." An isotopically labeled mixture of polyaromatic hydrocarbons (PAH) was added as internal standards. Each sample was prepared individually, and a pooled mixture of all samples was prepared as a quality control. Five replicates of each sample were acquired (n = 180). An Agilent 7890/7250 GC-QTOF system was used for all analysis. The GC-QTOF was equipped with a split/split-less inlet operated at 280 °C and mid-column backflush where column 1 and column 2 were Agilent J&W DB-35 ms Ultra Inert (30 m x 250 μm x 0.25 μm) operated at 1.2 mL/min and 1.4 mL/min, respectively. A 1.0 μL injection was made in split mode with a split ratio of 10:1. The oven program was 60 °C (1.5 min), then 30 °C/min to 180 °C (0 min), then 15 °C/min to 255 °C (0 min), then 10 °C/min to 320 °C (6.5 min). The transfer line temperature was 320 °C. The QTOF mass spectrometer was operated in scan mode over a range of 45–650 m/z using electron ionization (EI). Two EI modes were used: (1) standard EI where the source temperature was 300 °C, the electron energy was 70 eV, and the emission current was 5 μA, and (2) Low EI mode where the source temperature was 200 °C, the electron energy was 14 eV, and the emission current was 0.8 μA. Chemical features from the raw data were identified using Mass Hunter Unknowns Analysis software. Annotation was performed by searching commercially available spectral databases. Mass Profiler Professional software was used for covariate analysis.

The results of the pet supplement study "identified fatty acids, fatty acid esters, di- and triglycerides, terpenoids, cannabinoids, tocopherols, steroids and sterols, essential oils, fragrances, alkanes, alkenes, and alcohols" (Curtis et al. 2019). Principle component analysis (PCA) revealed 4/6 products occupying unique regions in the PCA space, and 2/6 products that were very similar to one another (made by the same manufacturer). THC was also identified in each product with a relative abundance indicative of exceeding 0.3% in 2/6 products. Figure 9 illustrates the relative abundance of THC in each product. This non-targeted workflow is

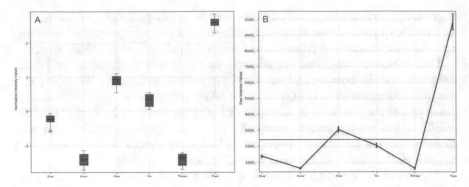

Fig. 9 (**a**) Box-Whisker plot of the normalized THC abundance in the 6 CBD pet supplements. (**b**) Raw THC intensity values. Black horizontal line estimates the 0.3% THC cut off

easily translated to cannabinomics, terpenomics, and plant metabolomics in discovery-based studies.

General Screening of Cannabis and Cannabinoid Products in Regulatory Labs Using TOF Technologies

In a presentation at Pittcon in 2018, Macherone proposed a comprehensive testing approach for cannabinoid quantitation and other chromatographic based assays common to the regulatory testing lab [17]. The workflow confirmed that TOF-based testing for regulated cannabinoids is overkill and that traditional HPLC with UV detection is adequate for that purpose. However, a semi-targeted analysis with HRAM systems offers comprehensive approach for simultaneous quantitative potency determinations, qualitative profiling for other cannabinoids, and characterization of flavonoids, phenols, alkaloids, amides, fatty acids, and other endogenous chemicals. This approach may also facilitate strain verification and product quality control based on chemical profiles. Perhaps a more important role for TOF assays would be screening for adulterants like synthetic cannabinoids, cathinones (bath salts), and designer drugs, and characterization of extractable or leachable compounds tainting cannabinoid products through packaging and inks. In this model, all samples received in a lab are screened with LC-TOF and GC-TOF systems. Cannabinoid and terpene concentrations would be quantified. Only a subset of samples that are presumptive positives for pesticides would require confirmation and quantitation using targeted methodologies. An example of this model was developed by Wylie et al. (2020) [18]. In that work, 1000 pesticides were screened using GC-QTOF and a commercially available HRAM Personal Compound Database and Library (PCDL). This proof-of-principle study demonstrated the viability of non-targeted suspect screening of pesticides in cannabis inflorescence.

Summary Discussion

There are at least 70 known cannabinoids that can be attributed to *Cannabis spp.* [19, 20] yet state and country regulatory entities have focused on only a handful (THC, THCA, CBD, CBDA, CBN, and CBG). From a product safety perspective, is this enough? Oxidation, irradiation, photochemical conversion converts cannabinoids into non-natural chemicals. For example, aerobic oxidation of CBD yields cannabidiolquinone (CBQ, HU-331) [21]. Originally synthesized from CBD as an anti-cancer drug, HU-331 inhibits DNA topoisomerase II [22] but development was abandoned due to the production of reactive oxygen species, disruption of the mitochondria transmembrane potential, and cytotoxicity in primary and transformed cells in vitro [23]. Other questions include: should synthetic cannabinoids be on these lists, and what other chemicals should be monitored to ensure the safety of the products? Hitherto, only the degradation product CBN is regulated and only in one state.

This chapter focused on cannabinoid quantitation using HPLC, LC-MS, GC, GC-MS, and TOF mass spectrometry. Sample preparation methods were briefly discussed and potential pitfalls that may cause CBD conversion to THC were illustrated as well as caveats to solvent selection and extraction temperatures. For the most part, HPLC testing is adequate for compliance in regulatory testing labs. HPLC testing is simple and does not require derivatization to measure phytocannabinoid acids, decarboxylation compounds like THC and CBD, and degradation products like CBN. Cannabinoid testing by LC-MS was also discussed and caveats discussed concerning in situ decarboxylation of phytocannabinoid acids in ESI-mode. Gas phase methodologies with or without derivatization can also be used for the quantification of cannabinoids but may offer a limited set of information without derivatization and are more complex with derivatization. A comprehensive approach to cannabis screening was discussed with respect to omics type discovery studies and as an option for screening of cannabis samples. Some of the questions posed above like: 'are we testing for enough cannabinoids or looking for adulterants' may be addressed through these more comprehensive approaches. Although this chapter focused on cannabis, these methodologies are equally applied to hemp and hemp products where warranted.

Acknowledgements The author would like to thank Anastasia Andrianova, Matthew Curtis, Sue D'Antonio, Eric Fausett, Wendi A. Hale, Terry Harper, Jeffery S. Hollis, Nikolas C. Lau, Bruce Quimby, Peter J.W. Stone, Christy Storm, Jessica Westland, Michael Zumwalt of Agilent Technologies, and Julie Kowalski of JA Science and Support.

References

1. Thomas BF, Elsohly M (2016) Biosynthesis and pharmacology of phytocannabinoids and related chemical constituents. In: Thomas BF (ed) The analytical chemistry of cannabis: quality assessment, assurance, and regulation of medicinal marijuana and cannabinoid preparations. Elsevier, Amsterdam, pp 27–41
2. Lewis MM et al (2017) Chemical profiling of medical cannabis extracts. ACS Omega 2:6091–6103
3. Drug Scheduling. Drug Schedules. https://www.dea.gov/drug-scheduling. Accessed 14 April 2020
4. Macherone A (2019) Hemp in the United States: an opinion. Cannabis Industry Journal. https://cannabisindustryjournal.com/column/hemp-in-the-united-states-an-opinion/. Accessed 6 April 2020
5. Federal Register Rules and Regulations (2019) Establishment of a domestic. Hemp Production Program 84(211):58522–58564
6. Bureau of Cannabis Control Text of Regulations (2019) California Code of Regulations Title 16 Division 42. https://www.bcc.ca.gov/law_regs/cannabis_order_of_adoption.pdf. Accessed 19 December 2019
7. Storm C, Zumwalt M, Macherone A (2109) A dedicated cannabinoid potency testing in cannabis or hemp products using the agilent 1220 infinity II LC system, Application note 5991-9285, Agilent Technologies, Inc., Santa Clara, CA
8. Kiselak TD, Koerber R, Verbeck GF (2020) Synthetic route sourcing of illicit at home cannabidiol (CBD) isomerization to psychoactive cannabinoids using ion mobility-coupled-LC–MS/MS. Forensic Sci Int 308:1–8
9. Andrews R, Paterson S (2012) Production of identical retention times and mass spectra for D9-tetrahydrocannabinol and cannabidiol following derivatization with trifluoracetic anhydride with 1,1,1,3,3,3-hexafluoroisopropanol. J Anal Toxicol 36:61–65
10. Dussy FE et al (2005) Isolation of Δ9-THCA-A from hemp and analytical aspects concerning the determination of D9-THC in cannabis products. Forensic Sci Int 149:3–10
11. D'Antonio S, Li G, Macherone A (2020) Quantitation of phytocannabinoid oils using the Agilent Infinity II 1260 Prime/InfinityLab LC/MSD iQ LC/MS system, Application note 5994-1706. Agilent Technologies, Inc., Santa Clara, CA
12. Stone PJW et al (2020) The stability of acid phytocannabinoids using electrospray ionization LC–MS in positive and negative modes. Cannabis Sci Technol 3(3):34–40
13. Baselt RC (2004) Disposition of toxic drugs and chemicals in man. Biomedical Publications, Foster City
14. Macherone A (2013) The future of GC/Q-TOF in environmental analysis. In: Ferrer I, Thurman EM (eds) Comprehensive analytical chemistry, vol 61. Elsevier, Amsterdam, pp 471–490
15. Broadrup RL et al (2019) Honey bee (Apis mellifera) exposomes and dysregulated metabolic pathways associated with Nosema ceranae infection. PLoS One 14(3):e0213249. https://doi.org/10.1371/journal.pone.0215166
16. Curtis M, D'Antonio S, Macherone A (2019) GC-TOF discovery-based profiling of CBD oil pet supplements. Cannabis Sci Technol 2(6):6–11
17. Macherone A Cannabinoid quantitation using HPLC-UV & Multi-platform, Comprehensive Quality Testing. In: PittCon Conference & Expo. Orlando, February 26 - March 1, 2018
18. Wylie PL et al (2020) Screening for more than 1,000 pesticides and environmental contaminants in cannabis by GC/Q-TOF. Med Cannabis Cannabinoids 3(1):1–1. https://doi.org/10.1159/000504391
19. ElSohly MA, Slade D (2005) Chemical constituents of marijuana: the complex mixture of natural cannabinoids. Life Sci 78:39–548
20. Marilyn A, Huestis MA (2007) Human cannabinoid pharmacokinetics. Chem Biodivers 4(8):1770–1804

21. Caprioglio D et al (2020) The oxidation of phytocannabinoids to cannabinoquinoids. J Nat Prod 83(5):1711–1715. https://doi.org/10.1021/acs.jnatprod.9b01284
22. Natalya M et al (2007) HU-331, a novel cannabinoid-based anticancer topoisomerase II inhibitor. Mol Cancer Ther 6(1):173–183
23. del Rio C et al (2016) The cannabinoid quinol VCE-004.8 alleviates bleomycin-induced scleroderma and exerts potent antifibrotic effects through peroxisome proliferator-activated receptor-γ and CB2 pathways. Sci Rep 18(6):21703

Utilizing GC-MS and GC Instrumentation for Residual Solvents in Cannabis and Hemp

Bob Clifford, Nicole Lock, Richard Karbowski, Vikki Johnson, Andy Sandy, Leyda Lugo-Morales, Alan Owens, and Christine Sheehan

Abstract This chapter provides an overview of residual solvents analysis in cannabis flower and related products, such as waxes, hash and hemp oil, butters and shatter. The chapter provides background on why it is important to test for residual solvents as well as information on sources of contamination, classification of the compounds, and state, federal and international regulations. Beyond these general discussion points, the chapter will offer details on head space GC-MS and GC-FID, the most widely used analytical instruments for residual solvents testing. Topics will include system blank vs. method blank, initial vs. continuing calibration curves, lab control samples, matrix spikes, and replicates samples. In addition, information about validation, installation qualification, operational qualification, and performance qualification will be discussed. Examples of cannabis sample data are also included. The chapter concludes with future direction and needs.

Background on Residual Solvents

Residual solvents are volatile organic chemicals that are left behind during the manufacturing process. The chemicals are used within the manufacturing process of concentrated cannabis products like oils and waxes. Solvents used for the manufacturing process are harmful to humans. Their toxicity is classified into three major categories: Class 1, Class 2, and Class 3 and these classifications are defined within United States Pharmacopeia (USP) Method <467> [1, 2]. Class 1 solvents are known carcinogens and are designated "To Be Avoided." Class 2 solvents are nongenotoxic animal carcinogens and are designated "To Be Limited." Class 3 solvents are considered the safest carcinogens and are designated "Low Toxic Potential."

B. Clifford (✉) · N. Lock · R. Karbowski · V. Johnson · A. Sandy · L. Lugo-Morales · A. Owens
Shimadzu Scientific Instruments, Columbia, MD, USA
e-mail: RHClifford@SHIMADZU.com

C. Sheehan
Albany Medical Center, Albany, NY, USA

© Springer Nature Switzerland AG 2021
S. R. Opie (ed.), *Cannabis Laboratory Fundamentals*,
https://doi.org/10.1007/978-3-030-62716-4_10

To understand where residual solvents originate, we need to understand the process in which the concentrates are made. There are many processes that can be done to create these concentrates, but an example would be to crush a cannabis sample in a solvent like isopropyl alcohol (isopropanol) to get the desirable cannabidiols (CBD) and terpenes out of the plant. The crushing and adding of a solvent may occur multiple times. The samples are then typically strained of plant material and the final liquid (which is primarily solvent) is heated to remove the solvent. The ultimate goal of heating this liquid is to obtain concentrated CBD. While this process may remove most of the solvent, there is a possibility for residual solvents after this process (and sometimes at a much higher concentration). Based on this process and the concentrate being made, different chemicals (solvents) might be used. Below are some examples of different chemicals that may be used in the creation of concentrates.

Chemical	Class number
Acetone	Class 3
Benzene	Class 1
Chloroform	Class 2
Ethanol	Class 3
Heptane	Class 3
Hexane	Class 2
Isopropanol	Class 3
Pentane	Class 3

While Class 3 chemicals are listed as "Low Toxic Potential," these chemicals still have the potential to be toxic to humans. Each state has different requirements regarding which classes are allowed for cannabinoid solvent use and which solvents should be screened for residual solvents. This information can be found in the next section.

Federal and State Legal Requirements

Cannabis is illegal on the federal level in the United States (US). For this reason, each state has determined its own list of solvents that can be used for extracting cannabinoids and terpenes, as well as the maximum residual solvent levels, also referred to as action levels, in the products. Table 1 provides an overview of the status of residual solvent testing in select states (due to space limitations). The lack of uniformity between the states' lists is a challenge, but as laws and regulations change, it remains the responsibility of the testing laboratory to confirm current testing requirements. In contrast, because Canada has legalized cannabis on the federal level, only one column in Table 1 is required for its list.

Table 1 Permitted solvents by select states and action levels utilized for extraction

State/solvent (μg/g)	CAS #	AK	AR	CA	FL	Hi	ND	OR	RI	Canada
Acetic Acid	64-19-7									5000
Acetone	67-64-1		5000	5000	750		5000	5000	5000	5000
Acetonitrile	75-05-8		410	410	60		410	410	410	
Anisole	100-66-3									5000
Benzene	71-43-2	1	2	1	1	1	2	2	2	
Butane	106-97-8	800	5000	5000	2000	800		5000	5000	5000
1-butanol	71-36-3		5000					5000	5000	5000
2-butanol	78-92-2		5000				5000	5000	5000	5000
2-butanone	78-93-3		5000					5000	5000	
Butyl acetate	123-86-4									5000
Tert-butylmethyl ether	1634-04-4									5000
Carbon tetrachloride	56-23-5								4	
Chloroform	67-66-3			1	2					
Cumene	98-82-8		70				70	70	70	
Cyclohexane	110-82-7		3880				3880	3880	3880	
1,2-dichloroethane	107-06-2			1	2				5	
1,1-dichloroethene	75-35-4				8				8	
1,2-dichloroethene	540-59-0								1870	
1,2-dimethoxyethane	110-71-4		100					100	100	
N,N-dimethylacetamide	127-19-5		1090					1090	1090	
1,2-dimethylbenzene	95-47-6								2170	
1,3-dimethylbenzene	108-38-3								2170	
1,4-dimethylbenzene	106-42-3								2170	
2,2-dimethylbutane	75-83-2		290					290	290	
2,3-dimethylbutane	79-29-8		290					290	290	
N,N-dimethylfromamide	68-12-2		880					880	880	
Dimethyl sulfoxide	67-68-5		5000					5000	5000	5000
1,4-dioxane	123-91-1		380				380	380	380	
Ethanol	64-17-5		5000	5000	5000			5000	5000	5000
2-ethoxyethanol	110-80-5		160				160	160	160	
Ethyl acetate	141-78-6		5000	5000	400		5000	5000	5000	5000
Ethylbenzene	100-41-4								70	
Ethyl ether	60-29-7		5000	5000	500		5000	5000	5000	5000
Ethyl Formate	109-94-4									5000
Ethylene glycol	107-21-1		620				620	620	620	
Ethylene oxide	75-21-8		50	1	5			50	50	
Formic acid	64-18-6									5000
Heptane	142-82-5	500	5000	5000	500	500	5000	5000	5000	5000
Hexane	110-54-3	10	290	290	250	10	290	290	290	
Isopropyl acetate	108-21-4		5000				5000	5000	5000	5000
Isopropyl alcohol (IPA)	67-63-0		5000	5000	500		5000	5000	5000	5000

(continued)

Table 1 (continued)

State/solvent (µg/g)	CAS #	AK	AR	CA	FL	Hi	ND	OR	RI	Canada
Isobutyl acetate	110-19-0									5000
Methanol	67-56-1		3000	5000	250		3000	3000	3000	
Methyl acetate	79-20-9									5000
2-methylbutane	78-78-4		5000					5000	5000	
3-methy-1-butanol	123-51-3									5000
Methylbutylketone	591-78-6								50	
Methylene chloride	75-09-2		600	1	125		600	600	600	
Methylcyclohexane	108-87-2								1180	
Methyl ethyl ketone	78-93-3									5000
Methylisobutylketone	108-10-1								5000	
2-methylpentane	107-83-5		290					290	290	
3-methylpentane	96-14-0		290					290	290	
Methylpropane	75-28-5		5000					5000	5000	
2-methyl-1-propanol	78-83-1								5000	5000
N-Methylpyrrolidone	872-50-4								530	
Nitromethane	72-52-5								50	
Pentane	109-66-0		5000	5000	750		5000	5000	5000	5000
1-pentanol	71-41-0		5000					5000	5000	5000
Propane	74-98-6		5000	5000	2100			5000	5000	5000
1-propanol	71-23-8		5000					5000	5000	5000
Propyl acetate	109-60-4								5000	5000
Pyridine	110-86-1		200					200	200	
Sulfolane	126-33-0		160					160	160	
Tetrahydrofuran	109-99-9		720				720	720	720	
Tetralin	119-64-2								100	
Toluene	108-88-3	1	890	890	150	1	890	890	890	
Total xylenes (o,m,p)	1330-20-7	1	2170	2170	150	1	2170	2170	2170	
1,1,1-trichloroethane	71-55-6								1500	
1,1,2-trichloroethylene	79-01-6			1	25				80	
Triethylamine	121-44-8									5000
Total requirements		6	41	20	21	6	21	41	59	28

When comparing solvents, it is important to reference the Chemical Abstracts Service (CAS) Registry Number since many compounds have alternative names in documentation. For example, methylene chloride and dicholormethane (CAS # 75-09-2) are synonyms. Similarly, isopropanol, isopropyl alcohol, and 2-propanol are also synonyms registered under CAS# 67-63-0. Each unique compound has a unique registry number.

In New York (NY), the only solvent that can be used for medical cannabis is carbon dioxide (CO_2), which is relatively nontoxic and not combustible [3]. Carbon dioxide is a food additive and generally recognized as safe (GRAS) according to the US Food and Drug Administration (FDA). Vermont (VT) takes it a step further by also permitting ethanol as an extraction solvent but prohibits petroleum solvent extraction [4]. Many states do not require any testing for solvents. However, other states do require testing of residual solvents but do not specify compounds or action levels. This may be the result of these states being in the infancy of cannabis rules and regulations. Therefore, those states are not listed as major change is expected and space is limited.

Rhode Island (RI) has the largest list of residual solvents to test at 59 [5]. Oregon (OR) and Arkansas (AR) each test for the same 41 residual solvents with the same action levels [6, 7]. This is an example of one state utilizing the criteria of another state (OR) that has a longer history of regulations. Another example of states with the same residual solvents and the same action levels are the two non-continental US states of Alaska (AK) and Hawaii (HI), which only list six solvents for use and testing [8, 9]. Louisiana (LA) utilizes those same six solvents at the same action levels but includes a seventh solvent with ethanol.

Florida (FL) and North Dakota (ND) each list 21 solvents, but the list and the action levels are quite different [10, 11]. For instance, Florida has 8 solvents not included on the ND list and ND has 8 compounds not included on the FL list. Action levels are listed in the FL documentation while ND references the International Conference on Harmonization (ICH) of Technical Requirements for Pharmaceuticals for Human Use guidance—Residual Solvents Q3C(R6). In comparison, all common compounds in the ND list are permitted to be 2–14.5 times higher than in the FL list.

Florida and California (CA) also have a similar list, but CA has one less solvent—1,1-dichloroethene (CAS # 75-35-4), putting the list at 20 compounds [12]. Comparing the overlapping 20 compounds, the CA list has action levels between 0.04 and 14.5 times lower or higher than FL, respectively, resulting in a range of 362 times the difference. On the low end, CA lists 1,1,2-trichloroethylene (CAS # 79-01-6) at a 1 µg/g action level while FL permits up to 25 µg/g. On the high end, CA has an action level for total xylenes (CAS # 1330-20-7) at 2170 µg/g while FL promotes an action level below 150 µg/g. It should be noted that most states report total xylenes (CAS # 1330-20-7) which include three isomers of ortho, meta and para, which are 1,2-dimethylbenzene (CAS # 95-47-6), 1,3-dimethylbenzene (CAS # 108-38-3) and 1,4-dimethylbenzene (CAS # 106-42-3), respectively.

Canada has 28 solvents on their list not including carbon dioxide, nitrogen and water [13]. All of the solvents listed have an action level of 5000 ppm (µg/g). Of further note, 13 of the solvents on the Canadian list are not listed on any of the states' lists described here.

Many of the state requirements reference organizations such as the FDA, ICH, USP, Association of Official Agricultural Chemists (AOAC), American Herbal Pharmacopeia (AHP), United States Department of Agriculture (USDA), and World Health Organization (WHO). Rhode Island states that Class 1 solvents may not be used in the production of any medical marijuana product. ICH Q3C(R6) uses a similar classification system as USP <467> with three classes of compounds [14]. If compounds exceed specified action levels, there are various ways to address the issues, one of which is to dispose of the product as it is not fit for consumption.

Since cannabis is illegal on the federal level in the US, a cultivator or processor cannot ship samples to out-of-state labs. Therefore, national entities must setup a laboratory in each state they wish to operate. Examples of national laboratories include Kaycha Labs, EVIO Labs, and Steep Hill. Federal oversight, with a uniform list of residual solvents and action levels, would certainly make the process much simpler, unlike the complex discussion above. Further, if cannabis becomes legal for international trade, the US and Canada would need to agree on which solvents to test and the action levels for each one.

Cannabis Matrices Tested for Residual Solvents

Analysis for residual solvents is typically performed on super-concentrated forms on cannabis. These matrices can be of several different types. Waxes, hash and hemp oil, butters and shatter are all produced by extracting cannabinoids and terpenoids from the physical plant material with a solvent. There is also a type of sample called a pre-roll, which consists of the cannabis flower mixed with some form of cannabis concentrate (such as oil, wax, or butter), and this also must be tested for residual solvents. It is important that the extracted products are free of solvents, hence the need for residual solvent testing. It is also possible to test finished products such as edibles and pills for residual solvents. Most solvents that are used are considered safe at levels below 0.5%, but the pervasive use of impure solvents requires that more hazardous solvent contaminants are also tested for, and at much lower concentration limits.

Cannabis Product Sampling

A representative sample must be comprised of several sample increments collected from a larger batch size for residual solvent testing. The amount and type of sample required for testing can vary by state. In California, regulatory compliance testing for residual solvents is required on cannabis concentrates (oils, shatter, waxes, etc.) and pre-rolls [11]. Pre-rolls contain the cannabis flower and are often infused with cannabis concentrates. Table 2 shows the number of samples required to be tested for per batch of cannabis products or pre-rolls.

A licensed testing laboratory employee performs the sampling of the cannabis products at the distributor site and transports the product back to the laboratory for testing. Pre-rolls must be rolled prior to sampling and testing. In order to ensure proper sample size, the process is often documented via video recording with the batch number, date, and time. The sample is transported back to the laboratory in a locked, fully enclosed container not visible to the outside of the vehicle. Once the sample arrives in the lab, it can be homogenized (pre-rolls) or directly placed into the headspace gas chromatograph mass spectrometer (HS GC-MS) vial for analysis. Chain of custody

Table 2 Batch and sample size requirements in CA

Cannabis product or pre-roll batch size (units)	Number of sample increments (per sample)
≤50	2
51–150	3
151–500	5
501–1200	8
1201–3200	13
3201–10,000	20
10,001–35,000	32
35,001–150,000	50

(COC) forms are completed at every step to document the chronological process of the sampling, handing, transport, storage, and eventual, destruction of the sample.

The above procedure is only required for regulatory compliance testing. Research and development (R&D) samples are not held to any sampling requirements. R&D samples can be dropped off directly at the testing laboratory and video documentation of sampling nor COC forms are required. R&D tests are typically requested by manufacturers to test their processing procedures.

Instrumentation

In the cannabis industry, a variety of solvents are used in the extraction process. Each solvent is classified based on the level of toxicity. Regulated action levels are varied based on the category classification of residual solvents, some requiring a high sensitivity method such as GC-MS for their low-level quantitation. Also, sample preparation and analysis is difficult with processed cannabis matrices due to their lack of solubility and the limitation of available non-interferent diluent solvents. The ideal method for testing processed cannabis matrices is direct analysis of the matrix in a headspace vial without any sample preparation [15, 16]. However, this process can be challenging due to high sample amount requirements required by some cannabis control regulatory agencies. Furthermore, other method development challenges exist, such as co-elution of several residual solvent analytes, or interference from the environment (presence of normal levels of mass-to-charge ions for water, air, and carbon dioxide). Finding an acceptable balance between proper separation for unequivocal identification and quantitation, and reasonable analysis throughput time can be complicated.

In this chapter, a GC-MS method using selected ion monitoring (SIM) with headspace injection developed for the analysis of cannabis concentrates is described. This includes the identification and use of appropriate standard mixes and diluent solvents for simple preparation of calibration curves, as well as method optimization to eliminate or minimize co-elution and ambient interferences for maximum sensitivity and accuracy in quantitation, all without compromising run time. Moreover, method parameters and hardware configuration changes performed to accommodate high sample size requirements for compliant testing are also discussed herein. These were conducted to avoid analytical column overloading and MS detector saturation, while directly testing sample without preparation steps.

The work described herein demonstrates the viability of HS-GC-MS with SIM mode for the analysis of twenty Category I and II residual solvents, including three soluble gases, which may be present in cannabis oil-based products.

Analysis

Sample Types

To ensure proper instrument calibration and method performance, calibration curves and quality control (QC) samples are regularly prepared and analyzed. Common QCs sample types are listed and defined below. Each state has their own requirements in terms of frequency and acceptance criteria for QC samples.

System Blank (SB)—Clean solvent used to evaluate instrument contamination.

Method (or Matrix) Blank (MB)—Analyte free matrix used to evaluate contamination from method preparation and matrix interferences.

Initial Calibration Verification (ICV)—A second source (different lot number or different vendor from the calibration standard) of each analyte used to verify the calibration curve. The ICV is injected immediately following the calibration curve.

Continuing Calibration Curve (CCV)—A reinjection of a calibration standard used to determine the validity of each analyte in the calibration curve.

Laboratory Control Sample (LCS)—A blank matrix that is spiked with a known concentration for all analytes.

Matrix Spike Sample (MS)—A matrix that is spiked with a known concentration for all analytes. Also known as Laboratory Fortified Sample Matrix.

Laboratory Replicate Sample (LRS)—A replicate of an unknown sample taken at random. The LRS ensures analytical precision of the method.

Instrument Validation

For an analytical instrument method to be of use, it must be capable of producing results with acceptable accuracy and precision. In addition, the method must have the sensitivity to measure and detect the target compound, selective for analyte comparison to other unknowns and be robust. The above parameters are pivotal in instrument validation.

Instrument validation is important in authenticating performance specifications given by the manufacturer of an instrument. Depending on the method, instrument validation criteria will vary. Instrument validation is conducted when an instrument is installed, in development of new methods, major part replacement, and/or instrument relocation.

The information presented in this section will only describe instrumentation validation from a method perspective. Installation Qualification (IQ), Operational Qualification (OQ) and Performance Qualification (PQ) will be presented elsewhere.

The validation of an instrument method should include the following tests:

1. Precision
2. Linearity
3. Accuracy
4. Sensitivity (Limit of Quantitation (LOQ) and Limit of Detection (LOD))
5. Specificity
6. Robustness
7. Range

Precision

The degree of repeatability among several independent measurements under the same conditions. Precision results are listed as coefficient of variation, standard deviation or percent relative standard deviation.

Linearity

The reportable range that analyte can be measured with a specified degree of confidence, such as regression value of 0.999. The reportable range is typically linear but can be polynomial.

Accuracy

An assessment of the measured value to a standard of known value. An accuracy experiment can be conducted by spiking a blank matrix with a known standard concentration and then analyzing the standard to determine its recovery.

Sensitivity

An absolute quantity, the smallest absolute amount that can be detected (LOD) or measured (LOQ) by an instrument. Depending on the method validation requirements of a regulatory agency, the procedure for calculating the LOD and LOQ may vary.

Specificity

The ability to unequivocally assess the analyte in the presence of other components.

Robustness

The ability of the method to remain unaffected by small changes in method conditions.

Range

An array of analyte values to measure an analyte without dilution, concentration or sample preparation. An acceptable accuracy and precision should be attained within the validated range.

Checks

Analytical quality control is important as it affects the quality of the data generated. Quality control includes the process of sample collection to data reporting. Checks on the analytical procedure should always be conducted to ensure that the data is precise and accurate. Moreover, daily system checks should be conducted to minimize random and systematic error; to ensure the method is good standing and is carrying out the functions of its validated purpose. There are several systematic daily checks that should be conducted on GC-MS instruments. These checks consist of laboratory reagent blanks, laboratory fortified blank, quality control sample, laboratory fortified matrix and continuing calibration verification.

Laboratory Reagent Blanks

Laboratory reagent blanks (LRB) should be analyzed before sample analysis to ensure the laboratory, including the instrument, is free of contamination that may inadvertently affect the measuring or identification of an analyte. Sources of background contamination may stem from glassware, reagent water, instrument, etc. Ideally background noise should be about 3–5 below the Method Detection Limit (MDL) of the analyte.

Laboratory Fortified Blank

Laboratory fortified blank (LFB) is a method check to determine the accuracy and precision of the instrument and operatory while handling samples in real-time. In this process, a blank sample is spiked with a known concentration of each analyte, in addition to internal standards and surrogates. These blanks are typically analyzed with each sample batch.

Quality Control Sample

A quality control sample (QCS) from an external source can be measured periodically to determine the accuracy of the laboratory. If accuracy is unacceptable, a check of the entire system must be done to rectify the problem.

Laboratory Fortified Matrix

Matrix effects are indicated by significant changes in internal standard responses in the sample. If matrix effects are observed during sample analysis, a laboratory fortified matrix (LFM) must be used. With matrix effects, sample results should be documented and the LFM must be reported.

Continuing Calibration Verification

An acceptable initial calibration is required for most methods. Calibration curve criteria will depend on the QC criteria of the method. Criteria include the use of internal vs. external standards, the use of linear or second order regression calibration curves or an acceptable percent relative standard deviation of the response factor. However, despite the demonstration of acceptable initial calibration, satisfactory performance must be attained occasionally throughout the analysis using a continuing calibration check. Continuing calibration verification (CCV) can be conducted as deemed by the QC criterion.

Instrument Tuning

The GC-MS system requires fine tune adjustment of various parameters for peak operating performance. Since this tuning is often beyond the scope of an operator, the GCMS manufacturer has software to make these complicated adjustments automatically.

Evaluating a GC-MS autotune can help determine if a system is in satisfactory operating condition or to diagnose potential problems. It is important to keep in mind that while this is a very important tool, the autotune report alone cannot diagnose all GC-MS problems. Furthermore, failure to reach one or more of these criteria does require the system to be out of service for sample analysis. The number of discrepancies in the autotune report, the severity and type of the discrepancies should be considered.

Depending on the vendor, the contents of the autotune report may be different. The remaining of this instrument tune section will focus on autotune reports generated by Shimadzu GC-MS instruments.

An autotune report generated by Shimadzu GC-MS consists of three sections: Peak Profile, Spectrum Data and Hardware Values. The overall tune can be evaluated by reviewing several key indicators in each of these sections.

Peak Shape

The peak should have a Gaussian shape as shown in Fig. 1. Note that the shape is not a perfect triangle, but it is symmetrical.

Full Width at Half Maximum

Full width at half maximum (FWHM) is a criterion that further defines the shape of the peak. A width measurement is taken across the peak at its halfway point. The software performs this calculation and reports out an expression of atomic mass units (amu). Criterion for this measurement is twofold. First, all three peaks displayed in the autotune report should be 0.6 amu ± 0.1 amu. Second, there should be

Fig. 1 Gaussian shaped peak

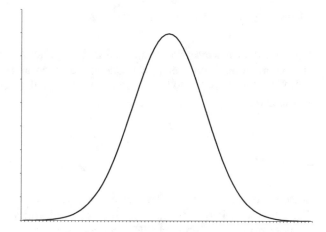

no more than 0.1 amu difference between all three peaks. Therefore, if an autotune report displays three peaks with FWHM values of 0.54, 0.6 and 0.67, the criterion has not been met. This is due to a greater than 0.1 amu difference between 0.67 and 0.54.

502–503 Resolution

Examining resolution on a GC-MS is similar to evaluating resolution on a GC chromatogram. The difference is that when speaking of MS resolution, an attempt is being made to determine the separation of one mass from another in the mass spectrometer, instead of separating one compound from another on the GC column. The criterion for resolution is examined at 502 and 503 m/z (mass to charge). The height of the valley between 502 and 503 m/z should be less than 50% of 503 peak top. This criterion also applies to 69–70 and 219–220 m/z ratios.

Base Peak of the Spectrum

In mass spectrometry, the tallest peak of the spectrum is called the base peak. When evaluating a GC-MS autotune report, it is expected that 69 m/z should be the base peak. If an autotune is run shortly after the completion of maintenance procedures, 18 or 28 m/z ions may be the base peak. This is an indicator that more pumping time is required.

Mass Accuracy

Mass accuracy is a term that simply refers to whether the mass peaks are displayed at the correct mass number (m/z). Check the seven largest peaks in the spectrum (69, 131, 219, 264, 414, 502, and 614 m/z) to ensure that they are ±0.1 amu from the expected mass.

28–69 Ratio

Since air is approximately 80% nitrogen, the presence of nitrogen (28 m/z) in a GC-MS system is often used as an indication of an air leak. The tuning solution contains a tuning compound, PFTBA, that is released at a constant rate into the GC-MS during the tuning process. The 69 m/z peak is almost always at a constant height from one tune to the next. It has been determined that if the 28 m/z is less than 20% of 69 m/z, the amount of air in the system is low enough to allow operation.

Lens 1 Voltage

The Shimadzu GCMS-QP2010 series employs four lenses that repel and focus ion beams from the ion box (source) to the pre-rods. Lens 1 is also known as the repeller. Ideally, lens 1 voltage should have a value near 0 V, indicating a clean system. If the voltages increase, this indicates that the system is has contaminants. Voltages of +3 V and above indicate that the system requires maintenance cleaning procedures.

RF Gain and RF Offset

RF Gain and RF Offset are adjustments that are made to the direct current (DC) voltages on the quadropoles to improve resolution. Values are not displayed in any meaningful unit of measure, but the value are still important. When the RF Offset is adjusted upward, all mass peaks across the spectrum improve in resolution at approximately the same rate. When the RF Gain is adjusted, mass peaks of higher mass improve in resolution more quickly than lower mass peaks. Therefore, the resultant tune values can give an indication of the condition of the rods and pre-rods. RF Gain should be 5000 ± 200 and RF Offset should be 4900 ± 100.

Pre-rod Voltage

Pre-rod voltage has a bias voltage that is placed to attract ions as they leave lens 4. Adjusting this voltage will influence ion travel speed towards the pre-rods and will consequently impact mass peak resolution and size. The pre-rod voltage should always tune at -3.5 V in electron impact (EI) mode (-2.5 for chemical ionization (CI or PCI) and 2.5 for negative chemical ionization (NCI)). Any deviation from this value usually indicates system contamination, damage, or improper assembly.

Detector Voltage

Detector voltage is simply the amount of voltage that is placed on the detector to amplify the signal to a desired level. As with all amplifiers, there is a point at which the noise begins to be amplified at a greater rate than the desired signal. The result is that the amplifier may have increased the signal by a factor of 5, but the noise may be increased by a factor of 10, thus actually reducing the signal to noise ratio. To avoid operating the detector in this range, the detector voltage should be less than 1.8 kV.

502 Intensity

Poor intensity of high mass ions can be an indication of system contamination. The tuning solution (PFTBA) produces ions of known masses and relative intensities. The autotune report generates a mass table that lists several ions, raw abundance, and m/z ratio to 69 (the base peak). The intensity of the 502 ion should be at least 2% of 69 (at least 1% 69 on the 2010S).

IQ/OQ/PQ

IQ/OQ/PQ processes are extremely common in the cannabis market and with the use of any and all analytical instrumentation. These processes are implemented when something new, or even something that has been altered in the system. IQ/OQ/PQ processes are often provided by the instrument manufacturer as each piece of equipment require or follow different qualification tests.

These qualifications are in reference to the type of instrument(s) used in cannabis production. It is important to understand design specifications, as provide insight into the exact materials and processes involved in cannabis production, but also the type of power source and procedures used in their manufacture.

Installation Qualification

To validate whether new equipment can deliver the desired results, design qualification (DQ), is conducted. However, in real-world applications, the performance of the equipment depends on the installation process. IQ authenticates whether the equipment or instrument being qualified, along with all the ancillary and sub-systems, have been delivered, installed and commissioned as per manufacturer instruction. Any current good manufacturing practice (cGMP) requirements that are relevant to the IQ and the approach employed for conducting the IQ are well-documented in a validation master plan (VMP).

To pass the IQ successfully, installation must fall in sync with the manufacturer requirements per the following:

- Location and floor space
- Power, gas supply and other energy sources
- Operating and environmental conditions
- Unpacking instruments and checking for damage
- Equipment drawings (if any)
- Expendables and consumables location
- Documenting computer-controlled instrumentation
- Verifying communication and connections with peripheral units
- Lists of tests and special tools included
- Keeping a record of validation and calibration dates used for IQ
- Preserving certificates and manuals of conformity
- Operational Qualification (OQ)

Operational Qualification

IQ is typically followed by OQ. Operational qualification tests equipment for performance as per manufacturer requirements and within the operating range specified by the vendor. One of the prerequisites for technical acceptance of the equipment and the facility is that all the parts listed in the test plan should be assessed individually and their performance be documented. OQ can only be conducted only after a successful IQ.

Requalification is also important, especially after any services are performed on the equipment, post any major maintenance work, or as a part of the quality assurance schedule. Basically, OQ serves the purpose of identifying and assessing the various features of equipment listed below that contribute towards final quality of the product:

- Signaling LEDs and display units
- Temperature fluctuations and controls
- Low-temperature and overheating protection systems and alarms

- Transfer communication test
- Leak and pressurization tests

Performance Qualification

Performance qualification is the last phase of the qualification process and comprises verifying and documenting the performance of the equipment. It is conducted to assess whether the instrument is functioning properly within the desired working specifications. Based on the process description, an in-depth assessment plan is created before conducting PQ. The FDA has conditions pertaining to the manufacturing aspect of the instrument such as equipment limits, operating parameters and component inputs that are a part of the PQ protocols. They are as follows:

- Specific catalog of data which should be recorded or analyzed while conducting tests, calibration and validation
- Certain tests that should be performed to ascertain consistent quality during different steps of production
- Sampling plan that summarizes the methods used for sampling, during and in between the batches of production
- Methodology for analysis that makes data, scientific and risk-oriented decisions based on statistical data
- Establishing contingency plans and variability limits to handle non-compliance

Tests that are included range from repeatability and carry-over, sensitivity, and the instrument(s) detection limit.

The system of GMP necessitates the practices of IQ, OQ and PQ for equipment qualification process. It helps manufacturers ascertain a consistent quality delivery from the equipment. Additionally, these practices considerably minimize errors so that product quality can be maintained in accordance with the relevant regulations and industry standards. Equipment qualification can be conducted in-house or by equipment validation service professionals.

A 6-point calibration curve was created using certified reference material (CRM) standards acquired from CPI International (Category I, Z-G34-115300-03 and II, Z-G34-115301-02 standards). An aliquot of 150 μL was placed on a 20 mL headspace vial and capped using a vial crimper. Butyl acetate was used as diluent solvent for the preparation of calibration curve standards by serial dilution. Concentration ranges and calculated weight sample amounts are listed in Table 3 below. Analytical conditions and system configuration are listed in Table 4.

Quality control standards were prepared by weighing approximately 250 mg of methyl cellulose (substitute matrix) and spiked with residual solvents standards to obtain a Level 5 concentration in the QC samples. The QC samples consisted of an initial calibration verification, a continuing calibration verification, a laboratory control sample, and a matrix spike sample (Fig. 2).

Precautions during preparation of standards for calibration curve:

Table 3 Calculated amount of residual solvent calibration curve standards

Calibration curve standards	Concentration (µg/mL)	Volume in vial (mL)	Amount (µg)
Category I residual solvent			
Calibrator level 1	0.781	0.150	0.117
Calibrator level 2	1.563	0.150	0.234
Calibrator level 3	3.125	0.150	0.469
Calibrator level 4	6.25	0.150	0.938
Calibrator level 5	12.5	0.150	1.875
Calibrator level 6	25.0	0.150	3.75
Category II residual solvent			
Calibrator level 1	312.5	0.150	46.875
Calibrator level 2	625	0.150	93.75
Calibrator level 3	1250	0.150	187.5
Calibrator level 4	2500	0.150	375
Calibrator level 5	5000	0.150	750
Calibrator level 6	10000	0.150	1500

Table 4 System configuration and instrument parameters

Headspace	**HS-20 loop model**
Operation mode	Static headspace with loop
Sample	150 µL sample volume 20-mL headspace vial
Equilibration	15.00 min at 120 °C
Sample loop	0.2-mL loop Vial pressure 350 kP, pressurizing time-1.50 min Loop load time 0.20 min. Equilibration 0.20 min Injection time 0.20 min
Sample line temperature	150 °C
Transfer line temperature	150 °C
Gas chromatograph	**GC-2010 plus or 2030 NX**
Injection	Split injection from HS-20. wi1h 50:1 split ratio
Column	Rxi-624 Sil MS 30.0 m × 0.25 mm × 1.40 µm Helium carrier gas Constant linear velocity. 39.9 cm/s Column Flow 1.24 mL/min Purge flow 0.0 mL/min
Oven program	30 °C, hold 3.0 min 10 °C/min to 140 °C, hold 0.0 min 45 °C/min to 200 °C, hold 1.0 min Total GC run time 16.33 min Total GC cycle time: 25.00 min
Detector	**GCMS-QP2010 SE or 2020 NX**
Operation mode	Selected ion monitoring mode (SIM)
Ion source	200 °C. El mode. 70 eV
Solvent cut time	0.1 min
MS interface	300 °C

Fig. 2 Shimadzu GCMS-QP2020 NX with HS-20 autosampler

- Keep samples cold during the duration of the sample preparation
- Important to prime pipette tip or gas-tight syringe with sample before sampling
- Minimize exposure of sample to environment by replacing cap quickly after sampling
- Vortex sample after preparation and before sampling to mix contents in headspace of vial back into solution
- Use vials with minimum headspace over solution
- Ampules are considered by vendor single use only
- Use one pipette tip per sample or rinse the gas-tight syringe between sample to prevent cross-contamination

Figure 3 shows representative SIM chromatograms for each of the 20 residual solvents analyzed, with target and reference m/z ions (left panel). The calibration curve for each is also presented (right panel). The curve fit type used is Quadratic with a 1/C weighting regression.

Figures 4 and 5 show the total ion chromatogram (TIC) with the identified peaks for the 20 residual solvent standards required in California, and a magnified version to show lower concentration solvent peaks.

Tables 5 and 6 summarize the calibration curve and QC results, respectively. Results showed good coefficients of determination and percent of accuracy for calibrator standards. All QC samples (ICV, CCV, LCS, and MS) showed good percent of recovery.

Figure 6 shows a representative picture of some of the different types of cannabis concentrate samples, and pre-roll tested for residual solvents in a cannabis laboratory using a GC-MS or GC-FID instrument with headspace autosampler.

Cannabis concentrates can be produced in numerous ways. One of the most common procedures employed to extract therapeutic compounds found in cannabis plant (cannabinoids and terpenes) is by performing a solvent extraction with butane, which is a soluble gas [17]. After the extraction of the therapeutic compounds, the extraction solvent is then removed by evaporation via heat or vacuum. This extraction method allows the terpenes and cannabinoids to be concentrated in the extract, and the toxic organic solvent is removed. However, residual amounts of solvent may persist in the extracted sample. Therefore, these residual amounts must be checked for action levels as specified by regulatory agencies. These results are reported in a

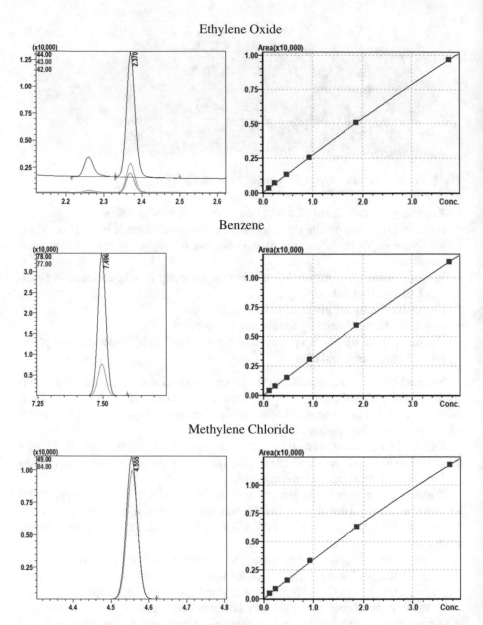

Fig. 3 Right side—Calibration curves for Category I (ethylene oxide, benzene, methylene chloride, 1,2-dichloroethane, chloroform, trichloroethylene), and Category II (propane, isopropanol, butane, acetonitrile, acetone, total xylenes, pentane, methanol, ethanol, ethyl acetate, hexane, ethyl ether, heptane, toluene) residual solvents required by the state of California. Left side—Representative SIM chromatograms for highest calibrator level for each solvent

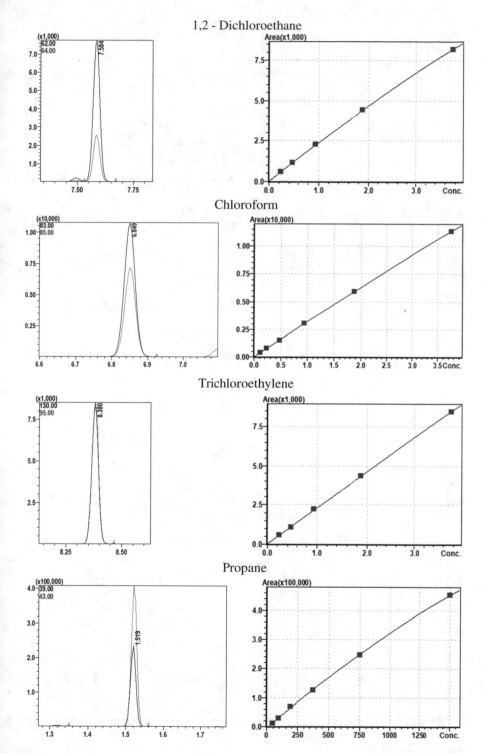

Fig. 3 (continued)

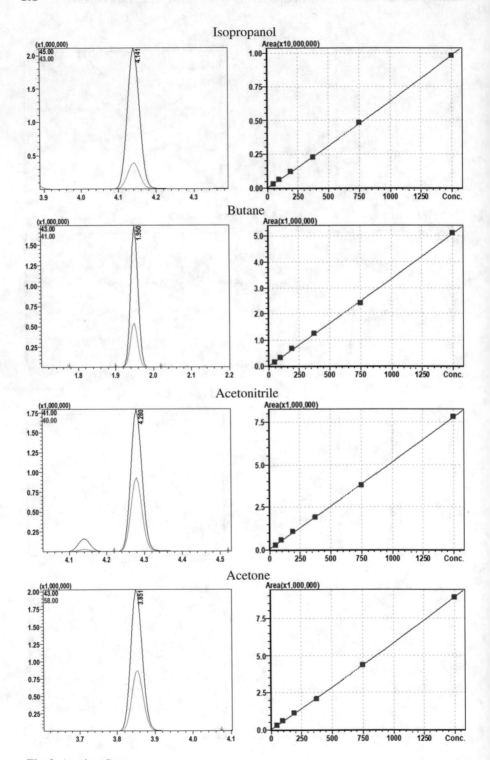

Fig. 3 (continued)

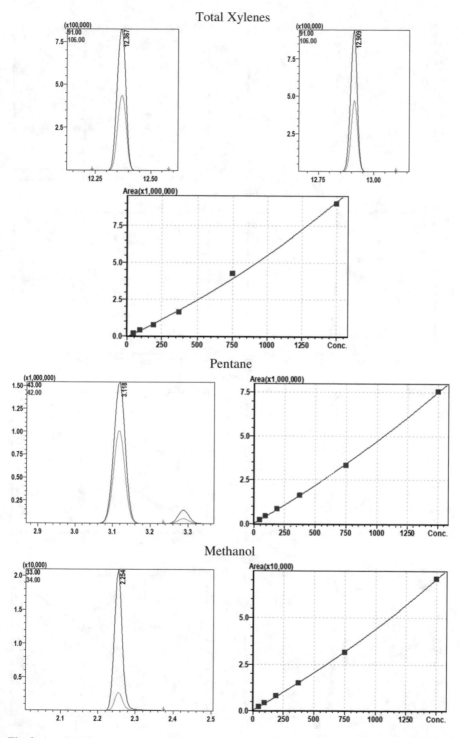

Fig. 3 (continued)

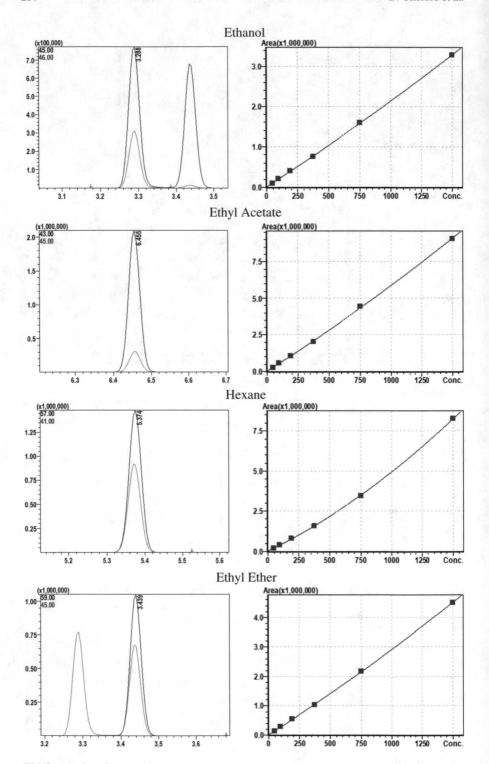

Fig. 3 (continued)

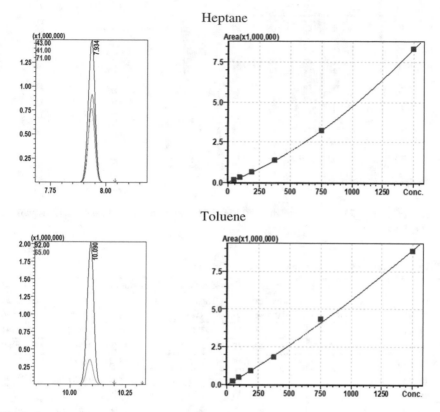

Fig. 3 (continued)

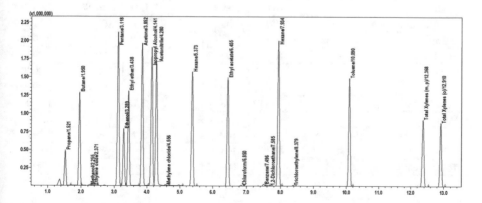

Fig. 4 TIC chromatogram of 20 Residual Solvent standards (required in CA)

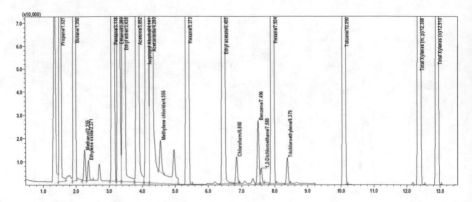

Fig. 5 Magnified TIC chromatogram of 20 Residual Solvent standards to show solvent peaks at lower concentration

Table 5 Accuracy % and R^2 results for calibration curves

| Residual solvents calibration curve standards | Accuracy % (Criteria: 80–120% for standards) | | | | R^2 |
	Lowest level in Cal curve (µg)	Result (%)	Highest level in Cal curve (µg)	Result (%)	(Criteria: ≥0.99)
Propane	46.875	97.46	1500	100.40	0.9991
Butane	46.875	91.54	1500	100.61	0.9989
Methanol	46.875	92.07	1500	100.24	0.9996
Ethylene oxide	0.117	93.56	3.75	100.15	0.9996
Pentane	46.875	91.19	1500	100.43	0.9993
Ethanol	46.875	94.71	1500	100.03	0.9997
Ethyl ether	46.875	96.21	1500	99.95	0.9997
Acetone	46.875	94.54	1500	100.03	0.9996
Isopropanol	46.875	96.44	1500	99.85	0.9996
Acetonitrile	46.875	89.69	1500	100.47	0.9990
Methylene chloride	0.117	99.04	3.75	100.07	0.9998
Hexane	46.875	91.67	1500	100.27	0.9996
Ethyl acetate	46.875	97.59	1500	99.73	0.9994
Chloroform	0.117	100.12	3.75	100.00	1.0000
Bentene	0.117	99.58	3.75	99.99	0.9999
1,2-Dichloroethane	0.117	98.64	3.75	99.98	1.0000
Heptane	46.875	94.04	1500	100.10	0.9999
Trichloroethylene	0.117	100.67	3.75	99.93	0.9997
Toluene	46.875	104.01	1500	99.32	0.9984
Total xylenes (m, p, and o)	46.875	109.97	1500	99.02	0.9964

Table 6 Recovery % results for spiked QC samples

Residual solvents quality control standards	Recovery % (Criteria: 70–130% for spiked QC's)			
	ICV (level 5; %)	CCV (level 5, %)	LCS (level 5; %)	MS (level 5; %)
Propane	109.27	106.68	105.23	101.03
Butane	109.68	106.53	105.83	101.22
Methanol	109.11	105.78	88.00	83.33
Ethylene oxide	106.61	101.09	94.18	87.46
Pentane	107.83	104.79	104.84	100.93
Ethanol	107.19	104.95	88.56	83.77
Ethyl ether	104.80	101.74	100.72	97.26
Acetone	104.63	103.22	97.54	93.28
Isopropanol	104.34	102.80	92.49	88.08
Acetonitrile	108.30	107.54	93.25	88.30
Methylene chloride	102.65	99.70	94.46	90.08
Hexane	106.91	104.14	106.12	102.74
Ethyl acetate	100.91	100.51	100.27	96.56
Chloroform	100.92	99.77	96.66	93.66
Benzene	100.25	99.27	98.73	95.10
1,2-dichloroethane	98.80	97.66	96.95	92.56
Heptane	102.68	101.70	108.45	105.40
Trichloroethylene	101.75	99.05	100.95	97.05
Toluene	95.80	96.28	99.92	96.64
Total Xylenes (m, p, and o)	91.29	92.71	98.08	95.36

Fig. 6 Shimadzu 10 mL headspace vials containing no less than 250 mg of cannabis concentrate samples and pre-roll for analysis of residual solvents using either GCMS or GC-FID with headspace analysis

certificate of analysis (COA) which contains identifier information about the sample. This COA is sent to the regulatory agency. If the sample fails to identify a residual solvent quantitated at the action level or above, then the product cannot be released for retail sale. If the product cannot be remediated, then it shall be destroyed per the regulatory agency requirements.

A representative result from a GC-MS, and from a GC-FID acquired from the analysis of residual solvents in a cannabis concentrate sample is shown in Fig. 7. In this example, a concentration of butane was found at a level much higher than the action level per regulations in CA [12]. The action level for butane is CA is of 5000 µg/g (per regulations effective January 2019); and the concentration of butane found in the cannabis concentrate sample was of 15,809 µg/g. This cannabis sample failed the residual solvents test.

The additional pictures shown in Fig. 8 provide another visual example of the many varying types of concentrate and pre-roll samples that are received in the laboratory for the state-required residual solvent analysis. Headspace analysis is ideal for these types of samples because of the high volatility of residual solvents and these sample types can be weighted and analyzed directly in a headspace vial. Therefore, minimizing the amount of manual sample handling required and time spent in sample preparation.

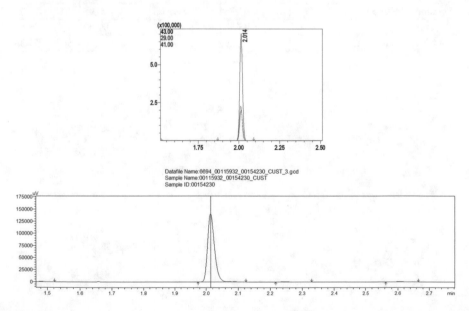

Fig. 7 Top—Representative GCMS SIM chromatograms for the Category II Residual Solvent Butane found in a cannabis concentrate sample at a level of 15,809 µg/g, which is about 3.2 times higher the action level for Butane in California, failing the residual solvent test. Bottom—Representative GC-FID result showing butane chromatogram peak for the same cannabis sample Photos and GCMS/GC-FID data courtesy of Cannalysis Cannabis Testing Laboratory in South California

A representative portion of at least 250 mg in sample size taken from the pre-roll sample shown in Fig. 8 was weighted in a 20 mL headspace vial for residual solvent analysis using headspace analysis with GC-MS only. The SIM results from the analysis are shown in Fig. 9. Ethanol, methanol, acetone, isopropanol, and acetonitrile were identified in the sample but with a quantitation result below the regulated action level for each of these residual solvents. The results were below the action limits, therefore the sample passed the criteria for residual solvent analysis.

Data is processed using Shimadzu GCMSsolutions (or LabSolutions GC if using GC-FID) and LabSolutions Insight GCMS software. Data files can be set to be exported as .csv or .txt files prior to acquisition, or also in post-run after acquisition.

Fig. 8 Picture of the variety of concentrate and pre-roll samples received in cannabis testing laboratories for the analysis of residual solvents

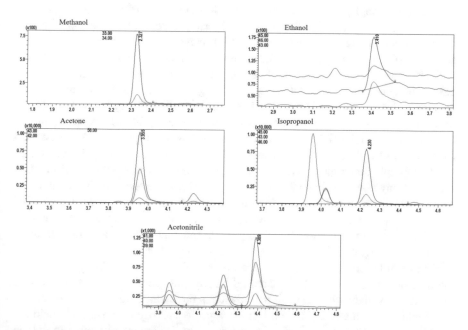

Fig. 9 Representative GC-MS SIM chromatograms for the category II residual solvents methanol, ethanol, acetone, isopropanol, and acetonitrile found in an aliquot sample from the pre-roll pictured in Fig. 8. These residual solvents were quantitated each one at a level below the action level required by the State of CA, indicating that this sample passed the test for residual solvents
Photos and GC-MS data courtesy of Caligreen Cannabis Testing Laboratory in South California

These types of files can then be set to be used by the laboratory information management system (LIMS) to generate COA report.

In GCMSsolutions (and LabSolutions GC) software, a dilution factor and sample amount (in mg) can be included in the acquisition batch. The software automatically uses these values in the calculation of the concentration result. For the quantitation of residual solvents with isomer peaks, such as in the case of xylene, which is reported as Total Xylenes (ortho-, para-, and meta isomers), a group calibration can be set to automatically quantitate Total Xylenes. Otherwise, the alternate option is to use the isomer peak ratio (% of isomer proportion) reported on the CRM Residual Solvent Standard mix COA provided by the vendor. This value is then used by the LIMS to output Total Xylenes concentration.

For laboratories that require to be compliant with regulations and guidance, such as FDA 21 CFR Part 11, and need to properly and efficiently manage and control multiple analytical instruments, Shimadzu offers LabSolutions DB and CS to meet these needs.

Conclusion and Future Directions

Gas chromatography-mass spectrometry (GC-MS) is currently the analytical solution for the analysis of residual solvents in cannabis material; however, gas chromatography-flame ionization detectors (GC-FID) can also be used for this testing (as seen in USP methodologies). Results obtained from the use of a Shimadzu GC-MS with HS showed good coefficient of determination and accuracy for calibrator standards; and good recovery for QC samples. Changes are made as the cannabis market changes to be able to meet the analysis demands. As the industry grows and laboratories need to run more samples, one could see where the GC-MS could be substituted out for a simpler GC-FID. This would allow the mass spectrometers to be used for pesticides (triple-quadrupole being standard) and terpenes (can be done by either single- or triple-quadrupole). Another change that the cannabis industry will be facing, and many non-cannabis laboratories nationwide have been dealing with, is the shortage of high-purity helium; which is the most common carrier gas used in GC chromatography. Helium supplies are running critically low and prices are increasing. Fortunately, there is another carrier gas that can generate the same fast, accurate results you see with helium at a lower cost. Hydrogen is an alternate carrier gas that achieves similar chromatographic separation to helium at higher flow rates, resulting in faster analysis times and increased throughput. There are many easy-to-use online tools available that serve as a guidance in the method transfer for the laboratory looking into make the switch from helium to hydrogen. These tools help the laboratory shorten their method development time. One challenge this presents in many facilities is the question of safety. Safety is improved by the availability of hydrogen sensor that can be added to most manufacturers GC units. An important point to note however, hydrogen **cannot** be used as the gas for vial pressurization in headspace analysis safety reasons. In this case, the non-flammable gas nitrogen can be used instead of helium or hydrogen. Finally, unification of these methods is critical. While no two states are the same, the solvents used in all areas are. Seeing more unified guidelines can help ensure the safety of all from processing cannabis start to finish. Understanding the seriousness of using residual solvents in the creation of things like concentrates not only helps prepare laboratories better tracking the use of them in manufacturing, but also makes the consumer products safer for the general public to use.

References

1. United States Pharmacopeia (USP) Method <467> Residual Solvents (2019). https://www.uspnf.com/sites/default/files/usp_pdf/EN/USPNF/revisions/gc-467-residual-solvents-ira-20190927.pdf. Accessed 18 Mar 2020
2. Analysis of Residual Solvents – Class 1, Class 2A, Class 2B in Pharmaceuticals Using Headspace-GC/MS (2014) Shimadzu's application note. https://solutions.shimadzu.co.jp/an/n/en/gcms/jpo214097.pdf?_ga=2.157293946.1518677979.1584484027-1959975583.1528907761. Accessed 18 Mar 2020
3. New York Medical Marijuana Program Regulations. https://www.health.ny.gov/regulations/medical_marijuana/docs/regulations.pdf. Accessed 18 Mar 2020
4. Hreeman K, McHenry M, Cats-Baril W, Grace T (2016) Cannabis Testing for Public Safety – Best Practices for Vermont Analytical Laboratories. PhytoScience Institute. https://legislature.vermont.gov/Documents/2016/WorkGroups/Senate%20Health%20and%20Welfare/Marijuana/W~Dr.%20Kalev%20Freeman~Canabis%20Testing%20for%20Public%20Safety%20-%20Best%20Practices%20for%20Vermont~1-20-2016.pdf. Accessed 18 Mar 2020
5. Title 216, Chapter 60, Subchapter 05, Part 6 (2018) Licensing Analytical Laboratories for Sampling and Testing Medical Marijuana. https://risos-apa-production-public.s3.amazonaws.com/DOH/REG_10098_20180811132502.pdf. Accessed 18 Mar 2020
6. David F (2015) Technical Report: Oregon Health Authority's Process to Determine Which Types of Contaminants to Test for in Cannabis Product. Oregon Health Authority. https://www.oregon.gov/oha/ph/preventionwellness/marijuana/documents/oha-8964-technical-report-marijuana-contaminant-testing.pdf. Accessed 18 Mar 2020
7. Rules and Regulations Governing Medical Marijuana Registration, Testing, and Labeling in Arkansas (2017). https://www.healthy.arkansas.gov/images/uploads/rules/Medical_Marijuana_Emergency_Rule_4-27-2017_(Signed).pdf. Accessed 18 Mar 2020
8. 3 AAC 306 Regulations for the Marijuana Control Board (2019). https://www.commerce.alaska.gov/web/Portals/9/pub/MCB/StatutesAndRegulations/3AAC306%209.18.19.pdf. Accessed 18 Mar 2020
9. H.B. No. 2534 Relating to Medical Cannabis (2017). https://www.capitol.hawaii.gov/session2018/bills/HB2534_.HTM Accessed 18 Mar 2020
10. Rule 64-4.307 Standard Operating Procedures. https://s27415.pcdn.co/wp-content/uploads/2019/09/DRAFT-Standards-for-Laboratories.pdf. Accessed 18 Mar 2020
11. Article 33–44 Medical Marijuana Chapter 33-44-01 Medical Marijuana. 01 April 2010. https://www.legis.nd.gov/information/acdata/pdf/33-44-01.pdf. Accessed 18 Mar 2020
12. Bureau of Cannabis Control, Text of Regulations, California Code of Regulations, Tittle 16 Division 42 (2019). https://www.bcc.ca.gov/law_regs/cannabis_order_of_adoption.pdf. Accessed 18 Mar 2020

13. Limits for Residual Solvents in Cannabis Products (2019). https://www.canada.ca/en/health-canada/services/drugs-medication/cannabis/laws-regulations/regulations-support-cannabis-act/limits-residual-solvents.html. Accessed 18 Mar 2020
14. ICH guideline Q3C (R6) on impurities: guideline for residual solvents (2019). https://www.ema.europa.eu/en/documents/scientific-guideline/international-conference-harmonisation-technical-requirements-registration-pharmaceuticals-human-use_en-33.pdf. Accessed 18 Mar 2020
15. Hilliard C, Amanda R; William S; Christi S, Theo F. A fast, simple FET headspace GC-FID technique for determining residual solvents in cannabis concentrates. Foods, Flavors & Fragrances Applications. Restek Corporation. https://pdf2.chromtech.net.au/A%20Fast,%20Simple%20FET%20Headspace%20GC-FID_FFAN2009A-UNV.pdf. Accessed 18 Mar 2020
16. Amanda R (2016) Protocol for quantitative determination of residual solvents in cannabis concentrates. Emerald Scientific. https://cdn.shopify.com/s/files/1/0355/4493/files/091616_Protocol_for_Cannabis_Residual_Solvents_Analysis.pdf?3351052693274207236. Accessed 18 March 2020
17. Beadle A (2019) Advances in cannabis extraction techniques. Analytical Cannabis Articles. https://www.analyticalcannabis.com/articles/advances-in-cannabis-extraction-techniques-311772. Accessed 18 Mar 2020

Elemental Analysis of Cannabis and Hemp: Regulations, Instrumentation, and Best Practices

Bob Clifford, Andrew Fornadel, and Jon Peters

Abstract The legality of cannabis and derivative products such as edibles, oils, extracts, and topical ointments for medicinal and recreational use has been expanding in recent years in multiple countries worldwide. As with any product destined for human consumption, assessing their chemistry ensures safety for the consumer. The focus of this chapter is the assessment of elemental contamination in cannabis, which is of concern due to the negative health effects of certain metals when inhaled, ingested, or otherwise absorbed by the human body.

In this chapter, we address the current regulatory framework for elemental contaminants in cannabis products, sample preparation, analytical instrumentation and workflows, as well as associated challenges. Particular attention is paid to instrument calibration, method validation, installation, operational, and performance qualifications, as well as standard operating procedures and best practices, with example data provided. We conclude with future avenues for investigation and market needs for elemental analysis.

Background

Metallic elements are found throughout the environment and within biological materials, including the tissues of plants and animals [1]. Chemical analysis for heavy metals has an extensive history and been performed on products intended for human consumption, such as drinking water, foods and beverages, and pharmaceutical products. This is due to these metals having both beneficial and deleterious biological effects. For example, elements such as iron or calcium are known for their beneficial biological and metabolic effects at appropriate concentrations. On the contrary, elements such as mercury (Hg) and lead (Pb) have long been understood to have negative health impacts, even at low concentrations, when ingested or otherwise incorporated into the body [2]. Metals with toxic effects are commonly referred to as "heavy metals," which is somewhat of a misnomer as some lighter

B. Clifford (✉) · A. Fornadel · J. Peters
Shimadzu Scientific Instruments, Columbia, MD, USA
e-mail: RHClifford@SHIMADZU.com

© Springer Nature Switzerland AG 2021
S. R. Opie (ed.), *Cannabis Laboratory Fundamentals*,
https://doi.org/10.1007/978-3-030-62716-4_11

metallic elements also are known to be toxic at sufficient concentrations (e.g., beryllium). As a result of the negative health effects of some metals, elemental analysis is critical to ensuring consumer health and safety.

In general, the toxicity of most metals in the human body arises from their substitution into proteins at sites in which other biologically-necessary metals, called cofactors, would bind [3]. This can inhibit the normal function of the protein, or due to the chemistry of the toxic metal, might distort the physical structure of the enzyme. Long-term exposure to low concentrations or short-term exposure to high concentrations of toxic metals can cause adverse effects on otherwise healthy proteins and enzymes. Moreover, these toxic metals can lead to mutagenic, carcinogenic, and/or neurodegenerative effects within the body, among other negative outcomes.

As the market for cannabis increases in legitimacy, various regulatory bodies have recognized the need for assessing the metallic content of cannabis and derivative products to ensure consumer safety. As of now, in the United States (US), regulations are imposed at the state level as cannabis remains a Schedule I drug at the federal level. Various voluntary, consensus-based standards bodies such as ASTM International and AOAC International are also actively developing analytical methodologies and regulatory limits that may experience widespread adoption. Other countries with fewer restrictions on cannabis typically regulate analytical testing requirements at the federal level (e.g., Canada).

Common modes of delivery for cannabis and related products include inhalation, ingestion, and topical application. Metals are bioavailable to humans through all three common modes of delivery. The recognition of this risk has led to organizations such as the United States Pharmacopeia (USP) and Japan Pharmacopeia (JP) to establish analytical methodologies and regulatory limits for metals in pharmaceutical products that are administered through the same routes [4, 5]. Therefore, cannabis and its end-use products have the potential to expose consumers to toxic metals during typical routes of administration and analytical testing is necessary to ensure safety.

Although minimizing metal content in cannabis products is beneficial for ensuring the safety of all consumers, it is particularly important for medicinal cannabis users whose bodies and immune systems may not be as tolerant to ingestion, inhalation, or otherwise consumption of metal-laden products when compared to healthy individuals. Furthermore, medical cannabis consumers may use cannabis products more regularly than recreational users to alleviate their symptoms, giving rise to concern for chronic, consistent exposure of toxic metals.

Sources of Heavy Metals in Cannabis and Derivative Products

Metallic elements in cannabis and associated cannabis-derived products originate from a variety of sources [1]. Primary sources of metals stem from the soil and water in which cannabis plants grow, wherein metals are mineralogically bound and freed during natural weathering processes into the groundwater [6]. Once solubilized, metals are incorporated into the plant as it draws water from the soil, where they are bound into plant proteins or stored in the intracellular space. *Cannabis sativa* is effective at sequestering metals from its environment, termed a "hyperaccumulator," and its utility as a bioremediation agent for metal-contaminated sites has been explored [7]. Secondary sources of metals include water used for irrigation, pesticides and fertilizers or other chemicals applied to enhance crop yield [8], or metals transferred to the plant during harvest and processing (e.g., from cutting tools, grinding, or storage).

Tertiary sources of metals include the consumption of cannabis products via a specific delivery vehicle. As consumers are opting away from traditional smoking vehicles for the inhalational delivery of cannabis due to negative health effects, many are turning to vaping. The vaping process works by heating cannabis flower or derivative waxes and oils to the point that the psychoactive compounds vaporize but do not combust, allowing the vapors to be inhaled while minimizing or eliminating smoke inhalation. This process is perceived as more health-conscious, however recent studies have shown that repeated heating cycles on the metallic filaments within vaporizers may induce the sputtering of metals into the vapor, which are then inhaled by the consumer [9].

Further adding to concern of metals contamination, is the growth in popularity of concentrated cannabis derivatives, such as extracts, waxes, oils, and "shatter," among others. In these products, the active compounds in cannabis are extracted from the plant and concentrated to change their mode of consumption and/or enhance their effects. Although effective at concentrating desirable psychoactive and non-psychoactive compounds, these processes raise the likelihood of concentrating unwanted compounds, such as pesticides or heavy metals. In turn, the consumer may experience elevated risks associated with consumption of these concentrated products.

Irrespective of the source of metallic elements in cannabis and cannabis products, their presence can pose health concerns for the consumer and should be accounted for as part of a general quality assurance and quality control program for cannabis growers, processors, and suppliers.

Federal Requirements vs. State Requirements

Cannabis is federally legal in Canada, therefore cannabis testing regulations are determined by a single governmental organization. In contrast, currently, cannabis is federally illegal in the United States, thus each state is responsible for determining contaminant elements and compounds as well as setting action levels because cannabis cannot legally be transported across state lines. Contaminants include heavy metals, the focus of this section, as well as residual pesticides, residual solvents, mycotoxins, yeast, mold, and microorganisms.

It should be noted that the data included here is current at the time of writing, but due to additional and changing regulations, the reader should confirm the information herein is current.

Table 1 shows elemental state testing requirements for the US, Washington DC, and Canada. Action levels listed in Table 1 have all been converted to micrograms/gram ($\mu g/g$), which, for the purposes of this volume, is considered the same as part-per-million (ppm). For example, Arkansas (AR), Florida (FL), Massachusetts (MA), New Hampshire (NH), and Rhode Island (RI) list action levels as microgram/kilogram ($\mu g/kg$) or part-per-billion (ppb), so values have been divided by 1000 to provide results as $\mu g/g$. Regulatory requirements for some states outline different action levels specific to different routes of ingestion such as inhalable products, ingestible topicals, etc. The table has a column labeled "Code" where "L" is for metals with specific action limits, "NC" for states with no cannabis (or hemp) such as South Dakota (SD), "NT" for states that don't require testing of heavy metals, "TNE" for states that require testing of heavy metals but don't list testing elements or limits, and "TNL," which indicates that testing of specific elements is required but action levels are not listed. TNL elements are listed as a check mark since action levels are not provided by the states. In many states cannabis is illegal, but some states allow the growing of hemp for cannabidiol (CBD), thus the code of "CBD." Arizona (AZ) has proposed testing to be implemented in November 2020, as indicated by a code "P." Maryland (MD), Montana (MT), and Washington (WA) have an asterisk after the state name indicating the values are in microgram per daily dose of 5 g of product for the permitted daily exposure (PDE).

Table 1 prioritizes the "big four" elements of arsenic (As), cadmium (Cd), lead (Pb), and mercury (Hg) in alphabetical order by symbol, followed by the less commonly tested elements. Twenty states list action limits for the big four elements while five require testing but do not specify action levels. For the five states without action limits, New Mexico (NM), Ohio (OH), and West Virginia (WV) only require testing for the big four. New York (NY) and New Jersey (NJ) include additional elements for testing with no action levels, bringing their totals to 9 and 10 elements, respectively. Of the additional elements, these states have chromium (Cr), nickel (Ni), and zinc (Zn) in common. New Jersey also requires iron (Fe), manganese (Mn), and selenium (Se), whereas NY includes copper (Cu) and antimony (Sb). There are only three other states, Maryland (MD), Michigan (MI) and Missouri (MO), that require additional elemental testing and provide action levels. MO requires testing for chromium with an action limit of 0.6 ppm. In addition to testing for chromium at the same level, MD also requires testing for silver (Ag), selenium (Se), and barium (Ba), making it the only state to test for barium. MI requires copper and nickel testing, which is required in other states, although MI is the only state to list action levels.

The big four elements have various action levels. For example, the action level for arsenic in Connecticut (CT) is 0.1 ppm while the action level is 100 times higher at 10 ppm in Hawaii (HI), Louisiana (LA), and Washington (WA). Similarly, the action level for cadmium in Connecticut (CT) in 0.1 ppm while the action level is 41 times higher at 4.1 ppm in Louisiana (LA), Montana (MT), and Washington

Table 1 Heavy metal testing in cannabis by state and country

State/Country	Element Code	As	Cd	Hg	Pb	Ag	Ba	Cr	Cu	Fe	Mn	Ni	Sb	Se	Zn
		Part per million (ppm) = microgram/gram (ug/g)													
Canada	L	1.5	1.0	0.1	5.0										
Alabama	CBD														
Alaska	TNE														
Arizona	P														
Arkansas	L	0.2	0.2	0.1	0.2										
California	L	0.2	0.2	0.1	0.5										
Colorado	L	0.4	0.4	0.2	1.0										
Connecticut	L	0.1	0.1	0.3	0.3										
District of Columbia	NT														
Delaware	NT														
Florida	L	1.5	0.5	3.0	0.5										
Georgia	CBD														
Hawaii	L	10.0	4.0	2.0	6.0										
Idaho	CBD														
Illinois	NT														
Indiana	CBD														
Iowa	CBD														
Kansas	CBD														
Kentucky	CBD														
Louisiana	L	10.0	4.1	2.0	10.0										
Maine	TNE														
Maryland[a]	L	0.4	0.4	0.2	1.0	1.4	60.0	0.6						26.0	
Massachusetts	L	0.2	0.2	0.1	0.5										
Michigan	L	0.98	0.63	2.00	2.00				3.0			0.5			

(continued)

Table 1 (continued)

State/Country	Element Code	As	Cd	Hg	Pb	Ag	Ba	Cr	Cu	Fe	Mn	Ni	Sb	Se	Zn
		Part per million (ppm) = microgram/gram (ug/g)													
Minnesota	NT														
Mississippi	CBD														
Missouri	L	0.2	0.2	0.1	0.5			0.6							
Montana[a]	L	10.0	4.1	2.0	6.0										
Nebraska	CBD														
Nevada	L	2.00	0.82	0.40	1.20										
New Hampshire	L	4.206	2.704	8.712	8.712										
New Jersey	TNL	✓	✓	✓	✓			✓			✓	✓			✓
New Mexico	TNL	✓	✓	✓	✓									✓	
New York	TNL	✓	✓	✓	✓			✓	✓	✓		✓	✓		✓
North Carolina	CBD														
North Dakota	NT														
Ohio	TNL	✓	✓	✓	✓										
Oklahoma	L	0.4	0.4	0.2	1.0										
Oregon	NT														
Pennsylvania	L	0.4	0.3	0.2	1.0										
Rhode Island	L	0.2	0.2	0.1	0.5										
South Carolina	CBD														
South Dakota	NC														
Tennessee	CBD														
Texas	CBD														
Utah	TNE														
Vermont	NT														
Virginia	CBD														
Washington[a]	L	10.0	4.1	2.0	5.0										

State/Country	Element	As	Cd	Hg	Pb	Ag	Ba	Cr	Cu	Fe	Mn	Ni	Sb	Se	Zn
	Code	Part per million (ppm) = microgram/gram (ug/g)													
West Virginia	TNL	✓	✓	✓	✓										
Wisconsin	CBD														
Wyoming	CBD														

Key: *CBD* CBD only, *L* limits for metals, *NC* no cannabis/hemp, *NT* no testing, *P* proposed Nov 2020, *TNE* testing, no elements, *TNL* testing, no limit
[a]μg/daily does (5 g)

(WA). For mercury and lead, the lowest action levels are 0.1 ppm and 0.2 ppm, respectively. The highest action levels are 8.7 ppm and 10 ppm, or 87 and 50 times higher, respectively.

Sample Types

Cannabis can be administered and consumed in a variety of forms. Plant material, such as leaves, buds, and flowers can be smoked, or added directly to baked food products. Extracted forms of cannabis include edible waxes, tetrahydrocannabinol (THC) extracted into butter and used in cooking, or as a liquid tincture used for direct ingestion of the oil product.

Even as new technologies allow manufacturers to create numerous cannabis products, traditional flower smoking remains the most common and preferred way of consumption. Flowers, also called bud(s), refers to the smokable, trichome-covered part of a female cannabis plant. The cannabis flower continues to be a popular choice for its versatility and highest content of THC. When a female flower is denied male pollen, it produces a higher volume of cannabinoids and a terpene-rich resin to attract insects for possible pollination. The abundant resin renders the unpollinated female buds as the most coveted part of the cannabis plant. In comparison, male cannabis plants are typically discarded so their flowers never have the chance to inadvertently pollinate the female flower. The female flower is then harvested, dried, and cured before it is sold on the market.

Outside of the cannabis flower, various cannabis concentrates are available. Concentrates refer to purified products that have been processed for the most desirable portion of the plant compounds (primarily cannabinoids and terpenes). Purification of these products remove excess plant material and other impurities. Cannabis concentrates are diverse and present as a wide range of products with varying levels of THC concentration. CBD oils are extracts of the cannabis plant that is formulated to contain high levels of CBD, and exceedingly low concentrations of THC. CBD oils are non-intoxicating unlike THC-rich products which cause the high commonly associated with cannabis. CBD oils have been purported to have myriad medical benefits, including pain relief, cancer treatment, anxiety and depression relief, and insomnia management, among other conditions. THC oils, in comparison, are intoxicating oils that contain high concentrations of THC. THC oils come in many forms, but the most popular are solids that can be vaporized (called dabs), tinctures, wax, and captures.

Another widely available cannabis-derived product is edibles. Cannabis edibles refers to food-based products that have been infused with cannabis, THC, CBD, and/or other cannabinoids. Generally, edibles are found in a variety of foods such as brownies, chocolates, cookies, gummies, butter, beef jerky, candies, marshmallows, truffles, and beverages, among others.

New cannabis products and modes of consumption are introduced regularly in the rapidly evolving industry. Topical delivery is one of the more widely used modes of consumption and experiencing a growth in consumer interest. Topical products include cannabis-infused lotions, balms, and oils that are absorbed through the skin for localized relief of pain, soreness, and inflammation. Cannabis topicals do not cause a psychoactive effect, therefore they can be easily incorporated into most daily routines.

Sample Size Selection Criteria

The objective of cannabis sampling analysis is no different than the goal of any other sample. The goal is to collect a small portion of the material that can be conveniently transported, handled, and analyzed, while at the same time accurately representing the chemistry of the bulk material. Unless the correct sampling procedure is used, analytical data regarding heavy metals can be biased and not representative of the original bulk material.

Random sampling is the most basic type of sampling to ensure collection of representative samples. The principle of random sampling is that every portion of the original bulk material has the same probability of being chosen. By selecting multiple portions of the original bulk material, the goal is that the sample being chosen accurately represents the bulk. An increased sample size increases the likelihood that any variation in composition in the bulk material will be captured, thereby providing analytical results that are representative of the whole. The sampling size, however, is limited by economy, time, and personnel, among other constraints. In sampling methods in which the sample being tested is a high-value commodity, sample size is critical. A sample size that is too small risks collecting nonrepresentative samples. In some states, cannabis facilities are required to provide pounds of material for testing, after which the unused portion is returned to the provider. This procedure raises not only economic concerns for the provider, but also issues of homogeneity, and finally of contamination risk during transportation. In general, three samples is the minimum sample size that should be collected and analyzed to obtain statistically representative data.

Representative sampling is more difficult with a solid cannabis plant because of the heterogeneity of the plant. By its very nature, dried cannabis plant material is not homogeneous. A cannabis flower product may contain stems, leaves, seeds, and sometimes foreign matter, all of which can have different metallic content. Furthermore, different buds in the same product may contain different levels of heavy metals, making sampling the cannabis flower even more complex.

Instruments used to analyze heavy metals are so sensitive that only small amounts of sample are required for analysis. For example, analysis of the "big four" elements using inductively coupled plasma mass spectrometry (ICP-MS) requires approximately 3 mL of aqueous sample. Dilution factors are accounted for during sample preparation, therefore approximately 6 mg of raw material is needed. A sample size smaller than the recommendation, however, can increase bias and error unless it can be reasonably assumed that the sample is homogeneous.

It is critical for the analyst to analyze homogenous samples that are representative of the portion of the plant or material destined for consumption. This is relatively simple for extracts and derivative products, as they these products are homogenized during extraction and concentration. However, it is critical for plant material, as the stems, leaves, and flower buds of the cannabis plant may be chemically heterogenous, which has been demonstrated in other plant species [10].

The most common method to obtain homogeneous solid cannabis plant material is grinding to reduce the particle size of the material. Reducing particle size allows for a smaller sample of dry material to more accurately represent the entire batch. In addition to homogenization, grinding samples has other benefits for sample preparation because the process increases surface area, which improves dissolution or extraction efficiency.

However, there are disadvantages with sample preparation via grinding, including potential contamination from the grinding mechanism and evaporation of volatile metal compounds. The process of grinding a sample with a mortar and pestle or ball mill and passing it through a metallic sieve can be a major source of contamination. This can occur from the remnants of a previous sample that had been prepared earlier or from the materials used for grinding and sieving equipment. To minimize contamination, plastic sieves are often used for sample preparation and it is good practice to discard the first batch of the sample when using grinding equipment.

There are also challenges involved with grinding cannabis plant material when preparing samples. Cannabis, by nature, is a very fibrous and tough plant that may resist grinding unless sufficient force is applied. Furthermore, heat generated due to shear stresses can cause melting and mobilization of oily resins within the plant, clogging or otherwise sticking to the grinding mechanism. Among the most difficult samples to homogenize by grinding are so-called "gummies," which deform elastically rather than in a brittle fashion and are resistant to shear stresses. One method of overcoming this elastic nature is cryo-milling or freezing the samples prior to grinding.

Liquid nitrogen is the most popular cryogenic liquid using in grinding because of its wide availability and relatively low cost. Liquid nitrogen causes rapid freezing and has a boiling point of -195.79 °C. Lowering samples to this temperature causes the sample to become extremely brittle. This allows for quick pulverization, an increase in throughput and efficiency, and particle size reduction. Therefore, difficult-to-process samples can be more efficiently processed before analysis. Higher efficiency in a grinding system also subjects the mechanical portions of the grinder to less wear, thereby transferring fewer metals into the samples. The reduction in temperature of a material also minimizes loss of potentially volatile metals like mercury.

Sampling liquid cannabis material is far simpler compared to solid samples. The homogeneity of liquid samples depends on the viscosity of the liquids and any concerns about liquid inhomogeneity can be overcome by mixing or stirring the sample. Liquid cannabis samples should be stored in thoroughly clean containers. For analysis of liquid cannabis samples that are long-term storage, it is essential that the analytes stay in solution in a preservative such as a diluted acid. Long-term storage in an acidic solution will assist the analyte from being adsorbed into the storage vessel. It is also important to keep the samples as cool as possible to avoid losses through evaporation, though this is of minor concern compared to samples stored in more volatile solvents.

Sample Preparation

Sample preparation is a very important aspect of characterizing heavy metals in cannabis products via graphite furnace atomic absorption (GFAA) or ICP-MS instruments. Regardless of the analytical method used, samples must be dissolved into solution prior to analysis. Reliable characterization data cannot be generated unless the material is completely in solution when introduced into the instrument for analysis. Sample dissolution is generally required to decompose various forms of cannabis products for release of heavy metals into aqueous solution. The solution must be clear and colorless with no obvious particulate matter with all organic and inorganic components of the sample dissolved into an aqueous solution. Common dissolution techniques include hot plate (block) or microwave digestion. These techniques use concentrated acids and oxidizing agents, such as nitric acid, perchloric acid, hydrofluoric acid, *aqua regia*, hydrogen peroxide, or a combination thereof. Hot plate (block) digestion involves placing a sample in a vessel containing concentrated acids and/or oxidizing agents. The mixture is then heated inside a fume hood for a predefined time period via a hot plate (block) or heating mantle. Microwave digestion is usually accomplished by exposing a sample to concentrated acids and/ or oxidizing agents in a closed vessel and raising the temperature and pressure through microwave irradiation. These conditions increase both the speed of thermal decomposition and the solubility of heavy metals in the sample mixture. A typical microwave digester can hold 12 or more samples and completely digest them in 30–40 min. For reference, hot plate (block) digestions often take 12–24 h to complete. The resulting digested sample is then ready for analysis by GFAA or ICP-MS, however is typically diluted further with water to minimize the acidity and to dilute other constituents of the sample matrix, and adhere to the instrument manufacturer's sample matrix recommendations.

Closed-vessel microwave digestion offer several advantages over open-vessel hot plate techniques, including:

1. Minimization of volatile metals loss (e.g. mercury and lead).
2. Dissolution temperature above the boiling point selected solvent(s).
3. Oxidation potential of concentrated acids and oxidizing agents higher at elevated temperatures for efficient and complete digestion.
4. Lower acid consumption.
5. Microwave dissolution conditions and parameters allow for reproducibility, ensuring consistent dissolution and results.
6. Reduced exposure to hazardous fumes from corrosive agents.

7. Temperature monitoring is automated and individualized, a variety of sample types can be digested in the same batch.
8. High sample throughput.
9. Contamination risk minimized in a closed system.

Despite numerous advantages, microwave digestion processes can suffer from cracking, breaking, or explosion of digestion tubes due to the simultaneous buildup of pressure. This can be mitigated via vessel pressure relief valves and microwave sensors that will limit microwave power to avoid over pressuring the vessels. In the event of a vessel failure, solvents are safely contained within the housing of the microwave digester.

A typical microwave digestion procedure involves:

1. Weighing the required amount of sample, typically 0.1–0.5 g depending on the method and regulation. Placing sample into a plastic vessel, along with the appropriate acids.
2. Sealing the vessel with a tight-fitting cap to create a pressurized environment.
3. Transferring the vessel to the microwave digestion system and digest the sample using the appropriate program.
4. After sample digestion, which takes 10–40 min depending on the matrix, transferring the resulting liquid to a volumetric flask and diluting the sample to the desired volume using high purity water.

The order of addition between acid, water, and sample may need experimentation for optimum digestion results. For example, if the sample floats on top of the acid or sticks to the side of the vessel, some of the sample may not be completely digested in the acid mixture. Therefore, off gassing is required as the sample reacts with the acid material. This may take up to 15 min.

Acid selection for sample preparation is important, especially when ICP-MS is used to characterize heavy metals because of polyatomic interferences. Although this is not strictly a contamination problem, it can significantly impact data if not taken into consideration. For example, if vanadium or arsenic is the target element, hydrochloric acid (HCl) or perchloric acid ($HClO_4$) should not be used if possible, due to the generation of polyatomic ions such as $^{35}Cl^{16}O^+$ and $^{40}Ar^{35}Cl^+$. These ions interfere with $^{51}V^+$ and $^{75}As^+$, respectively. Sulfuric acid (H_2SO_4) and phosphoric acid (H_3PO_4) are also acids that should be avoided, if possible, because they can lead to the generation of sulfur- and phosphorus-based polyatomic interferences. Nitric acid (HNO_3) generates the fewest polyatomic interference among those acids and is generally preferred. If the sample cannot be completely digested using only HNO_3, a combination of HNO_3 and hydrogen peroxide (H_2O_2) can be used to increase the oxidative potential of the solvent. Other acids can be added if a combination of HNO_3 and H_2O_2 does not work. In some cases, hydrofluoric acid (HF) may be needed to dissolve hard-to-digest silicate-based components common to plant materials. In cases where HF is required, specialized polytetrafluoroethylene (PTFE) sample introduction components should be used. Hydrofluoric acid is highly corrosive and a powerful contact poison and should be used with extreme caution. Silicates can also be removed using 0.45 µm PTFE syringe filters. Polyatomic interferences can also be removed using collision cell technology. However, increased analysis time is required for collision cell technology if a multi-element method is used to determine analytes that require both cell and no-cell conditions. Instrument sensitivity may also be compromised if collision cell technology is used.

Contamination issues are probably the greatest concern during sample preparation because of the high sensitivity of ICP-MS and other elemental analysis techniques. Reagent grade trace metals and high purity water should be used in the preparation and dilution of the sample. Trace metal acids should have a certificate of analysis indicating heavy metal content in the parts per billion (ppb) concentration range or less. Using high purity water for the cleaning of vessels and containers is critical. Commercially available water purification systems use combinations of filters, ion exchange cartridges, and/or reverse osmosis systems to remove particulates, organic matter, and trace metal contaminants. Ultra-high-purity water systems typically produce water with a resistance ≥ 18 mΩ cm^{-1}, which is a common benchmark value for water used in ICP-MS analysis.

Calibration standards can be an additional source of preparation error and lead to analytical uncertainty. It is typically recommended to use elemental standards that are specified for inductively coupled plasma optical emission spectroscopy (ICP-OES) or ICP-MS use, as opposed to atomic absorption (AA) standards. Both standard types are certified to contain the target analyte at a specific concentration, ICP-OES and ICP-MS standards are also certified to not contain any other trace elements. Care must be taken when mixing single element standards to make a multi-element stock solution. Certain elements and the acids in which they are solubilized may not be compatible with one another.

In order to save time, effort, and minimize the potential for error, it is advised to purchase multi-element standards from a reputable vendor to ensure their purity, concentration, and stability. Many standards vendors will provide options for purchasing custom blends of elements and concentrations to meet analysis need. It should also be noted that elemental standards typically have a certified shelf-life of 18–24 months after date of manufacturing. Additionally, lower concentration standards have a shorter shelf life, therefore it is recommended to remake analytical standards at least weekly.

The vessels and containers used for the preparation, dilution, storage, and introduction of the sample are another source of potential contamination. For example, glass made from soda lime contains high levels of silicon, sodium, calcium, magnesium, and aluminum. Borosilicate glass contains high levels of boron. Besides these major elements, glass might also contain minor concentrations of zirconium, lithium, barium, iron, potassium, and manganese. If the sample solution is highly acidic, there is a possibility that these elements can be leached or exchanged with the elements present in the sample solution. Moreover, analytes in the sample solution can be adsorbed into the walls of glassware. If the sample or standard is to be stored for extended period, especially if the analyte concentrations are extremely low, adsorption of analytes can significantly bias analytical results. Therefore, plastic vessels and containers are recommended for sample preparation.

If sample vessels are to be reused, the best practice is to soak them in a bath of dilute (1–5%) HNO_3 at 40–50 °C overnight. The vessel should be triple rinsed with ultrapure water before drying and storage. If possible, disposable plasticware can be used to minimize cross-contamination and generation of hazardous acid waste.

Extra care should be taken when preparing and analyzing mercury in samples due to high volatility. Mercury is also noted for its tendency to adhere to walls of vessels, containers, glassware, and tubing. Thereby, accurate and precise determination of mercury can be extremely difficult. In order to minimize this issue, hydrochloric acid or gold can be added into sample to stabilize mercury in solution. Other organic reagents, like ethylenediaminetetraacetic acid (EDTA), 2-mercaptoeothnol, and L-cysteine can be also used successfully, avoiding the difficulties of using HCl.

Samples that are ready for heavy metal analysis should be clear and colorless or a pale-yellow aqueous solution with no particulate matter. If particulates are present, the sample should be filtered with a 0.45 PTFE filter with 0.45 μm pore size is recommended for this purpose. Particulates will clog the nebulizer, torch, and skimmer and sampling cones. If these parts are clogged, the system will need to be back flushed, cleaned, or replaced before acceptable data can be acquired. If the sample is measured via ICP-MS, sample dilution is required to lower acid concentration. Introduction of a sample with a high acid concentration will corrode the nickel or copper formed sampling and skimmer cones. High concentration of acids in the sample solution can also degrade the detector and other instrument hardware quickly.

Instrumentation and Methodology

There are a variety of analytical instruments and methodologies that are employed for elemental analysis of cannabis and derivative products, all of which have historically been widely used in analysis of consumer products. Although there are other elemental analytical techniques available, such as inductively coupled plasma atomic emission spectroscopy (ICP-AES) or ICP-OES and X-ray fluorescence (XRF), these techniques have not proven to be sensitive enough for most cannabis elemental analyses. Here we will focus on the two most common techniques—ICP-MS and GFAA. In general, ICP-MS is the preferred and most common method for analysis whereas GFAA is applicable, but rarely used for reasons explored below.

Analytical Theory

Graphite Furnace Atomic Absorption

The theory and function of atomic absorption spectrometers, including GFAA, is detailed in Welz and Sperling [11] and an overview is provided below. A GFAA consists of a light source, a graphite furnace, an optical bench, and a detector. For analysis by GFAA, a graphite tube is housed within the optical path of the spectrometer such that light can be shone through the tube from end-to-end. A small aliquot of sample is placed into the graphite tube, typically by an automated sampling arm. An electrical current is passed through the electrically-conductive graphite, causing it to heat up, evaporating the solvent and then atomizing the elements contained within the sample.

Simultaneously, a beam of light is passed through the atomized sample from a hollow cathode light source in which the cathode is constructed of the target element. This emits a wavelength of light characteristic to the target element, which is then absorbed by the atomized sample. The amount of light absorbed is directly correlated to the concentration of the target analyte in the sample in accordance with the Beer-Lambert Law. The most common method of detection in a GFAA is a photomultiplier tube (PMT), which converts incident photons into an electrical potential by generating cascading electrons which is interpreted by software as intensity and converted to a concentration.

This process is typically repeated several times. First, most analysts will perform two to three repeats for each element to ensure repeatability. Then, the process is repeated for each individual element for which data is desired, either using different hollow cathode lamps for each or using a single lamp whose cathode consists of multiple elements.

The graphite tube atomizer within a GFAA is capable of reaching very high temperatures, allowing for efficient atomization of samples. Furthermore, the tubular shape serves to contain the atomized sample, as opposed to allowing it to disperse, increasing light absorption even for small concentrations. As a result, typical detection limits are usually in the lower ppb range, which tends to be acceptable to meet regulatory limits for cannabis samples. The primary advantages of using a GFAA is its robustness, ease of use, low cost to purchase, and low cost of operation.

Inductively Coupled Plasma Mass Spectrometry

The analytical theory behind ICP-MS analyses is well-described by Thomas [12] and is outlined here. During analysis by ICP-MS, liquid sample is aspirated into the ion source, called a torch, as a fine, aerosolized mist transported in argon carrier gas. Within the torch, a high-energy argon plasma is maintained by applying a time-varying magnetic field to the argon through an induction coil. Temperatures within the plasma are typically between 6000 and 10,000 K. When injected into the plasma, the aerosol rapidly undergoes several processes: desolvation in which the liquids are evaporated, atomization in which the bonds holding together molecules are broken, and ionization in which an electron is stripped from the individual atoms. The ions generated are then pulled into the mass analyzer section of the instrument by differential vacuum; the ion source is at atmospheric pressure, whereas the mass analyzer is under high vacuum, typically in the range of 1×10^{-5} Torr (~1.3×10^{-3} Pa).

In the mass analyzer, the ion beam is focused through a series of electronic lenses. Magnetic fields are applied to a series of electrically-charged rods, typically in a quadrupole configuration of four opposing rods, in which ions of certain mass to charge (m/z) ratio either resonate in-phase and are transmitted down the length of the quadrupole to the detector, or resonate out of phase, in which they are ejected from the quadrupoles. The electrical fields that are applied to the quadrupole can be rapidly altered on the order of milliseconds, thus allowing for fast focusing on target ions of a certain m/z and then moving to the next target m/z ratio. Such fast switching between transmitted ions gives rise to the pseudo-simultaneous nature of ICP-MS.

The detector on most modern ICP-MS instruments is a discrete dynode electron multiplier. The detector consists of a series of negatively-charged dynodes coated in an ablative semi-conductor material. When an ion impacts the first dynode, the coating ejects a series of electrons, which are attracted to the next dynode, where they eject more electrons in a cascading effect. The detector registers the electrical potential generated by an ion of given m/z, which is proportional to the concentration of that particular element in the starting material and is compared to a known calibration curve to calculate a concentration.

Detection limits for ICP-MS are commonly in the low parts per trillion (ppt) range for most elements, with some instrument vendors even boasting upper parts per quadrillion (ppq) sensitivity for some instruments because of the ICP-MS utilizes an energetic source that efficiently ionizes most elements. This level of sensitivity is more than sufficient to meet the regulatory specification for metals in cannabis products and leaves room for further sample dilution without sacrificing its ability to detect target analytes.

Challenges and Limitations

Along with the adoption of any analytical technique comes the potential for challenges and limitations to the instrument's ability. Both GFAA and ICP-MS are no exception to this rule. Here, we will provide a broad overview of the limitations of each technique particularly as it pertains to analysis in the cannabis laboratory.

The primary limitation of GFAA is sample throughput. Each burn of the graphite furnace yields results only for one element and typical results are comprised of two to three burns for each element. After the first result is obtained, the hollow cathode lamp is changed, and another sample is analyzed for another element. To obtain results for four metals for an individual sample, at least eight injections and burns must be completed. The first limitation to this is time, with each analysis lasting 2–3 min, and analysts usually performing two to three repetitions to ensure accuracy, a complete assay for a single metal would require 4–9 min. A complete assay for four metals for one sample would require 16–36 min.

The second limitation is that the number of lamps housed within an instrument is limited. Low-end instruments only hold a single hollow cathode lamp, while higher-end instruments may have an automated carousel to change lamps and contain between four and eight lamps. Because a unique lamp is used for each element, this limits the number of elements that can be analyzed without the user manually changing the lamps. Many state regulations include eight or more elements, which may immediately limit the appeal of GFAA for analysis due to time constraints or the capacity of the instrument for hollow cathode lamps.

Because of these fundamental limitations, the utility and appeal of GFAA as an elemental analytical technique diminishes with an increasing number of target analytes. Laboratories interested that perform a minimal number of elemental analyses, only are interested in quantifying a small number of elements, and/or those interested in minimizing costs are laboratories that may be interested in GFAA for analysis.

Use of ICP-MS is common for cannabis analysis because it overcomes many of the challenges associated with GFAA, particularly as it pertains to sample throughput. The duration of typical metal analysis by ICP-MS is on the order of 3–5 min and may be made even shorter by utilizing front-end peripherals that rapidly inject samples and rinse solutions into the instrument. However, that is not to say that ICP-MS analyses are without limitation.

The first major challenge with ICP-MS analyses is the overall knowledge base of the typical laboratory analyst. Particularly in cannabis labs, in which the majority of analyses are related to gas or liquid chromatography techniques, most users are not as familiar with ICP-MS and the results that are generated. Much of this limitation is alleviated by software and automation that has made ICP-MS as a technique more accessible for the common user.

Coupled with a generally limited knowledge base is the existence of spectral interferences on certain analytes, including polyatomic, isobaric, and physical spectral interferences. These interferences may negatively impact detection limits, may generate false-positive results, or make analysis of a given analyte impossible. Perhaps the most common example of such is the polyatomic interference on m/z 75 of $^{40}Ar^{35}Cl$ with ^{75}As when analyzing arsenic, exacerbated by the fact that arsenic only has one stable isotopic mass to analyze. When properly understood and accounted for, these spectral interferences can be compensated or corrected for using a collision or reaction cell to minimize interference, proper selection of target masses for the sample and matrix, and mathematical compensation to subtract the portion of a signal generated by interference. Nonetheless, the potential for spectral interferences must be understood and accounted for whenever performing ICP-MS analysis.

Another challenge to obtaining accurate and precise ICP-MS results is the challenges with the instrument and sample matrices. First, matrices are typically strong acids to ensure complete digestion, this requires a level of safety during sample preparation and handling. Furthermore, these matrices can be corrosive to the instrument hardware, even during normal operation. Additionally, because of differences in ionization efficiency, sample solubility, and viscosity, among other factors, it is crucial that all samples and external samples be contained in the same sample matrix to eliminate result variability based on changing the sample matrix.

Finally, and perhaps above all, is the reality that ICP-MS instruments have a higher initial cost-to-purchase as well as ongoing operating costs compared to GFAA. A typical collision cell equipped, single quadrupole ICP-MS may cost

$150,000 or more to purchase whereas high-end triple quadrupole ICP-MS/MS units can cost $250,000 or more. The primary recurring costs during usage are argon gas to generate the plasma at 10–20 liters per minute during operation, trace metal grade acids for sample preparation, and sample introduction consumables within the instrument.

Despite the costs and other challenges associated with ICP-MS analysis, most cannabis laboratories employ ICP-MS for their elemental analysis needs. This is primarily due to high sample throughput, with many states and jurisdictions requiring tests for at least four metals, with many requiring more. Analysis by ICP-MS is typically rapid enough that the limiting factor for laboratory throughput is sample preparation.

Analysis

Analysis of samples involves a multi-step process which begins after a complete validation of all steps to ensure the quality and integrity of results. A standard operating procedure (SOP) must be validated using standard criteria and techniques derived from the published guidelines of the governing regulatory agency. For example, pharmaceutical industry SOPs are validated using the U.S. Food and Drug Administration (FDA) method validation guidelines. In the absence of the regulatory oversight of a single enforcement agency, the method validation guidelines of the FDA have been used as a benchmark(s). Parameters of method validation include range of linearity, interferences, robustness, limits of detection and limits of quantitation. Once the instrument and the SOP are validated, routine analyses can be conducted in a manner that is easily backed up by an audit trail that supports the data. Some of the components of a successful analysis protocol are detailed in the following sections.

Instrument Calibration

Before starting any type of analysis with ICP-MS, key operational parameters of the ICP-MS components should be adjusted using a specific tuning solution. Tuning an ICP-MS should not be confused with "calibrating" the instrument, the latter of which usually refers to generating calibration curves. A tuning solution is typically a vendor-specific, unique blend of several elements at known concentrations. The solution is typically purchased from an instrument vendor, a standards vendor, or can be prepared by mixing and diluting standard solutions.

The purpose of tuning the instrument is to optimize a variety of physical and electronic parameters to generate the highest signal intensities, a minimal amount of electronic signal noise, and acceptable mass resolution. The parameters that undergo regular tuning include the torch position and the voltages applied to electromagnetic lenses, quadrupole mass filter and detector. Tuning the ICP-MS ensures consistent day to day performance. Additionally, instrument calibration results would be stored by software and can be used for tracking the instrument performance over time, allowing the user to identify slow drift or acute faults within the instrument. This start up procedure usually includes following parameters, with some model-specific variables:

1. **Torch Position:** Adjustment of the position of the torch to achieve highest intensity.
2. **Gain Voltage:** Optimization of the voltage of the detector (analog/pulse). The detector is a consumable device and it will degrade over time, by recording the gain voltage, the deterioration of the detector can be monitored, and the user alerted when the detector is no longer operating efficiently.
3. **Lens Voltages:** Optimization of the voltages applied to electromagnetic focusing lenses in the ion transport system to achieve desired sensitivity and resolution.
4. **Mass Axis and Resolution:** Calibration of the mass axis and resolution of the mass analyzer. Usually peak width at 10% of the peak height is used for calibration.
5. **Pulse/Analog Calibration:** A discrete dynode electron multiplier is used as a detector in most of the ICP-MS instruments to the impact of analyte ions into an electric signal. The detector can operate in two different modes: pulse mode for low concentrations and analog mode for high concentrations. It can rapidly switch between these two modes and protect the dynodes from being exposed to high ion fluxes in high concentration samples. Pulse/analog calibration determines and optimizes the factor for converting the analog count into a pulse count.
6. **Database Calibration:** Measures the intensity for calibration sample elements and determines the calibration factor to compensate the drift in intensities of the database and those of the current conditions.
7. **RF Power:** The amount of power applied to the RF generator affects the plasma and is sometimes part of routine ICP-MS tuning. Higher power will create a stronger signal but will also increase the abundance of oxides and double charged interferences. Optimizing this ratio is the goal of tuning. Many ICP-MS manufacturers recommend keeping a constant power setting and not tuning this parameter.

Instrument Validation

Although adjustment of ICP-MS key parameters helps to optimize the instrument performance, instrument performance could be tested and validated at any time using the vendor recommended tuning solution. The following parameters are usually required to be evaluated for this purpose.

1. **Repeatability:** Relative standard deviation (RSD) is used to evaluate repeatability.

2. **Mass Resolution:** Commercially available ICP-MS instruments can provide a mass resolution below 1 atomic mass units (amu).

3. **Oxide Formation:** Polyatomic species are ions made up of multiple atoms that have the same total mass as the analyte of interest. Many of these can be removed using the collision cell technology. However, plasma conditions should be adjusted so that the formation of polyatomic species minimized. Oxides are the most common because of the oxygen in air, but chlorides and sulfides can also be formed depending on the matrix. For example, $^{100}Mo^{16}O^+$ (m/z = 116) has the same m/z as $^{116}Cd^+$.

4. **Divalent Formation:** The first ionization potential for argon is higher than most of the elements in the periodic table which makes argon a great plasma source for generating univalent cations. Argon plasma allows for ionization of most of the elements in the periodic table enabling a wide mass range typically between 5 and 260 m/z for ICP-MS analysis. However, the second ionization potential for some elements, notably the rare earth elements (REE), is also lower than the ionization potential of argon, leading to the formation of undesired divalents (*i.e.*, M^{2+}). The fundamentals of quadrupole mass filtering are based around electromagnetic focusing based on mass to charge ratio (m/z). A divalent cation may have the same m/z as a univalent cation, leading to spectral interference. For example, $^{88}Sr^{2+}$ has the same m/z as $^{44}Ca^+$ and is functionally indistinguishable by the quadrupole and detector. Oxide formation should be monitored for one or more isotope to ensure it is less than a specific standard ratio (usually less than 2% for CeO/Ce). Cerium (Ce) is monitored as it formed oxides more readily than any other element and the logic is that if cerium oxide (CeO) levels are limited, so too must be those of other polyatomic oxides.

5. **Background Signal:** Background intensity is monitored at several masses to assure it is less than a specific standard value.

Installation Qualification, Operational Qualification, Performance Qualification

Installation Qualification (IQ), Operational Qualification (OQ), and Performance Qualification (PQ) are protocols designed to ensure the instrument has been installed according to the manufacturer recommendations. IQ/OQ is usually performed by the ICP-MS vendor as part of an instrument purchase agreement.

An IQ report contains documentation of installation location and environment, vendor and customer contact information, equipment documentation resources (hardware and software manuals, qualification certificates, and certificates of analysis), and delivered consumables and accessories. It should be noted that because of the modular nature of ICP-MS, each unit and accessory needs to be qualified and documented separately according to recommendations by the manufacturer and detailed on the pre-installation checklist and IQ/OQ report template. In addition to the hardware, the version number for the software and firmware is also typically included in the installation qualification report. The installation location should meet the requirements for environmental parameters such as temperature range, temperature fluctuations, relative humidity, dust, corrosive gases or liquids, external magnetic field, bench-top vibration, bench weight tolerance and level, and exposure to direct sun light. All necessary components of a successful installation are verified and documented for future audits.

An OQ report documents that all hardware operations are functioning to specification. This is required for each component, and for the integrated system. While the specific procedures depend on the vendor and hardware configuration, typical components include confirmation of firmware and software versions, gas flowrates (argon gas including carrier, plasma, and auxiliary gas, argon/oxygen mixture if applicable, and helium gas), vacuum level, and results of the instrument calibration and validation as mentioned above. Operational qualification is also usually performed as part of an instrument installation agreement by the instrument manufacturer. The purpose is to support the lab in any on-site audit by an overseeing regulatory agency like the FDA.

Qualitative and Quantitative Analysis Modes (Blanks and Calibration Curves)

ICP-MS is one of the most sensitive techniques for qualitative and quantitative elemental analysis with the analytical range of nine orders of magnitude or more. Elemental measurements can be done in qualitative and/or quantitative mode.

In qualitative measurement mode, calibration curves are not generated. Rather, approximate results are estimated using a series of expected signal intensities built into the software. Qualitative data generally requires less collection time than quantitative results but the results are not accurate enough for stringent reporting purposes. As an approximation, qualitative analysis will yield a concentration for an element that is within an order of magnitude of the true value. Qualitative analyses are primarily used as a quick screening tool to assess the elemental content of a sample or matrix, to determine the appropriate calibration ranges for quantitative analysis, and determine the potential for spectral interferences during analysis.

Analyte concentration can also be measured in quantitative measurement mode. In this method, instrument response is calibrated with respect to the analyte concentration using a blank and several standard solutions of known concentrations. Calibration solutions can be prepared by dilution of commercially available stock standard solutions and blank samples can be prepared by acidifying deionized water. Notably, the same acid matrix should be used in the preparation of the blank sample, calibration standards, sample digestion, and dilution. Using the same acid matrix will minimize the matrix effects generated from different viscosities, ionization potential, and aerosol efficiency effects of different sample matrices.

Instrument performance evaluation prior to analysis of unknown samples requires the determination of instrument detection limits (IDL), method detection limits (MDL), and linear dynamic ranges (LDR) for each analyte. As there is no widely accepted methodology for these measurements, Environmental Protection Agency (EPA) Method 200.8 has been adopted by different labs and industries [13].

The IDL is the lowest concentration of analyte that is distinguishable but not necessarily quantifiable from the blank sample (noise level). Based on the EPA method, the IDL is determined as the mean concentration plus three times the standard deviation (3σ) of replicate measurements of a calibration blank, usually six or more.

The MDL is the minimum concentration of an analyte that can be reported with 99% confidence level. The MDL is calculated by the following equation where the DI water spiked at a concentration 2 to 5 times of the IDL and 7 replicate measurements were run using this solution.

$$MDL = t \times S.$$

Where, t is t-distribution (99% confidence level which is 3.14 for a sample size of seven) and S is the standard deviation for 7 replicated measurements.

The LDR is the range of concentrations where the detected intensity (signal) is proportional to the concentration of the analyte. The LDR should be measured after the calibration of the instrument(s) using the blank and standard samples and fitting the calibration curve, a series of samples of increasing concentrations are run beyond the range of calibration curve. LDR is the concentration at which calculated concentration from extrapolated calibration curve falls within ±10% of their actual concentration.

Instrument Performance Validation

After finishing the initial calibration and IDL, MDL, and LDR determinations, calibration curve validity should be checked by preparing a test sample, usually at half the upper limit concentration of the calibration curve – termed the initial calibration verification (ICV) sample. The sample should be prepared from a second source (different from calibration standard source) and the results should fall within ±10% of the known values for all analytes. If not, the source of problem should be identified and the ICV should be remeasured.

Verification of analytical calibration within the analysis sequence requires a calibration blank and at least a middle-range calibration standard to be run as surrogate sample after every 10 samples, at least. This process should also be completed at the beginning and the end of each batch of samples. The recovery criteria between 90–110% of known value and concentration below 5 times of MDL is desired for continuing calibration verification (CCV) and continuing calibration blank (CCB) samples, respectively. If the initial calibration is not validated, a new calibration is required, and last 10 samples should be remeasured by new calibration curve.

The "spike and recovery" method is another way to validate analytical results and instrument performance. While the matrix plays a key role in acquiring accurate results, analysis of fortified (spiked) samples can be used to assess the effect of the sample matrix. In this method, one sample from each batch of samples or one per 10 or 20 samples are spiked with a specified concentration of the target analyte(s) and analyzed as an unknown sample. The spike recovery is calculated by the following equation.

$$Percent\ Recovery = \frac{C_s - C}{s} \times 100$$

Where C and C_s are concentration of sample and spiked samples respectively and s is the concentration of added analyte to spiked sample. Ideally, the recovery should be fallen within 90–110% of the target value. If the recovery falls out of this range while the other control measurements are successfully passed, it indicates the problem in matrix such as failure in matrix effect correction or heterogeneous sample.

The intensities recorded for internal standards are also required to be monitored for all internal standard elements. While there is not any established criteria, the EPA suggests that the intensity for internal standards should fall within the range of 60–125% when it compared to the original internal standard intensities in blank solution (calibration blank). If the internal standard intensity falls out of this range, the instrument should be flushed with rinse solution and the intensity of the internal standards for calibration blank tested again. If it is back to the normal range, the further dilution of samples can solve the problem. If the drifting problem persist, cleaning the sampler cone or tuning the instrument parameters using the tuning solution can be applied. Checking the sample uptake rate and nebulizer efficiency is necessary to eliminate particulate clogs, worn peristaltic pump tubing, leaks, and other causes for reduced sample in the plasma, which leads to decreased signal intensities and may result in poor analytical results.

Analysis of the "Big Four" Elements

Mercury is a major concern for analysis due to its volatility. Although nitric acid can be used for the digestion of cannabis flower, edibles, and oils, mercury is not stable in nitric acid matrix. Mercury can adsorb onto plastic containers or tubing through binding to the active sites of the plastic. Due to the binding of mercury to active sites, this often causes a so-called carryover effect that may lead to the poor linearity of a calibration curve, long rinsing times, and poor repeatability of analysis. To address this issue, EPA guidelines suggest that 1 ppm gold trichloride ($AuCl_3$) should be added to samples and standards to help stabilize mercuric ions in solution. $AuCl_3$ can also increase the solubility of the silver in nitric acid matrix [14]. Alternatively, L-cysteine is an affinitive reagent to mercury and can stabilize mercury ions in the nitric acid matrix. Notably, continuous introduction of L-cysteine into an ICP-MS can lead to the deposition of carbon on the interface cones and require more frequent cleaning and maintenance [15]. Adding hydrochloric acid to the nitric acid matrix can effectively enhance stability and minimize the carryover effect of mercury, although this step can contribute to polyatomic chlorine-based spectral interferences. Mercury also readily forms volatile hydrides and must be contained in a closed vessel during the microwave digestion. If uncontained, mercury can be removed from a sample in the vapor phase, leading to anomalously low concentration results.

Cadmium has eight different stable isotopes in nature. However, potential isobaric and polyatomic interferences should be considered during method development. The isobaric interference arises in analysis of ^{112}Cd, ^{114}Cd, and ^{116}Cd in the presence of tin (Sn) and similarly there is an interference in the measurement of ^{113}Cd in the presence of In. There is also isobaric interference for ^{106}Cd, ^{108}Cd, ^{110}Cd in the presence of palladium (Pd). If there is molybdenum (Mo) in the sample solution, potential polyatomic interface may arise in analysis of ^{108}Cd, ^{110}Cd, ^{111}Cd, ^{112}Cd, ^{113}Cd, ^{114}Cd and ^{116}Cd as a result of the formation of molybdenum oxide (MoO) ions in plasma. The collision cell can be used to address the issue of polyatomic interferences and suppress the signals from MoO.

Arsenic measurement is critical because it has just one isotope at mass 75. If chlorine (Cl) ions are present in the sample solution, it can prompt a false positive count through the formation of polyatomic ^{40}Ar^{35}Cl, which has the same mass-to-charge ratio as the ^{75}As signal. The collision cell can be used to suppress false positive signal from ^{40}Ar^{35}Cl and facilitate highly accurate measurement. If the concentration of ^{150}Sm and ^{150}Nd is high enough in the sample to generate false positive signal larger than the detection limit of As, an equation correction can be applied to ^{75}As to compensate for the divalent interference. However, high concentration of samarium (Sm), neodymium (Nd), and other rare earth elements are exceedingly rare in cannabis samples.

For lead analysis, ^{208}Pb can be used for the ICP-MS quantitative analysis as the most abundant isotope. However, because of the variations of lead isotopes ratio in nature that can cause isotopic variations between samples and standards, the correction equation can be applied to the ^{208}Pb analysis and the sum of the signals from three Pb isotopes (^{206}Pb, ^{207}Pb, and ^{208}Pb) is used for the Pb measurement. This approach has long been applied to analysis of lead in environmental samples, for example, using EPA Method 200.8 [13].

Typical SOP Components and Representative Data

Elements are selected for analysis, each with the specific mass of highest intensity and lowest potential for interference. Both isobaric and polyatomic interferences should be considered, along with the abundances of each isotope. Reference charts with this information are available, and data is often included in the ICP-MS software as well.

Internal standards are selected and assigned to each analyte as shown in Fig. 1. Internal standards should be within ±30 amu of the analyte. Internal standards are selected from elements that are both rare in the environment and unlikely to be found as contamination, and that have few isotopes which could cause isobaric interferences with other analytes.

Most method conditions (Fig. 2) are standard, but they can be adjusted and optimized for each individual method requirement and/or sample matrix. Rinse and sample uptake times are adjusted to balance speed and throughput against carryover. Increased rinse time, particularly at a high speed, will result in lower carryover but there will be an increase in time and rinse solution used. Chamber temperature, "low" pump speed during sample analysis, and carrier gas adjustments can greatly affect signal, as all contribute to the efficiency of the aerosol produced by the nebulizer.

Peak profiles are shown in Fig. 3, 4, 5, and 6 for the four major elements in a typical cannabis analysis. As expected, intensities increase proportionally to the concentration in the matrix blank and standards. The pink underlined peaks are isotopes not selected for analysis in the representative method by Shimadzu software. They show the signal strength for those isotopes, which can be checked against the ratios expected by published isotopic abundance charts for the selected element. If an unexpectedly large amount of a particular isotope m/z signal is present, it could signify an interference.

Examples of calibration curves for the As, Hg, Cd, and Pb are shown in Figs. 7, 8, 9, and 10, respectively. Standards of 0, 10, 100, 250, 500, and 1000 ppb were selected for each element. The correlation coefficient shown in Fig. 11 for each calibration curve is 0.9995 or above for all the elements where a value r = 1 means a perfect positive correlation. Also shown are the limits of detection (3 s) and limits of quantitation (10 s) calculated by Shimadzu's LabSolutions ICPMS software from the calibration curve.

Shown in Fig. 12 are the analysis of 10 medium-chain triglyceride (MCT) oils with results reported in μg/L. A value of "n.d." means it was none detected as results were below the level of detection limit for the ICP-MS. In order to determine the instrumental detection limits (IDL) for each element shown in Fig. 13 are examples of the measured blank solutions concentration following EPA guidelines.

Shown in Fig. 14 is the Shimadzu ICPMS-2030 used for the analysis of the data described in this chapter.

	Elem	Mass	Type	Cond.	Cell Gas	Integration Time	Integration Mode	IntStd		Excl
1	As	75	QUANT	DB	ON	2.0	Auto	Y	89 (DBG)	☐
2	Cd	111	QUANT	DB	ON	2.0	Auto	Y	89 (DBG)	☐
3	Hg	202	QUANT	DB	ON	2.0	Auto	Tl	205 (DBG	☐
4	Pb	208	QUANT	DB	ON	2.0	Auto	Tl	205 (DBG	☐
5	Tl	205	INTERNAL	DB	ON	2.0	Auto			—
6	Y	89	INTERNAL	DB	ON	2.0	Auto			—

Fig. 1 Element and mass selections with internal standards

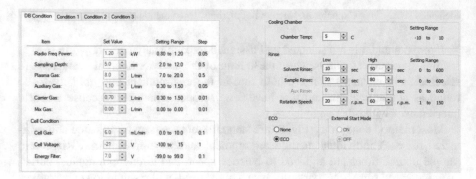

Fig. 2 Representative ICPMS method conditions

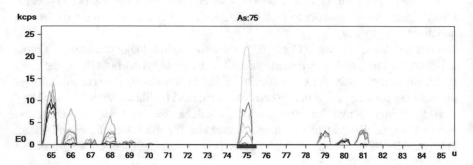

Fig. 3 Peak Profiles, Arsenic: 0, 10, 100, 250, 500, and 1000 ppb

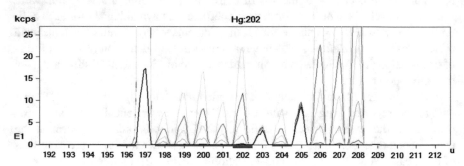

Fig. 4 Peak Profiles, Mercury: 0, 10, 100, 250, 500, and 1000 ppb

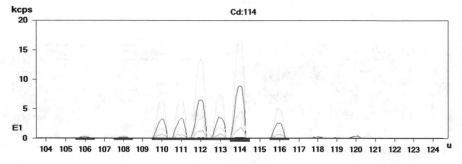

Fig. 5 Peak Profiles, Cadmium: 0, 10, 100, 250, 500, and 1000 ppb

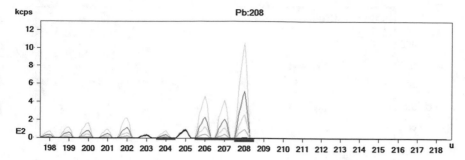

Fig. 6 Peak Profiles, Lead: 0, 10, 100, 250, 500, and 1000 ppb

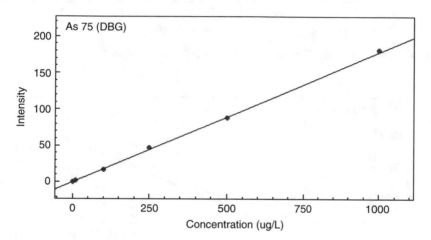

Fig. 7 Calibration Curve, Arsenic: 0, 10, 100, 250, 500, and 1000 ppb

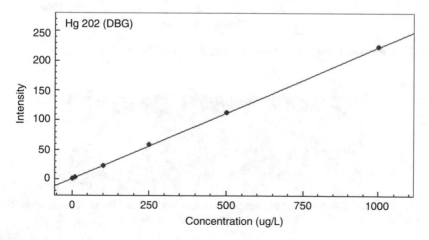

Fig. 8 Calibration Curve, Mercury: 0, 10, 100, 250, 500, and 1000 ppb

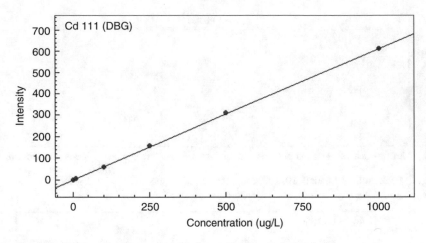

Fig. 9 Calibration Curve, Cadmium: 0, 10, 100, 250, 500, and 1000 ppb

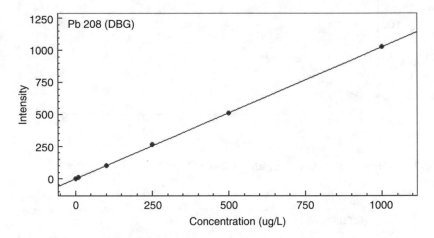

Fig. 10 Calibration Curve, Lead: 0, 10, 100, 250, 500, and 1000 ppb

Element	m/z	r	3s (µg/L)	10s (µg/L)
As	55	0.99962	0.5105	1.7017
Cd	98	0.99974	0.02282	0.0761
Hg	60	0.99947	0.3846	1.2822
Pb	207	0.99984	0.0547	0.1823

Fig. 11 Example Calibration Curves, with limits of detection (3 s) and limits of quantitation (10 s) calculated by LabSolutions ICPMS software from the calibration curve

Sample	As (µg/L)	Cd (µg/L)	Hg (µg/L)	Pb (µg/L)
Blank 1	0.0087	0.0098	0.0115	0.0095
Blank 2	0.009	0.0098	0.109	0.0097
Blank 3	0.0087	0.0099	0.0102	0.0096
Blank 4	0.0084	0.0095	0.0092	0.0095
Blank 5	0.0087	0.0098	0.0092	0.0096
Blank 6	0.0089	0.0098	0.0089	0.0095
Blank 7	0.0081	0.0093	0.0081	0.0093
Mean Conc.	0.0086	0.0097	0.0237	0.0095
3x Std Dev. (3σ)	0.0009	0.0006	0.1128	0.0004
IDL	0.0096	0.0103	0.1366	0.0099

Fig. 12 Example data: customer laboratory data, MCT Oil samples. Note, reporting "n.d.," none detected, is appropriate when the measured value is less than the detection limit

Sample	As (µg/L)	Cd (µg/L)	Hg (µg/L)	Pb (µg/L)
MCT Oil 001	n.d.	0.0009	n.d.	0.0105
MCT Oil 002	n.d.	0.0011	n.d.	0.0109
MCT Oil 003	n.d.	0.0021	n.d.	7.7100
MCT Oil 004	n.d.	n.d.	n.d.	0.0009
MCT Oil 005	n.d.	n.d.	n.d.	n.d.
MCT Oil 006	n.d.	n.d.	n.d.	0.0008
MCT Oil 007	n.d.	0.0001	n.d.	0.0144
MCT Oil 008	n.d.	n.d.	n.d.	n.d.
MCT Oil 009	n.d.	n.d.	n.d.	n.d.
MCT Oil 010	n.d.	0.0225	n.d.	0.0234

Fig. 13 Example IDL data: customer validation data, blank MCT oil samples to determine detection limits per EPA method. IDL is calculated as three times the standard deviation plus the mean concentration of a set of blank samples

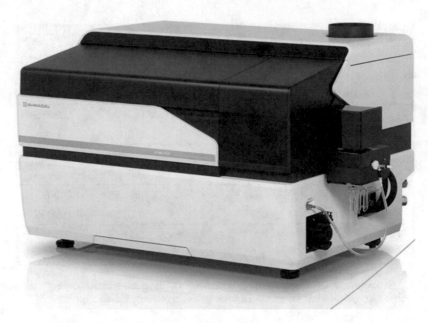

Fig. 14 Shimadzu's ICPMS-2030 used to analyze the MCT oils

Conclusion

Inductively coupled plasma mass spectrometry is the most common analytical technique for the analysis of metals in cannabis and cannabis-related products. The ability to measure a wide range of sample matrices, as well as pseudo-simultaneous analysis of a multitude of elements sets this technology apart from other elemental analytical tools. The ability to detect elements at low parts per trillion levels makes the ICP-MS ideal for measurement of contaminates in pharmaceutical, cannabis, and other applications affecting human health. Results obtained from the Shimadzu ICPMS-2030 showed good linearity, detection limits, and spike recoveries for QC samples for the big four elements – arsenic, cadmium, mercury, and lead. Expected changes in the cannabis industry will include standardization of testing methods and protocols for metals analysis. Currently similar industry-specific methods are published by the FDA, USP, EPA, United States Department of Agriculture (USDA), ASTM International, AOAC International, and others. A cannabis-specific method will include a standard list of elements and detection limits for different types of samples, and for different methods of ingestion and type of patient or consumer. Standardization of the requirements will likely accompany standardization of the sample preparation process, including the microwave digestion program for varying matrices. Other method parameters such as the technique for stabilizing mercury could become standardized as well. It is expected that additions to the regulations will include heavy metals testing for other parts of the manufacturing and growing process, including soils and nutrients. Required testing may also be expected for packaging, components of nebulizers and vape pens, and other ancillary products which are currently unregulated. Additionally, improvements in instrumentation technologies, software, and sample handling techniques will accompany changes in analytical testing methodologies in this growing market.

References

1. Lisk DJ (1972) Trace metals in soils, plants, and animals. Adv Agron 24:267–325
2. Tchounwou PB, Yedjou CG, Patlolla AK, Sutton DJ (2012) Heavy metal toxicity and the environment. In: Luch A (ed) Molecular, clinical and environmental toxicology, Experientia Supplementum, vol 101. Springer, Basel
3. Luckey TD, Venugopal B (1977) Metal toxicity in mammals. Volume 1. Physiologic and chemical basis for metal toxicity. Plenum, New York, p 238
4. <232> Elemental Impurities – Limits, in: United States Pharmacopeia National Formulary (USP 38-NF 33), United States Pharmacopeia Convention, 2012, p. 245–248.
5. <233> Elemental Impurities – Procedures, in: Second Supplement to United States Pharmacopeia National Formulary (USP 38-NF 33), United States Pharmacopeia Convention, 2012, p. 243–244.
6. Chojnacka K, Chojnacka A, Górecka H, Górecki H (2005) Bioavailability of heavy metals from polluted soils to plants. Sci Total Environ 337(1-3):172–185
7. Girdhar M, Sharma NR, Rehman H, Kumar A, Mohan A (2014) Comparative assessment for hyperaccumulatory and phytoremediation capability of three wild weeds. 3 Biotech 4(6):579–589
8. Gimeno-Garcia E, Andreu V, Boluda R (1996) Heavy metals incidence in the application of inorganic fertilizers and pesticides to rice farming soils. Environ Pollut 92(1):19–25
9. Olmedo P, Goessler W, Tanda S, Grau-Perez M, Jarmul S, Aherrera A, Chen R, Hilpert M, Cohen JE, Navas-Acien A, Rule AM (2018) Metal concentrations in e-cigarette liquid and aerosol samples: the contribution of metallic coils. Environ Health Perspect 126(2):027010-1–012010-11
10. Wei S, Zhou Q, Koval PV (2006) Flowering stage characteristics of cadmium hyperaccumulator Solanum nigrum L. and their significance to phytoremediation. Sci Total Environ 369:441–446
11. Welz B, Sperling M (1999) Atomic absorption spectrometry, 3rd edn. Wiley, New York, p 941
12. Thomas R (2013) Practical guide to ICP-MS: a tutorial for beginners, 3rd edn. CRC Press, Boca Raton, FL, p 409
13. Method 200.8, Determination of trace elements in waters and wastes by inductively coupled plasma-mass spectroscopy, Revision 5.5, EMMC Version
14. Mercury Preservation Techniques, US Environmental Protection Agency.
15. Li Y, Chen C, Li B, Sun J, Wang J, Gao Y, Zhao Y, Chai Z (2006) Elimination efficiency of different reagents for the memory effect of mercury using ICP-MS. J Anal At Spectrom 21(1):94–96

Quantitative Terpene Profiling from Cannabis Samples

Adam Floyd and Mike Tunis

Abstract Terpenes are naturally occurring compounds that are present throughout nature. Modern interest in terpenes and terpene analysis has grown significantly as the recreational and medicinal use of cannabis has expanded across the globe. Terpenes are the aromatic compounds that give cannabis its unique smell and flavor and are a critical aspect of the cannabis experience. They are built from repeating five carbon units called isoprene and their distinction as monoterpenes, sesquiterpenes, diterpenes, etc. is dependent on the number of isoprene units utilized in their biosynthesis. The objective of this chapter was to produce a broad overview of terpene analysis in cannabis. Several methodologies will be discussed in this chapter and it should provide a baseline for analyzing terpenes in cannabis flower and cannabis products as well as an overview of the role of gas chromatography in the cannabis laboratory.

Introduction

Terpenes are a classification of aromatic compounds that are nearly ubiquitous throughout nature. They are primarily found in plants but can also be observed in marine organisms, insects, and, to a lesser extent, higher-order animals. In addition to being responsible for the characteristic fragrances of many flowers, fruits, and trees, terpenes also have a multitude of commercial uses in cleaning products, essential oils, coffee, tea, and beer. At the commercial scale, terpenes are both synthesized and extracted from biological sources containing large concentrations of these molecules. Due to their extensive natural abundance and vast commercial use, there is no doubt that terpenes impact our lives daily.

Chemically, terpenes are comprised of repeating five-carbon molecules called isoprene units and, depending on the number of isoprene units in a terpene's biochemical synthesis, are further categorized as monoterpenes (two isoprene units), sesquiterpenes (three isoprene units), diterpenes (four isoprene units), etc. [1] Biochemically, terpenes are utilized by insects to metabolize growth hormones and

A. Floyd (✉) · M. Tunis
Think20 Labs, Irvine, CA, USA
e-mail: adam.floyd@think20labs.com

© Springer Nature Switzerland AG 2021
S. R. Opie (ed.), *Cannabis Laboratory Fundamentals*,
https://doi.org/10.1007/978-3-030-62716-4_12

pheromones [2]. While it is not completely understood yet, it is known that in plants terpenes play a significant role acting as crucial plant growth regulators (PGRs), attracting/repelling insects, preventing fungal growth, and signaling. Currently, there is still much research to be done on how terpenes are fully utilized in both plants and animals.

Terpene Biosynthesis

The biosynthesis of terpenes can occur through two pathways: the mevalonate (MVA) pathway and the non-mevalonate (MEP) pathway. In most eukaryotes, the mevalonate pathway is used exclusively while in most eubacteria, the non-mevalonate pathway is used. In plants, both pathways can operate simultaneously with biosynthesis via the MVA pathway taking place in the cytosol and the MEP pathway taking place in the plastids. Sesquiterpene and triterpene synthesis occurs in the cytosol while monoterpene, diterpene, and tetraterpene synthesis occurs in the plastids [3].

Terpenes are constructed from the precursor isopentenyl diphosphate (commonly known as IPP) and dimethylallyl diphosphate (DMAPP). In the mevalonate pathway, isopentenyl diphosphate is formed with the condensation of acetyl-CoA. In the non-mevalonate pathway, isopentenyl diphosphate is formed from pyruvate and glyceraldehydes 3- phosphate. In the next phase of biosynthesis, isopentenyl diphosphate and dimethylallyl diphosphate are utilized in condensation reactions to produce terpene precursors. The next phase produces monoterpenes, diterpenes, and sesquiterpenes [4, 5].

Terpenes in Cannabis

At this time, well over 150 different terpenes have been identified in cannabis with this number continually growing as research into cannabis becomes more prevalent. The number of terpenes endogenous to cannabis is impressive but they are a fraction of the over 30,000 different terpenes that have been discovered in nature [6]. Of the terpenes found in cannabis, the weight by weight majority is generally comprised of mono- and sesquiterpenes but several larger terpenes have also been observed. The most common terpenes endogenous to cannabis are myrcene, pinene, limonene, linalool, and caryophyllene.

In the cannabis user experience, terpenes are one of the primary distinguishing characteristics between strains. In fact, many cannabis strains are selectively bred by cultivators strictly for their unique terpene profiles. These profiles are responsible for the user's sensory experience of flavor and aroma and contribute significantly to how cannabis strains are ranked against each other in the cannabis community. Award-winning cannabis flowers tend to have diverse and strong terpene profiles while the tetrahydrocannabinol (THC) levels in these winning strains are typically very similar. There is significant research on the medical benefits of terpenes [7]. Future studies may link these compounds to some of the medical benefits of cannabis usage. For the manufacturing of concentrates for vape pens and dabs, conserving the terpene profile of the strain throughout the extraction process is considered critical as to not lose the "identity" of the cannabis flower. Terpenes are also considered to be integral in what is commonly referred to as the "entourage effect". The "entourage effect", currently based on anecdotal evidence, attributes novel and therapeutic experiences to synergistic interactions of cannabinoids and terpenes with the various biological systems of the brain and body [8].

Separation and Analysis

Liquid Chromatography

Liquid chromatography (LC) is a separation technique in which compounds present in liquid mixtures are separated through a chromatography column. In the column, compounds are separated from one other based on the degree to which they interact with the liquid mobile and solid stationary phases of the system. These interactions are typically governed by polarity, with nonpolar analytes interacting strongly with nonpolar phases and polar analytes interacting more strongly with polar phases. Analytes that readily partition in the column's solid stationary phase are retained longer than those that share the same polarity as the flowing liquid mobile phase. This is ultimately where LC gets its separatory power. Because liquid chromatography is a liquid based technique, it is most often used with non-volatile compounds.

There are LC applications for terpene analysis but as terpenes are volatile compounds, they are better suited to be separated using gas chromatography. That is not to say that LC is an inappropriate technique for terpene analysis. In fact, LC is widely used for the analysis of terpenes in academia and research-based settings. For high throughput cannabis laboratory settings, however, LC is not the most efficient choice as UV detection of terpenes is difficult in cannabis samples due to overwhelming concentrations of cannabinoids which also absorb light in the middle ultraviolet region.

Gas Chromatography

Much like in liquid chromatography, gas chromatography (GC) utilizes a chromatography column to separate compounds present within a mixture. However, as its name implies, samples undergoing GC analysis are first volatilized into a gas before being loaded onto the column for separation. The mobile phase in GC, referred to as the carrier gas, is most commonly either helium or hydrogen but nitrogen is also a viable choice for analyses requiring increased efficiency. As with LC, there are a wide variety of stationary phase choices that can be used based on the chemistry of the target compounds to be separated. Gas chromatography gets its separatory power through differences in volatility and polarity among target analytes with more volatile analytes being swept through the column at a quicker rate than less volatile ones or analytes with similar polarity to the column's stationary phase.

Due to its powerful separatory properties, gas chromatography is used ubiquitously for the analysis of volatile compounds and almost exclusively for the quantitation of terpenes in commercial analytical cannabis laboratories. Since the development of gas chromatography, it has been used for the analysis of terpenes in plants and fragrances. Thus, scientific literature is full of primary publications providing a wealth of information on the specifics of this analysis.

Multi-Dimensional Gas Chromatography

The concept of multi-dimensional gas chromatography (often denoted as GCxGC) is relatively straightforward, with two columns, typically of opposing polarities, used in series with one another. This allows for far greater separation and deeper understanding of complex sample matrices. As there are hundreds of compounds found in cannabis, GCxGC is a powerful tool for research and providing commercial customers with a deeper insight into the terpene profiles of their products. GCxGC excels at resolving co-elution problems that often occur in complex matrices such as cannabis. The unfortunate downside to GCxGC, however, is an extended run time in comparison to its single column counterpart. While the cost is greater than traditional GC, advances in technology continue to lower prices making GCxGC a more attractive technique.

Injector and Inlet Systems

There are a multitude of injector and inlet systems available for use with gas chromatography. However, most of them are not typically used for terpene analysis. The inlet system most used for the analysis of terpenes is the split/splitless inlet. This inlet, as its name implies, is a versatile inlet that can operate in either split or splitless mode (these topics are discussed in greater detail later in this chapter) depending on how concentrated the samples to be analyzed are and how the sample is introduced into the inlet.

Inlet Liners

Glass liners are an important part of the GC system as they are a vessel in which the sample can be heated, vaporized, and transferred to the head of the GC column. There are numerous variations of liner that are selected depending on what the analyst is trying to accomplish and the characteristics of the analytes that are being measured. Liners come in a variety of diameters and selecting the correct liner diameter allows for the proper expansion volume during the vaporization phase. If the liner is too small, there will be significant dead volume during the injection. The two types of liners typically used in terpene analysis are 1 mm deactivated split/splitless straight inlet liners and split/splitless liners with glass wool.

Direct Analysis via Headspace Injection

The simplest way to analyze a cannabis or cannabis product sample for terpene content is through direct analysis via headspace injection. This method is as simple as placing a small aliquot of the sample to be analyzed in a headspace vial and loading the sample vial into a headspace autosampler where it will be heated, pressurized, and subsequently injected in the GC system. Through the act of heating the vial in an oven, volatile terpenes present within the sample will move from the sample matrix into the "headspace" of the vial which will, in turn, be sampled from and sent to the GC system through a heated transfer line. The primary issue with this technique is that it is particularly susceptible to matrix effects when analyzing cannabis extracts such as distillates, butane hash oil (BHO), resins, and other viscous concentrates. These concentrates can cause significant matrix suppression resulting in incomplete recoveries. To combat this, some laboratories will prepare matrix matched calibrations where instruments are calibrated with standards prepared in the matrix being analyzed. Overall, this technique is not recommended for high throughput cannabis laboratories as each sample matrix requires a separate calibration to be run and with the number of cannabis containing sample matrices continually increasing, this would be an overly expensive and time consuming endeavor.

Total Evaporative Analysis Using Headspace

Total evaporative analysis (sometimes referred to as total evaporative technique (TET)) is a headspace analysis technique where the sample is extracted in a solvent and an aliquot (typically 10–40 uL) is injected into a headspace vial. This technique is preferred for cannabis analysis as cannabis containing products can vary significantly. Total evaporative analysis is an excellent way to overcome matrix effects which can be hugely problematic in cannabis analysis.

Liquid Injection Using Split/Splitless Vaporizing Injectors

Liquid injection is the most robust injection type and, unfortunately, also the dirtiest. Liners with glass wool must be used when performing liquid injection of cannabis extracts. The wool works to capture any solid particulates from the cannabis extract being injected into the system and depending on the sample preparation methods, these liners typically need to be changed often to prevent issues with carryover. This leads to increased operational cost and instrument downtime to conduct routine maintenance. The major advantage to liquid injection, however, is reproducibility. Figure 1 demonstrates the principles of split/splitless injection.

Split Injections

Split injections allow for a large volume to be injected while only a fraction makes it onto the column. Most of the injection is swept out of the split outlet. This technique is appropriate for terpene analysis and will most commonly be used with liquid injection. The advantages of this technique are high resolution and the ability to inject "dirty" matrices. This last advantage is key when analyzing cannabis as it tends to be a complex and dirty matrix. Split injection is typically done at a 20:1–50:1 for terpene analysis but 200:1–500:1 splits are not uncommon.

Splitless Injections

Most modern injection ports are split/splitless. Splitless injection works much like split injection, however, the split valve remains closed during the initial introduction of sample into the inlet. This allows for the entire volume of the injection to go onto the column. Splitless injections are primarily used for trace analysis as the entire

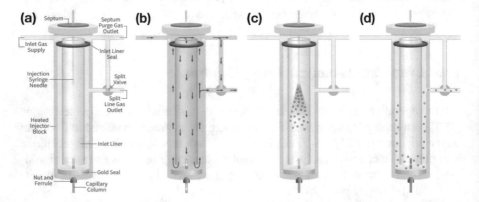

Fig. 1 (a) Schematic of a Split/Splitless Inlet. (b) The flow path of carrier gas through the inlet. (c) Sample being injected into the inlet liner. (d) Inlet operating in split mode

effluent volume enters the column. Splitless injections are commonly used in head-space analysis but are not recommended for use with liquid injection as column overload and backflash is likely to occur.

Autosamplers

Autosamplers for gas chromatography are included with nearly every modern instrument. The two autosampler configurations that are used in terpene analysis are headspace autosamplers and liquid injection autosamplers. Certain vendors offer combination units that can perform both types of injections. The use of autosam-plers is highly recommended for any sort of commercial or research analysis of terpenes.

Extraction and Sample Preparation

Quality sample preparation is critical for consistent and accurate analysis of ter-penes. Sample preparation is one of the large sources of error in the laboratory. The way samples are prepared depends on several factors. In commercial cannabis labs, a multitude of different matrices may need to be analyzed. In a research laboratory, only a single matrix type may need to be analyzed. In cannabis analysis, most sam-ples that are processed are solids and require solid-liquid extraction. Most liquid samples that arrive in the laboratory can be diluted directly.

Solid-Liquid Extraction

This is the most common extraction type used in commercial laboratory settings. The concept is relatively simple where a solid sample (in this case some type of cannabis or cannabis product) is extracted with a solvent that is conducive to solvat-ing the target analytes. Typically, between 0.25 g–1.0 g are used in the extraction process although this is dependent on state regulations. Isopropanol and methanol are commonly used solvents in this process. The sample is weighed out and extracted in a total solvent volume of 5–10 mL.

Solid Phase Extraction

Solid phase extraction is often used in a research-based setting when applied to terpenes. A sorbent material is placed near the sample (often the actual growing cannabis plant) and the volatile terpenes are adsorbed onto the material. Solid phase extraction has several distinct advantages: matrix effects are greatly mitigated as only the target analytes are introduced to the system.

Sample Matrices

Cannabis Flower

Cannabis flower is one of the easiest matrices to work with when analyzing terpenes. Most terpenes readily dissolve into methanol or other non-polar solvents. When using solid-liquid extraction, a representative sample is simply ground and extracted using solvent. The resulting liquid can be injected directly or used in headspace analysis. Cannabis flower can also be analyzed directly using headspace via full evaporation technique (FET). This is an ideal method as it does not require extraction and preserves the terpene profile of the plant.

Cannabis Extracts

Cannabis extracts pose more of an issue with sample preparation. The initial extraction process from the cannabis flower to the extract is a diverse process and the end material may vary from a crumble to oil. Direct analysis of cannabis extracts in headspace as previously discussed results in serious matrix suppression. Recoveries from direct analysis may be as low as ~20%. Standard addition or matrix matched calibrations may be required for accurate terpene analysis in cannabis extracts using direct analysis.

Tinctures

Tinctures are one of the easier matrices to extract for terpenes. Typically, dilution of the product is enough for analysis as tinctures are often made in ethanol. The terpene concentration of tinctures can vary significantly depending on the intentions of the producer. Certain companies want to retain the characteristic flavor of the strain while others choose to remove the terpenes completely.

Cannabis Edibles

Cannabis edibles rarely have significant terpene concentrations. In products such as mints or candies, terpenes can sometimes be found as flavoring agents. They are typically added during the product formulation and are not natural cannabis terpenes.

Columns for Gas Chromatography

Capillary Columns

The use of capillary columns is prevalent in modern day gas chromatography systems. The practical uses for packed columns are minimal and irrelevant to cannabis analysis and as such will not be discussed further in this text. Capillary columns (sometimes called open tubular columns) consist of a long thin tube typically made of fused silica. Fused silica is highly inert and flexible, it bonds well with silicon liquid phases resulting in efficient, long lasting columns. Fused silica columns have a protective coating of polyamide applied during the manufacturing process to prevent corrosion from atmospheric moisture. This coating, rather than the tubing itself sets an upper limit of roughly 400 °C on column temperature exposure. There are specialty columns and phases which can withstand higher temperatures for niche applications. For the purposes of terpene analysis, this is more than enough.

Column Characteristics

There are hundreds of commercially available capillary columns for thousands of applications.

GC columns are typically described with the following characteristics:

1. Stationary phase composition
2. Column length
3. Internal diameter (I.D.)
4. Film thickness
5. Maximum programmable temperature
6. Minimum bleed temperature

These characteristics allow the user to select the ideal column for their application.

Stationary Phase Composition

The stationary phase determines the chemical interactions between the target analytes and the columns. This will determine the elution order and speed in which analytes will elute from the column. For terpene analysis there are two types of columns typically used. If both residual solvent and terpene analysis are being performed on the same column, a mid-polarity siloxane column should be used. If terpene analysis is being done exclusively, a polar polyethylene glycol column can be used. The polyethylene glycol phase (sometimes called a wax column) provides excellent separation for terpenes but limits resolution and sensitivity for residual solvents. As most commercial cannabis laboratories do not have dedicated GCs for terpenes and residual solvents, a column that can reasonably resolve and retain both residual solvents and terpenes is ideal.

Column Length

The column length determines the efficiency of the chemical separation. The longer the column, the better the separation and resolution of the chromatography will be. The trade off for this is analysis time. Commercial capillary columns are typically sold in lengths between 15-105 m. Most routine terpene analysis can be done on a 30 m column although a 60 m or longer may be required for resolution of significant numbers of terpenes. Commercial laboratories typically choose a column length of 30 m as sample throughput is critical and instrument time is at a premium. When separating a large number of terpenes, a 60 m column can be used. The world record for longest column is 1300 m created by Jaap de Zeeuw.

Internal Diameter (I.D)

The internal diameter of the column determines the column's sample capacity and resolution. The internal diameters of modern-day capillary columns typically range from 100–530 μM. 250–320 μM is sufficient for most applications including terpene analysis. There should be no practical reason to stray outside of this internal diameter range.

Film Thickness

Film thickness is an indication of how much stationary phase is present in the column. Thick films allow for a high sample capacity on the column. Thin films allow for increased resolution in the chromatogram. For terpene analysis, a 0.25 μM film thickness is usually enough. It strikes a good balance between resolution and capacity as terpenes are typically found in significant concentrations.

Maximum Programmable Temperature

The maximum programmable temperature indicates the upper limit at which the temperature programmed GC cycle should reach. Generally, the temperature program should always end 10–20 °C below this value. This will help to prolong the life of the column and help to prevent column bleed.

Minimum Bleed Temperature

The minimum bleed temperature of the column indicates the highest temperature at which the column can be programmed to minimize the column bleed. Minimizing column bleed is critical to extending column life and maintaining a clean detector.

Column Conditioning

Column conditioning should be carried out whenever a new column is installed. Most modern capillary columns come from the factory pre-conditioned. It is good practice to condition a new column by running an initial temperature ramp of 2–4 °C/min up to approximately 10–15 °C below the maximum column temperature. Once this temperature is reached, it should be held for 30–45 minutes. Once the initial conditioning has been completed, it is good practice to tighten the nuts in the GC oven and run an additional quick conditioning method. The baseline and column bleed must be observed after conditioning the column. Once these have stabilized it should be ready for analysis. Air and water peaks need to be measured after column conditioning to ensure that there are no leaks in the system.

Detection

Now that we have covered all the topics necessary to successfully separate our target terpenes using gas chromatography, we must now discuss how to detect our analytes and generate a clean chromatogram. The two detection methods most relevant to terpene analysis are flame ionization detectors (FID) and mass spectrometers (MS).

Sensitivity

The sensitivity of both FID and MS detectors are sufficient for the majority of terpene applications. Overall, mass spectrometers will be the more sensitive detector with lower limits of detection than FIDs but flame ionization detectors will remain linear across a broader range than their MS counterparts.

Flame Ionization Detector (FID)

Flame ionization detectors are the most common GC detectors. These highly sensitive detectors burn column effluent in an oxy-hydrogen flame. Ions are created in this process and produce a current which becomes the signal. The ions are detected by an ammeter. A schematic of an FID is shown in Fig. 2. FID is appropriate for organic compounds; Therefore, the gasses used for this detector must be free from organic contaminants. The response from the detector is directly related to the number of carbons present in the analyte. Analytes containing no organic carbons will not produce a signal. Because terpenes are built from repeating units of isoprene this does not pose an issue when conducting terpene analysis using an FID. Nitrogen and helium are appropriate carrier gasses for FID and terpene analysis although a secondary source of hydrogen is still required. The major advantages of using an FID are as follows:

1. Low detection limits
2. Excellent linearity
3. Low noise levels
4. Simplicity
5. Ruggedness
6. Low cost

The primary issue that arises when using FID for terpene analysis is the requirement for baseline resolution among analytes. Depending on the number of terpenes that are targeted in the analysis, the run time of chromatographic separation can be significant. Due to the high number of terpenes found in cannabis and the similar chemical structures shared between them, co-elution is also a common issue in detection using an FID. They are significantly cheaper than using a mass spectrometer but lack the adaptability of mass selective detection. GC-FID is an excellent choice for laboratories that are budget conscious or not constrained by long analytical run times.

Ionization Sources for Mass Spectrometry

There are numerous ionization methods for mass spectrometry. For terpene analysis, using gas chromatography electron impact (EI) is the most common. The concept is quite simple; effluent from the GC flows into the source and is ionized using a filament. The source must be contained under a vacuum, in-source fragmentation occurs, and the resulting fragment ions are detected. This is considered a hard ionization method and the abundance of the parent ion is typically small. EI is appropriate for most organic compounds including terpenes. Chemical ionization is not a necessary methodology for terpene analysis. Electrospray ionization (ESI) is a common ionization method used in liquid chromatography.

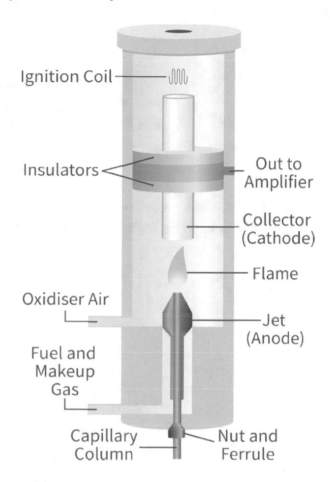

Fig. 2 A schematic of a flame ionization detector (FID)

Mass Spectrometry (MS)

Bench top mass spectrometers have become increasingly common as the price and size of the instruments has dropped significantly. Gas chromatographs coupled to mass spectrometers (GC/MS) allow for a more dynamic style of analysis compared to an FID. The major downside of purchasing a mass spectrometer over an FID is the cost. Most mass spectrometers used in commercial cannabis laboratories for terpene analysis are single quadrupole instruments. This is not always the case as interest in quantitative and qualitative information on extended terpene lists is growing. There are two major advantages in using mass spectrometry over FID in the analysis of terpenes. The first being detection of unknowns, with over 200 terpenes identified in cannabis. Having spectral data allows for untargeted terpene peaks to be identified. Current commercially available multi-component standards typically account for between 20–42 terpenes. An example chromatogram of a 22-component terpene standard is shown in Fig. 3. There will potentially be many untargeted

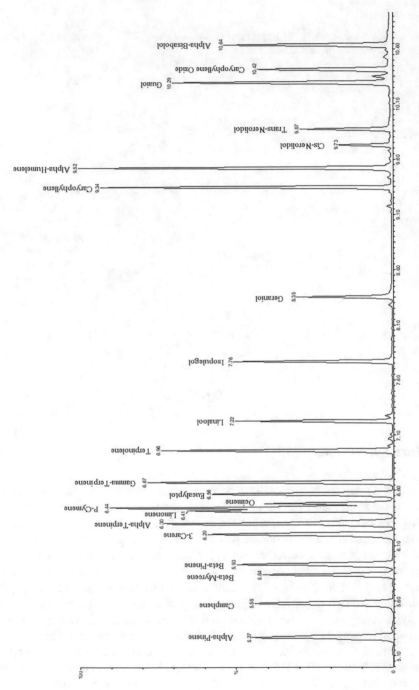

Fig. 3 Example of a terpene chromatogram. The visual appearance of the chromatogram will not differ significantly between FID and MS

terpenes present in the matrix. These untargeted compounds may co-elute with our target terpenes. These compounds can be resolved through mass separation using mass spectrometry. Due to the complex nature of cannabis samples, there will likely be unintended co-elution of non-target compounds. This can create constructive interference in the signal and provide an artificially high value for the target compound.

Mass Selection

Mass selection is particularly important when dealing with co-eluting peaks and is one of the primary reasons to use mass spectrometry over FID. One of the biggest challenges in mass spectral analysis of terpenes is the common ions that are present in many mass spectrums. When developing the initial GC methods, achieving baseline resolution for all analytes is ideal. This is not always an easily obtainable feat and separation by mass allows for a shortened run time. There are unique ions present in certain terpenes which makes separation by mass relatively easy. The co-eluting peaks can be deconvoluted using specific masses and mass ratios. This can help to dramatically reduce chromatographic run times. An example of co-eluting analytes is shown in Fig. 4 with the deconvolution shown in Fig. 5.

Tandem Mass Spectrometry

Tandem mass spectrometry (MS/MS) is a technique that uses multiple mass analyzers compared to the single mass analyzer present in single quadrupole mass spectrometry. Triple Quadrupole mass spectrometry combines three quadrupoles in tandem [9]. These instruments offer extraordinary quantitation/detection limits and are present in nearly every cannabis laboratory for the detection of pesticides. MS/MS applications for terpenes tend to be more research based. LC-MS/MS instruments are far more common than GC-MS/MS.

Database Identification of Unknown Compounds

Perhaps one of the more useful aspects of analyzing terpenes with a mass spectrometer is the identification of unknown analytes in a sample. It is unlikely that hundreds of terpenes standards will be used in instrument calibration. Database identification provides a versatile tool to identify untargeted terpenes present within samples. The concept is relatively straightforward, the mass spectrum that is experimentally produced is run against a database of mass spectra of known compounds. A probable match list is produced with top hits. Terpene database identification can

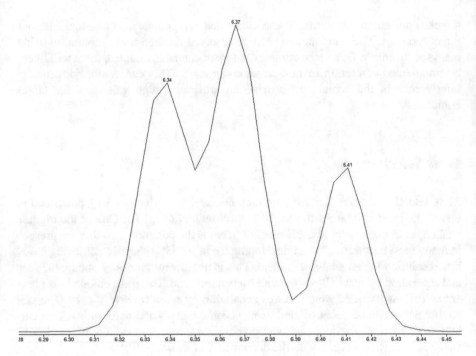

Fig. 4 Co-eluting terpene peaks. Limonene, p-cymene, and Ocimene all co-elute in this method. Separation is possible through mass selection

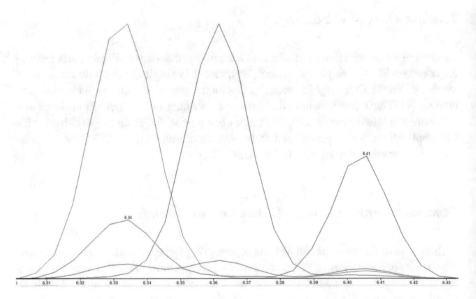

Fig. 5 Deconvolution of coeluting peaks by mass selection. Quantitation is possible using mass 68, 119, and 80 respectively. The green, purple, and black lines indicate the selected masses

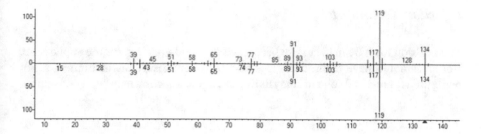

Fig. 6 Head to tail comparison of experimentally derived spectra versus database spectra of p-cymene. The mass spectrum in red in the experimental values. The blue mass spectrum is from the reference database. The x-axis is m/z and the y-axis is relative abundance

be somewhat difficult as terpenes share many structural similarities and thus have similar mass spectrums. However, this is a very useful tool for identifying unknown compounds. A head to tail database search result is shown in Fig. 6. Head to tail comparisons show how close an experimentally derived mass spectrum compares to a reference database. A match percentage is typically generated and the most likely hits are shown.

Method Development for Terpene Analysis [10]

Method development can be a long and time-consuming process. Following a logical pathway to accomplish the goals of the method is worth the time initially, it will undoubtedly save time in the long run. The most important aspect of method development is to clearly identify the goals for the method. Starting with literature references is a key step to developing a method in a reasonable amount of time. There is an abundance of methodology in the literature for terpene analysis as terpene analysis outside of cannabis has been developed for decades. There are growing numbers of cannabis specific terpene methodologies available for free. When developing a method, these can provide an excellent jumping off point.

Column Selection

Column selection is highly dependent on the goals of the method. A great place to start is literature reference. One of the largest factors in column selection for laboratories is the desire to run alternate analyses. Many laboratories run residual solvent analysis in addition to terpenes. Selecting a column that can accommodate both may be an important decision. Changing columns between analyses is not practical in most laboratories.

Injection Technique

The decision must be made to use headspace or liquid injection for the application. Both techniques have their advantages and disadvantages. If the instrument is not being used for residual solvent analysis as well, liquid injection may be a good option.

Temperature Program

Isothermal separations of terpenes are not practical. Consequently, a temperature program must be developed. Longer temperature programs tend to lead to better baseline resolution but at the cost of an extended run time. Identifying the goals of the method when developing the temperature program for terpene analysis is critical. A slow ramp speed allows for better analyte separation and may be critical if using an FID. Increasing the ramp speed can dramatically reduce the analysis time but may cause peaks to coelute. It is important to analyze both standards and samples when developing a temperature program. Baseline resolution may occur without issue in the standards but coeluting non-target terpenes may appear in the chromatography of samples. Matrix spiking is critical to overcoming this issue. An initial temperature of 60 °C is a good starting place for slower ramp speeds. An initial temperature of 100–120 °C is a good initial starting temperature for faster ramp speeds.

Mass Spectrometer Method

Terpenes in cannabis are typically 0.5–3% by weight, thus detection limits are rarely a problem barring analysis of minor terpenes. Most commercial analysis of terpenes is done with a single quadrupole mass spectrometer. There is rarely a need for running the mass spectrometer in single ion monitoring (SIM) mode for terpenes. The concentrations are significant enough to run full scan MS methods without issue. A m/z range of 35–300 should cover most cannabis terpenes as the majority are monoterpenes. If there is significant column bleed, a m/z range of 35–205 should provide cleaner chromatography and greater sensitivity.

Quantitation

Quantitating terpenes are a critical aspect of terpene analysis. Quantitation through external calibration is critical when reporting data in a commercial laboratory setting. In a research setting, relative comparison of peak areas between samples may be enough. Semi-quantitative analysis provides interesting information when comparing samples in a research-based setting. Semi-quantitative data is only marginally more interesting than a direct comparison of peak areas. Data quantitated through external calibration is critical. Terpenes are considered harmless and are not regulated beyond label claims that the manufacturer may print on their product label. Even this tends not to be strongly enforced. However, providing accurate data to consumers is of the utmost importance. The terpene profiles help guide the consumer to strains and products that they like. Many consumers have individual tastes and may prefer strains with specific dominant terpenes.

External Calibration

A multitude of commercial terpene mixtures are available for purchase. The simplest way to calibrate instrumentation is through the purchase of a commercial mixture of terpenes. As there are over 200+ terpenes found in cannabis, it is not a practical reality to calibrate and quantitate so many. Most laboratories calibrate for 20–40 terpenes that are most abundant in cannabis. Calibration curves are most easily made by serial dilutions of the commercial stock mixtures. Each analyte is most likely between 1000–2500 ug/mL in the commercial standard. Some scientists prefer perfectly round numbers in their calibration curves. This is acceptable however curves tend to have a better r^2 with serial dilutions. Individual terpene standards can be added if the desired list of analytes goes beyond what is present in a commercial mixture. The purity of commercial mixtures must be accounted for as many components are not pure.

Overcoming Matrix Issues in Cannabis Analysis

Internal Standards

Internal standards serve an important role in cannabis analysis. The concept behind internal standards is relatively simple, a known substance and concentration of a chemically similar analyte is added to the samples, calibration, laboratory quality control samples, and blanks. This serves as a marker and can help to correct for matrix effects and sample preparation issues. As cannabis is an extremely dirty matrix, ion suppression is a common theme. Ideal internal standards are deuterated or carbon-13 enriched versions of the target compounds. The logic behind this is that they will behave the same as the target analytes that are being measured.

Standard Addition

Standard addition is a methodology in which known amounts of analytes are added to directly to samples. This technique is excellent for understanding matrix effects and can be used to determine suppression of analytes.

Matrix Matched Calibrations

Matrix matched calibrations involve using the target matrix (e.g. oil) spiked with analytical standards. This is a labor-intensive process and not highly recommended for cannabis analysis unless a single type of matrix is being analyzed. The primary issue with this in cannabis analysis is the wide variety of matrices that are encountered during routine analysis.

Proficiency Tests

Proficiency tests provide an objective view of the accuracy and proficiency of a laboratory's ability to test for terpenes. There are commercially available proficiency tests in both hemp oil and hemp flower. There are known concentrations of target analytes and the tests provide an objective viewpoint of the laboratory's methods. Proficiency tests are critical in understanding errors in the laboratory and should be administered to every chemist running terpene analysis.

Terpene Analysis in a Commercial Laboratory Setting

Commercial throughput of terpene analysis has several factors that must be taken into consideration. The luxury of long chromatographic separations is not a reality in a commercial laboratory setting. Run time from sample to sample is a critical aspect that must be kept at the forefront when developing methodologies. A sixty plus minute run time severely impacts the daily sample throughput particularly when considering laboratory control samples. A target for total run time per sample should be 15–20 minutes. This allows for a total sample throughput of 72–96 samples per day (not including laboratory control samples).

Laboratory Quality Control Samples for Terpene Analysis

Requirements for laboratory control samples vary significantly from state to state. Laboratories may have their own internal regulations regarding laboratory controls samples as well. The basic idea of laboratory control samples is to ensure that day to day analysis is accurate and precise. They are typically analyzed during an analytical run with samples. These quality control checks allow for added confidence in the analytical values produced by instrumentation.

Calibration Curves

Terpene calibration curves tend to be linear for roughly 100x. Once the values increase beyond this, they tend to take on a quadratic fit. Both linear and quadratic curve fits are appropriate for most state regulations. Local regulations must be taken into consideration before beginning a method validation. If quadratic curve fits are not allowed, the range for the calibration curve must be reduced to have a linear fit.

Air Blank

The air blank is run to ensure that target compounds are not present (or at least present in low concentrations) in the background of the sample. The air blank is not used for liquid injection and is only relevant for headspace analysis. The air blank should be captured in the sample preparation area. As terpenes are volatile compounds they do occasionally appear in the background although rarely in significant concentrations.

Solvent Blank

The solvent blank is run to ensure that there are no target compounds present in the solvents used for extraction. For terpenes, this commonly occurs when a contaminated micropipette, pipette tip, or syringe is used to draw solvent from the container. Entire batches can potentially be ruined by using contaminated solvents.

Initial Calibration Verification

Initial calibration verifications (ICV) are typically run directly after the calibration curve. Their purpose is to ensure that the calibration curve can produce accurate values. They should be from a secondary source from the standards used to make the calibration curve. Typically, both a source from a secondary vendor or a secondary lot number from the one used to calibrate the instrument is appropriate.

Continuing calibration Verification

A continuing calibration verification (CCV) is a mid-point standard that is used to ensure that the calibration curve is still viable. CCVs are typically run every day prior to sample analysis. They ensure that the response from the instrument still provides viable data. Tolerances vary but are most often +/− 20–30% of the expected value.

Laboratory Control Sample

A laboratory control sample (LCS) is used to assess the laboratory's ability to recover analytes from a purified matrix. A good matrix to use for laboratory control samples is ground hemp seed or hemp seed oil. Running a laboratory control sample will help to identify if there are any issues in the extraction process. There should be minimal analyte loss for terpenes.

Matrix/Matrix Spike

Matrix/matrix spike samples are done to ensure that the target matrix does not significantly interfere with the analysis of the target analytes. This is typically done with hemp flower and hemp oil. Oil tends to be a more difficult matrix for recoveries.

Sample Duplicate

Sample duplicates can also be included in an analytical run. They showcase the laboratory's ability to consistently extract target analytes. The values for a sample duplicated should be +/− 20–30% of each other. If the values fall outside this range, the sample preparation process should be reviewed.

Method Validation for Terpene Analysis [11]

Licensing and good laboratory practices dictate that methods must be validated prior to analysis of commercial samples. Validation requirements vary from state to state and may even be federally regulated. Performing thorough method validations allows for added confidence in analytical values. The following serves as a general guide only and state and federal regulations should be consulted prior to method validation.

Accuracy

Accuracy is the closeness of agreement between a test result and an accepted reference value. If an in matrix certified reference material (CRM) is available, this can be used for an accuracy study. A source of terpenes from a secondary vendor or lot number is also appropriate for accuracy studies. This experiment illustrates the ability of the laboratory to accurately measure a value. State or internal acceptance ranges may vary but are typically +/− 20–30% from the accepted value. Data from a terpene accuracy study is present in Table 1.

Precision

Precision is the closeness of agreement between independent test results obtained under specified conditions. Typically, precision is done by repeated injections of the same sample. This experiment gives the laboratory an idea of how consistent their injections are. The precision values calculated from the following equation should be <10%.

Table 1 An example of a terpene accuracy study

Name	Expected Value (PPM)	Measured Value (PPM)	Recovery
a-Pinene	312.50	327.32	104.7%
Camphene	312.50	329.8	105.5%
b-Myrcene	312.50	336.86	107.8%
b-pinene	312.50	334.3	107.0%
3-Carene	312.50	330.61	105.8%
a-Terpinene	312.50	332.41	106.4%
Limonene	312.50	326.73	104.6%
p-Cymene	312.50	315.4	100.9%
Ocimene	312.50	323.96	103.7%
Eucalyptol	312.50	269.97	86.4%
y-Terpinene	312.50	332.88	106.5%
Terpinolene	312.50	335.56	107.4%
Linalool	312.50	334.18	106.9%
Isopulegol	312.50	333.38	106.7%
Geraniol	312.50	319.49	102.2%
b-Caryophyllene	312.50	331.06	105.9%
a-Humelene	312.50	323.95	103.7%
c-Nerolidol	121.88	125.2	102.7%
t-Nerolidol	190.63	200.13	105.0%
Guaiol	312.50	325.19	104.1%
Caryophyllene oxide	312.50	278.31	89.1%
a-Bisabolol	312.50	311.54	99.7%

$$Precision = \frac{STDEV * 100\%}{Average\,Value}$$

If precision values are greater the 10% the precision study should be repeated.

Limit of Detection/Limit of Quantitation

The limit of detection (LOD) describes the point at which an analyte can be distinguished from background noise. Limit of detection must be determined experimentally. A signal to noise (S/N) of 3:1 is considered acceptable for an LOD. The limit of quantitation (LOQ) is the level at which a signal can be distinguished from the background noise and quantitated. It is commonly defined as a signal to noise ratio of at least 10:1. LOD/LOQ can also be determined by using the slope of the calibration curve and the standard deviation of the response based on sample blanks. The LOD/LOQ are determined by multiplying 3.3 by the standard deviation divided by the slope for LOD and 10 times the standard deviation divided by the slop for LOQ. An example of an experimentally determined LOD is shown in Fig. 7.

Linearity

Linearity is the ability of a method, within a certain range, to provide an instrumental response or test results proportional to the quantity of analyte to be determined in the test sample [ref]. Terpenes are typically linear using a mass spectrometer across an order of magnitude of 100x. Regulatory bodies typically require an r^2 of between 0.99–0.995.

Selectivity

Selectivity is the extent to which a method can determine a particular analyte in a mixture or matrix without interferences from other components of similar behavior. Selectivity is much easier when using mass spectrometry rather than an FID for detection. FID relies only on retention time matching whereas mass spectrometry allows for ion ratio matching as well as database searching.

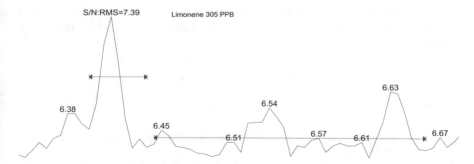

Fig. 7 An experimentally determined LoD for limonene at 305 parts per billion (ppb)

Recovery

The fraction of analyte remaining at the point of final determination after it is added to a specified amount of matrix and subjected to the entire analytical procedure. Matrix spike recoveries are performed to ensure there is limited interference from the matrix. This is done by spiking a known amount of each analyte into a matrix, then extracting the matrix according. The resulting data must be within 70–130% of our expected value. Spike recovery is typically expressed as a percentage.

Robustness

A measure of the capacity of an analytical procedure to remain unaffected by small but deliberate variations in method parameters and provides an indication of its reliability during normal usage.

Robustness can be determined by several different experiments. The FDA/ICH focuses more on physical parameters (such as the use of multiple columns from the same vendor) whereas some agencies prefer robustness be determined by multiple analysts or samples ran over multiple days. It is imperative that the requirements for robustness are understood before experimental design.

Conclusion

Terpenes are an amazing and diverse family of compounds that play an important role in everyday human life. They are an intricate part of the cannabis experience; award winning cannabis flower often has complex and diverse terpene profiles. Much like wine connoisseurs do not rate wines based on alcohol concentration but rather the characteristics of the wine. Terpene research in cannabis is still in its infancy and the relationship between the medical effects experienced from cannabis use and the terpenes present in the product is still a subject of great debate. There are still a great number of mysteries to be uncovered in the world of cannabis terpene research.

References

1. Singh B, Sharma RA (2015) Plant terpenes: defense responses, phylogenetic analysis, regulation and clinical applications. 3 Biotech 5(2):129–151
2. Stenerson KK, Halpenny MR (2017) Analysis of terpenes in Cannabis using headspace solid-phase microextraction and GC–MS. LCGC 35(5):27–32
3. Hanuš LO, Meyer SM, Muñoz E, Taglialatela-Scafati O, Appendino G (2016) Phytocannabinoids: a unified critical inventory. Nat Prod Rep 12:1357–1392
4. Cheng AX, Lou YG, Mao YB, Lu S, Wang LJ, Chen XY (2007) Plant terpenoids: Biosynthesis and ecological functions. J Integr Plant Biol 49(2):179–186
5. Gutzeit OH, Ludwig-Muller J (2014) Plant natural products: synthesis. In: Biological functions and practical applications. John Wiley & Sons, Hoboken, NJ
6. Newman JD, Chappell J (1999) Isoprenoid biosynthesis in plants: carbon partitioning within the cytoplasmic pathway. Crit Rev Biochem Mol Biol 34(2):95–106
7. Nuutinenab T (2018) Medicinal properties of terpenes found in Cannabis sativa and Humulus lupulus. Eur J Med Chem 157:198–228
8. Russo BE (2011) Taming THC: potential cannabis synergy and phytocannabinoid-terpenoid entourage effects. Br J Pharmacol 163:1344–1364
9. Mittal DR (2015) Tandem mass spectroscopy in diagnosis and clinical research. Indian J Clin Biochem 30(2):121–123
10. McNair MH, Miller MJ (2009) Basic gas chromatography, 2nd edn. John Wiley & Sons, Hoboken, NJ
11. Bhardwaja KS, Dwivedia K, Agarwala DD (2016) A Review: GC Method Development and validation. Int J Analyt Bioanalyt Chem 6(1):1–7
12. Breitmaier E (2006) Terpenes: flavors, fragrances, pharmaca, pheromones. John Wiley & Sons, Hoboken, NJ

Laboratory Safety and Compliance Testing for Microorganism Contamination in Marijuana

Patrick Bird, Nisha Corrigan, Renee Engle-Goodner, Benjamin A. Katchman, Jesse Miller, and Shaun R. Opie

Abstract Testing for the presence of microorganisms on marijuana is done for human safety. Pathogenic bacteria, mold, and fungi can colonize marijuana and, if present in sufficient quantity, can lead to illness or death following consumption. Marijuana safety testing is a relatively nascent field and laboratories should expect legislative changes, new quality standards, and innovative technologies to impact methodologies used, and validation and testing requirements. This chapter will provide an overview of the current understanding of health hazards, regulatory framework, and analytical methodology associated with marijuana safety testing.

P. Bird
PMB BioTek Consulting, LLC, Cincinnati, OH, USA
e-mail: consulting@pmbbiotek.com

N. Corrigan
Hygiena, Camarillo, CA, USA
e-mail: ncorrigan@hygiena.com

R. Engle-Goodner
REG Science, LLC, Sacramento, CA, USA

B. A. Katchman
PathogenDx, Scottsdale, AZ, USA
e-mail: bkatchman@pathogendex.com

J. Miller
Neogen Corporation, Lansing, MI, USA
e-mail: jmiller@neogen.com

S. R. Opie (✉)
E4 Bioscience, Charlevoix, MI, USA
e-mail: shaun@e4bioscience.com

© Springer Nature Switzerland AG 2021
S. R. Opie (ed.), *Cannabis Laboratory Fundamentals*,
https://doi.org/10.1007/978-3-030-62716-4_13

281

Introduction

Microorganisms, or microbes, are living organisms that are too small to be seen by the naked eye but are visible under a microscope. With the exception of artificially created sterile environments, microorganisms are found on all surfaces and in the air. Microbial detection, quantitation, and identification requires specialized equipment and training and is performed in a medical or analytical laboratory. Microbiological testing is performed regularly in a clinical setting and numerous reviews and textbooks that discuss the physical appearance, growth characteristics, and health related illnesses of these microorganisms are available [1, 2].

The type and number of microorganisms that are routinely encountered are harmless and do not cause any detectable symptoms or sickness for healthy individuals. However, pathogenic bacteria, yeasts, and molds exist, and under some conditions, they can cause severe health issues and even lead to death. A growing number of clinical reports directly link illnesses associated with contaminated marijuana including aspergillosis [3–12], salmonellosis [13] and meningitis [14]. Fungal growth and subsequent mycotoxin formation in marijuana has been reported [7, 9], although no reports of mycotoxin contaminated marijuana have been reported. While a healthy adult may fend off a potential microorganism colonization, an immunocompromised patient may not be able to effectively challenge a pathogenic microorganism. Therefore, the risk for acute infection is much higher for people with active HIV infection, taking immunosuppressive medications, post-organ transplant, post-chemotherapy, or being treated for hematological malignancy.

Because legalization of marijuana is relatively new and currently regulated at the state level, legislative guidelines are constantly changing. However, most states that have legalized either recreational or medical use require that marijuana and marijuana derived products undergo laboratory testing to determine if certain microorganisms are present. Testing for pathogenic microorganisms may seem like an obvious action, but the necessity of safety testing is often questioned. Common arguments to oppose microbiological testing include: (1) The cannabis plant has innate antibacterial properties resulting from naturally occurring terpenes, (2) Microorganisms are ubiquitous (3) Is there a true health benefit when very few documented cases of microorganism contaminated marijuana leading to new or exacerbated medical conditions have been published, (4) when cured and stored appropriately with a water activity level below <0.6 microbes should not grow (5) testing is a financial burden that reduces profitability.

With legalization and the loss of stigma, the number of marijuana users is expected to increase substantially and without testing and oversight, a corresponding increase in microbial contamination related adverse health effects might be expected. Fortunately, marijuana products that test positive for contamination are being removed from the consumer supply chain so microbial related health issues have not significantly increased. This is in contrast with a recently recognized serious health issue, e-cigarette or vaping product use-associated lung injury (EVALI), where additives that are not part of safety testing have caused lung damage and resulted in over 50 deaths and 2500 hospitalizations less than 4 months after the initial case [15]. Furthermore, the number of cases is likely to be much higher since the CDC only includes cases voluntarily and fully reported in their summary data which underrepresents the true number of cases.

Microbial Testing for the End User and Product Type

Unlike potency testing for product quality, microbial contamination testing is done for product safety. Most states recognize that microbial contamination is a natural by-product of the environment and that it can be introduced at all stages of the product supply chain. Irrespective of the product type, because microorganisms are present everywhere in the environment and because people handle the product at every processing step, microbial contamination testing should be included for all marijuana products.

Smoking the flower, or bud, has been the standard method to consume marijuana. Molds are common plant pathogens and when inhaled are known to cause a variety of immune lung disorders including asthma, allergic bronchopulmonary aspergillosis, and hypersensitivity pneumonitis to invasive fungal species in immunocompromised hosts. One study looking at the prevalence of mold in pre-rolled marijuana cigarettes found that 1 in 5 were contaminated with an estimated 270 viable fungal spores per cigarette [16]. Smoking a marijuana flower requires a heat source to ignite and burn and this level of heat will kill any microorganism. However, many users inhale from an unlit delivery tool (pipe/bowl) and particles and microorganisms can be transferred into the lung. Also, even with heat, a flower packed into a delivery tool does not burn evenly, and trace amounts can be inhaled.

Food contamination is a significant cause of morbidity and mortality in the US. The Centers for Disease Control estimates that eating contaminated food causes approximately 9.4 million illnesses, 56,000 hospitalizations, and 1300 deaths annually [17]. Cannabis Marijuana shares many growth properties of leafy green vegetables which are attributed to the highest percentage of illness than to any other commodity (22%). Further, illnesses associated with leafy vegetables were the second most frequent cause of hospitalizations (14%) and the fifth most frequent cause of death (6%) [18]. While there is limited data on illnesses associated with cannabis contamination, it is reasonable to expect that similar levels of contamination will be present due to similarities with the cultivation processes of leafy green vegetables and cannabis.

No injectable THC products have been developed for commercial sale. Whether intravenous, intramuscular, or subcutaneous, any injectable product significantly increases the risk of sepsis. Sepsis is a life-threatening condition that occurs when bacteria are introduced into the bloodstream. Standard microbial testing is wholly insufficient for this product type. Rather, injectables should follow United States Pharmacopeia (USP) requirements for sterile injectable products, which at this time is beyond the scope of safety compliance laboratories.

Upon initial legalization, states have commonly required a single set of tests for all product types. As the industry matures, legislators are recognizing that it is logical to pair specific testing requirements to potential safety hazards relevant to the different processing steps and the end user. For example, a baker who uses a concentrate to make cookies, would not be expected to introduce pesticides or solvents of any kind to a baked final product. So long as the concentrate had passed a pesticide and a residual solvent test before being purchased by the baker, there should be no reason to repeat this test. However, since a baker prepares and handles food, a microbial test should be performed on their product even if the infused marijuana product had previously tested negative. Using this risk-based approach and factoring in known routes of infection, it is logical to require testing for all products for microbial contamination. Certain products such as those that have undergone organic solvent-based extraction and subsequently packed under aseptic conditions, may carry a minimal risk and would not need to be evaluated as stringently.

Marijuana legalization usually starts with approval for medicinal use and, later, expanded to include recreational use creating two classes of marijuana. Should recreational marijuana have the same level of testing requirements as medical? A majority of states have not yet separated testing requirements for medicinal and recreational marijuana, and if regulations in similar industries are looked to for comparison, commodity specific regulations will become more common. Arguments can be made that routine testing of recreational marijuana is not necessary or can be reduced in frequency, but due to the risk of immunocompromised individuals any marijuana product designated for medical use should require microbial contamination testing after the final processing step before being sold to consumers.

Microbial Testing Regulatory Guidelines

In the absence of a federal framework of regulations, states have individually developed microbial testing requirements and set action limits for approving or rejecting batches. Not surprisingly, this has led to confusion throughout the industry as a result of the inconsistencies between state regulations. At the time of this publication, 46 states allow some form of marijuana or marijuana derivative to be manufactured and consumed. The type of legalization can be separated into 1 of 4 categories: Legal/decriminalized for medicinal and adult-use; legal/decriminalized for medicinal use only; legal/decriminalized for medicinal use with low levels of THC; Illegal in all forms. Figure 1 provides a breakdown of the current legal status by state as of July 2020.

While marijuana legalization can be broadly separated into 4 distinct categories, the microbial contaminant testing requirements within these states varies to a greater degree. These differences are minor in some instances but differ greatly in others. In an attempt to characterize these differences, the testing requirements have been placed into 7 different categories. It should be noted that although 7 categories are

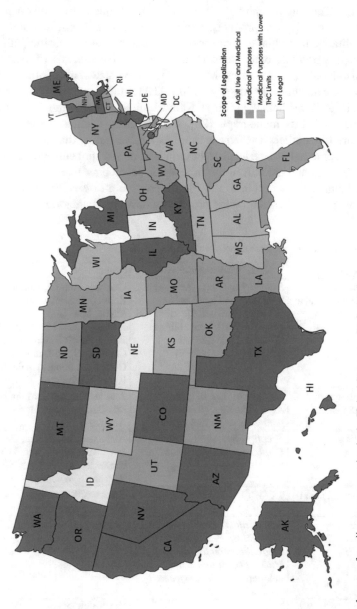

Fig. 1 Legal status of marijuana consumption in the United States as of July 2020

described, there does exist some minor variations between states within some categories, however, the overall scope of these states' regulations do align.

A majority of states follow the testing requirements as recommended by the American Herbal Pharmacopeia in their quality control monograph, *Cannabis inflorescence* [19]. This monograph outlines both a series of qualitative and quantitative requirements for marijuana products. Other states such as California and Alaska require qualitative microbial testing for only a small subset of specific pathogenic microorganisms or while other states require quantitative analysis for contamination organisms only (ex. Arkansas, Minnesota). Some states, a majority of which only allow for cannabis derivatives with low THC limits (essentially allowing the use of CBD as a medicinal product) follow guidance from the United States Pharmacopeia <1111> monograph [20]. Other states, such as New York and Delaware, are true outliers and require more expansive testing than all other states. Lastly, while some states have legalized marijuana, they do not currently have regulations established and are still in the process of developing them at the time this chapter is being written. Table 1 and Fig. 2 provide an overview of the testing requirements per state based on these previously described categories.

Table 1 Microbial testing requirements by state (April 2020)

Testing category	Types of analytes required	States
American herbal pharmacopeia or similar testing requirements	*Salmonella, E. coli* (pathogenic), total yeast and mold count (TYMC), total aerobic microbial count (TAMC), total coliform (TC), total bile tolerant Gram-negative (BTGN)	Colorado, Hawaii, Illinois, Kansas, Kentucky, Louisiana, Maine, Maryland, Massachusetts, Montana, New Hampshire, New Jersey, New Mexico, North Carolina, North Dakota, Ohio, Oklahoma, Pennsylvania, Rhode Island, Washington, Vermont
Specific pathogen and fungal contaminants	*Salmonella, E. coli* (pathogenic), *Aspergillus* spp. (*flavus, fumigatus, niger* and *terreus*)	Alaska, California, Michigan, Missouri, Nevada, Florida, Arizona
USP <1111>	*E. coli, S. aureus, P. aeruginosa, C. albicans,* TYMC, TAMC	Alabama, Connecticut, Georgia, Iowa, Mississippi, South Carolina, Tennessee, Texas, Utah, Virginia, Wisconsin, Wyoming
Minimal contamination organism testing	TYMC, TAMC, TC, BTGN	Arkansas, Minnesota, Oregon, West Virginia
Expansive testing requirements	*Salmonella, E. coli* (pathogenic), *Aspergillus, Pseudomonas, Klebsiella, Streptococcus, Mucor, Penicillium* and thermophilic *Actinomycetes* species	Delaware, New York
No current regulations or regulations in development	N/A	DC, South Dakota
Illegal	N/A	Idaho, Indiana, Nebraska

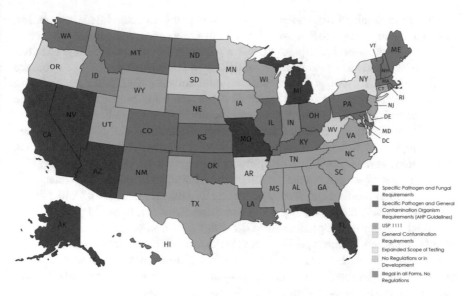

Fig. 2 General testing requirements of marijuana by state

Specific Microorganism Requirements: **Salmonella,** *STEC* *and* **Aspergillus**

On November 6, 2012, Colorado passed Amendment 64 to its state constitution, effectively became the first state to legalize marijuana for recreational adult use (Washington followed soon after) and set off a chain reaction of legislation for marijuana across the country [21]. As the first state to pass legislation, its regulations set the precedent for many future states. Drawing from a list of literature that existed on microbial analysis of marijuana as well as from regulations for similar type products (e.g. leafy greens), Colorado enacted microbiology guidelines requiring analysis for two bacterial pathogens of concern: *Salmonella* species and shiga-toxin producing *Escherichia coli* (STEC) as well as some specific fungal pathogens [22].

   Salmonella and STEC are known human pathogens that are found in the intestinal tracts of most mammals and birds and have been known to contaminate produce (leafy greens, romaine lettuce) similar to marijuana, through fecal contamination of irrigation water [23]. Their prevalence as contaminants in the food industry, as indicated by their ranking in the top 5 of foodborne bacterial illness, provides data driven support to include them as microbial contaminants in marijuana regulations [24].

   In addition to the bacterial pathogens, Colorado established requirements for plants and flowers requiring analysis for select molds (*Aspergillus, Penicillium* and *Mucor*), although further legislation removed these requirements for specific fungal pathogen analysis. Why did Colorado initially require testing for these three specific molds? Data obtained from analyzing marijuana plants indicated that these molds, which include species that are known human pathogens, were found frequently enough in unprocessed marijuana to pose a public health hazard [25–27]. The ability of these molds to cause invasive infections in humans, their ability to produce spores that can survive extraction with residual solvents [28], and their ability to produce mycotoxins, secondary metabolites that are toxic to humans [29] led the first state legislatures to identify them specifically in their regulations.

While it eventually modified its legislation, Colorado's early regulation were soon adopted (with slight modifications) by other states, particularly California, Alaska and Nevada [30–32]. When introducing regulations in these states, legislators identified *Aspergillus* as the highest concern based on the numerous metabolites the species can produce and focused solely on this fungal genus [33]. Unlike bacterial requirements, testing for *Aspergillus* in these states is a two-fold process. First an initial screen on the product for *Aspergillus* genus is performed. If a positive result is obtained, species specific testing is performed, but varies depending on the state. California and Nevada have identified four distinct species of *Aspergillus* that are included in its regulations (*flavus, fumigatus, niger, terreus*) while Alaska only identifies three as contaminants (*flavus, fumigatus* and *niger*).

These species have been singled out due to their ability to cause a range of human illnesses. Elected for varying reasons. *A. flavus* is known to produce Aflatoxins (a type of mycotoxin), including the most highly regulated Aflatoxin, B_1, a known human carcinogen [34]. *A. flavus* is the second leading cause of human mycosis and is a known allergen that can lead to allergic bronchopulmonary aspergillosis [34]. *A. fumigatus* has been classified as one of the most important fungal pathogens as it is the leading cause of fungal infections. *A. fumigatus* produces the most severe form of invasive aspergillosis and with that has a higher mortality rate than other *Aspergillus* species [35]. While important to the biotechnology industry for its ability to produce organic acids and extracellular enzymes, and its usefulness in bioremediation and waste pretreatment, *A. niger* is also known to produce several animal and human toxins including: Fumonisin and Ochratoxin [36]. *A. terreus*, while less commonly found than the other *Aspergillus* species, has one of the highest mortality rates and is a leading cause of death as a result of invasive aspergillosis [37].

Full or Partial Adoption of American Herbal Pharmacopeia Recommendations

While some states that were early adopters of legalizing marijuana chose to identify specific fungal pathogens in their regulations, the majority of states chose a different approach relying on guidance from a similar industry. These states incorporated some, or all of the recommendations outlined in the quality monograph developed by the American Herbal Pharmacopoeia (AHP). The AHP monograph outlines qualitative testing requirements (*Salmonella* and STEC) but also includes quantitative testing requirements. Due to the scope of quantitative tests outlined in the monograph, more variation is seen in states who have developed regulations using the AHP monograph as a basis. Within the AHP monograph, four main types of contamination tests are described: total aerobic microbial counts (TAMC), total yeast and mold counts (TYMC), total coliform counts (TC) and total bile-tolerant Gram-negative counts (BTGN). These tests are designed to help producers and manufacturers determine the overall cleanliness of the production facility or manufacturing conditions and are often employed in other industries (food, pharmaceutical or nutraceutical) as similar indicators. No universal action limits exist for these tests, as variations are seen within different industries based on the commodity, but most states have set their action levels as the following: $<10^4$ CFU/g for TAMC, $<10^3$ CFU/g for TYMC, $<10^2$ CFU/g for TC or BTGN.

While some states have adopted testing for each of these contaminants, the most common targets in regulations are yeast and molds. Yeasts and molds represent the two different forms of fungi that are found in the environment and would be enumerated following these regulations. These organisms are naturally found on cannabis plants and flowers, and at low concentrations do not typically pose a risk to humans, however, under the right conditions, these organisms are able to grow to levels where they can cause deleterious health effects for humans [38, 39] With an estimated five million different fungal species around the globe, a total yeast and mold count can provide a producer with valuable information on the total level of fungal contamination (viable organisms only) and determine if it meets a set action level threshold, often estimated at 10,000 colony forming units (CFU) [21, 32, 40]. However, for immunocompromised individuals, the concentration (CFU/g) at which a yeast or mold can cause infection is often much lower [41]. The FDA has published guidelines on limits of total yeast and mold counts in food commodities, ranging from as low as 10 CFU/g to 100,000 CFU/g [23]. While only a handful of species will cause illness in healthy humans, several hundred species have been known to cause illness in immunocompromised individuals, which constitutes a large portion of the cannabis market [41]. However, since many states initially legalized the use of marijuana for medicinal purposes only (with some later expanding to recreational use in some instance), with setting thresholds for TYMC at a level designed to protect the immunocompromised was key. These types of consumers are often at an increased risk for illness (i.e. immunocompromised). As states made laws legalizing marijuana recreationally, several states have begun to establish varying thresholds/action limits of acceptability based on the type of derivative being produced and its intended end user. One drawback of TYMC analysis is that Total yeast and mold enumerations will not specifically it does not identify the specific genus or species of fungal organisms present on a derivative but can provide guidance on the overall prevalence of fungal contamination. TYMC testing is designed to provide information to cultivators that can indicate issues exists within the cultivation or harvesting practices which can be altered to produce safer derivatives. Alternatively, assays designed to detect specific organisms (i.e. *Aspergillus)* provide a presence or absence results on for a particular pathogenic fungal strain but would not provide any information on the overall level of contamination in the product. A majority of states, including Colorado, which modified its original legis-lation, Nevada and Michigan, which require both specific *Aspergillus* species test as well as total fungal counts, as well as Canada which regulates cannabis at the fed-eral level, all established guidelines requiring cannabis marijuana derivative to be evaluated for total fungal contamination [21, 32].

USP <1111>

As mentioned previously, most states began to initially decriminalize the consump-tion of marijuana for medicinal purposes, however, some states proceeded down an even narrower path allowing for the consumption of only low THC (<0.3%), high CBD products., the first step most take is to allow it for medicinal purposes or for medicinal purposes with low levels of THC. In both instances, states typically regu-late these derivatives similar to other pharmaceutical products and have thus adopted regulations that mirror or reference guidelines for those commodities, specifically those outlined in the United States Pharmacopeia. While Even as these states looked to the USP for testing requirements, some found the monographs to be strict some states have a straightforward approach of using USP monographs as stated, others deemed them too strict and have modified them within their regulations to contain with less stringent limits [42].

While the USP has thousands of monographs, just a few monographs have been adopted as the basis for state regulations which adopt the These states have adopted the acceptance criteria outline in chapter <1111> with testing methodologies being performed according to the microbial limits chapter <61> for TAMC and TYMC, and chapter <62> for specific microorganisms [43, 44]. These guidelines typically require testing for TAMC, TYMC and then specific pathogens (*Escherichia coli* (generic not pathogenic), *Pseudomonas aeruginosa, Staphylococcus aureus,* or *Candida albicans*) based on the method of use/application (i.e. oral, mucosal, sup-pository, etc.). These recommendations align closely with guidance provided by the AHP guidelines (typically without *Salmonella* or STEC) as well as the World Health Organization and the European Pharmacopeia [42].

Minimal Contamination Organism Testing

A very small subset of states require limited analysis for microbial contaminants, specifically outlining requirements to test for contamination organisms only. These states (Arkansas, Minnesota, Oregon, and West Virginia) have limited scopes of analysis, mainly targeting organisms of the *Enterobacteriaceae* family (*E. coli,* TC or BTGN). These regulations are designed to specifically look for organisms that would be considered fecal contamination, an indication of poor sanitary conditions.

Expansive Testing Requirements

Almost all states, as well as Canada, that have legalized the use of marijuana in some form fall within the testing requirements as described in the previous four sections. However, there exist a few outlier states, New York and Delaware, that have set an expanded list of testing requirements far beyond the regulations of other states. In New York, in addition to traditional requirements for *Salmonella, E. coli* and *Aspergillus,* bacterial strains such as *Pseudomonas* (for vaporized cannabis derivatives only), *Klebsiella, Streptococcus* and fungal strains such as *Mucor, Penicillium* and thermophilic *Actinomycetes* species are required for analysis [45]. Delaware has similar requirements for *Salmonella, E. coli, Aspergillus*, TAMC, TYMC, *Pseudomonas, S. aureus* and *Streptococcus.*

In Development or No Regulations

As noted in the opening to this section, a few states (Idaho, Indiana, Nebraska and South Dakota) still outlaw marijuana in all forms while other states or areas (Florida, Washington D.C., etc..) that have recently passed legalization legalizing marijuana are still developing their microbial contaminant regulations as they solicit feedback from individuals in that state. Future versions of this text will be updated as legislation is adopted.

Cannabis Standards Organizations

CASP & ASTM D37

AOAC International, a non-profit organization that works with key leaders within industry, academia and government to develop consensus-based standards for method certification (i.e. approval for use) primarily for the food and dietary supplement industry, launched the Cannabis Analytical Science Program (CASP) in March of 2019. With no national (or globally) recognized standards for approving laboratory methods CASP was designed to serve as a forum for discussions to develop national standards and/or methods for cannabis analysis with initial programs for microbiological contaminants, potency and chemical contaminants [46].

Within CASP, three separate working groups were established. The Chemical Contaminants Working Group works to develop analytical chemistry standards for chemical residues (heavy metals, residual solvents, pesticides, etc.) that may be found in cannabis or cannabis derivatives. The Cannabinoids in Consumables Working Group works develop methods to accurately extract and determine cannabinoid concentration. The Microbial Contaminants Working Group is focused on developing standards for microbial contaminants, starting with *Aspergillus, Salmonella* and STEC. The objective of this group is to develop criteria for which methods can be certified as acceptable for use with an end goal of having harmonized methods. In September of 2019, the CASP community formally approved its first *Standard Method Performance Requirements* (SMPR) for the *Detection of Aspergillus in Cannabis and Cannabis Products* [47]. The scope of this SMPR is to provide a set of criteria for method developers to validate methods for the detection of *Aspergillus* species (including *flavus, fumigatus, niger* and *terreus*).

The American Society for Testing and Materials, now known as ASTM International, is another international standards organization that develops and publishes voluntary consensus based technical standards for a wide range of materials, products, systems and services. Within ASTM, over 12,000 standards have been developed and implemented. In 2017, ASTM launched a specific committee, D37, to develop standards for cannabis. Within the D37 committee, there are subcommittees that work to develop test methods and standards specific for quality assurance, laboratory design, sample packaging, security and practices and guidance for cultivation [48]. In the last 2 years, the 8 subcommittees of D37 have begun the development of 61 different standards.

While AOAC and ASTM are both voluntary consensus standard organizations, the scopes of these organizations diverge in the focus of the standards developed. The primary focus of the cannabis standards developed by AOAC are to be used to validate methods while ASTM standards have a wider focus, covering analytical methods as well as the processes surrounding cultivation and production [49].

Association of Public Health Laboratories

The Association of Public Health Laboratories (APHL) is comprised of local and state government health laboratories in the United States. APHL works with federal agencies to develop and execute national health initiates [50]. APHL has many areas of focus, including but not limited to laboratory science and training, emergency response and public health informatics. With medical cannabis being approved in nearly every state, APHL serves as a conduit for its member laboratories, hosting monthly teleconferences to discuss best practices, lessons learned and available resources. As states continued to develop their own unique guidelines, APHL developed a guidance document to harmonize testing for state medical cannabis testing programs [50]. The APHL guidance document provides recommendations for target analytes and action levels based on guidance from federal and state regulations. For microbial contaminants, APHL leans heavily on the guidance set for in the various chapters of the US Pharmacopeia.

Microbiological Analytical Methods

While certain cannabis safety testing is performed to determine what *is* in a sample, e.g. potency, THC %, microbial testing is conducted to determine what *is not* in a sample—it is a "rule out" test. Therefore, depending on state guidelines, a desirable test result is negative (non-detected) or below actionable limits. Marijuana and marijuana derivatives are not necessarily bacteria "free" or sterile, although some concentrated or processed products may have no detectable levels of organisms. Several questions should be asked when determining which methods are right for a laboratory.

- What are the target organisms required by the state the laboratory is located in?
- Which matrices will be evaluated in my laboratory? Flower and plant, infused edibles, infused non-edibles, concentrates or extracts?
- Has the method of choice been validated?

Unlike most industries, there are no cultural reference methods that have been developed, validated and adopted (ex: US FDA Bacteriological Manual (BAM) for food safety or the US Pharmacopeia for pharmaceutical raw materials and finished products) for the cannabis industry [20, 51]. Cultural methods can be employed to detect, enumerate, or confirm targets of concern; however, these methods are often time consuming and laborious to run. Due to this, most laboratories will look to either adopt a commercially available proprietary rapid test methods or develop their own in-house method.

Microbiological Culturing

The act of culturing microorganisms requires growing the microbes in a prepared medium at a specific temperature range and time point. Microbiological culturing is an essential process to test for microbial contaminants, because of testing detection and matrix limitations. It is of fundamental importance to isolate and proliferate microbes, if present, for the testing technology to work appropriately. Each US state that regulates cannabis testing has their own set of microbiological testing standards, encompassing qualitative and quantitative parameters. Specific microorganism requirements within this section will focus on bacteria, including *Salmonella* and Shiga toxin-producing *E. coli* (STEC), and yeasts and molds. Regardless of regulation, culturing is the first step in the microbiological testing process. Unfortunately, currently there are no reference methods for cannabis microbiological culturing. The cannabis industry often looks to food testing and pharmaceutical testing for method guidance, such as US FDA BAM, American Herbal Pharmacopeia and US Pharmacopeia [20, 51, 52].

Because of unique cannabis sample types, culturing is a necessity to ensure microbial proliferation in ever-changing cannabis matrix components, which may include waxes, fats, terpenoids, essential oils, solvents and unknown components.

These components can mask or overshadow low microbial counts so cultural growth will ensure detection during testing. Inhalable and non-inhalable cannabis products may require different microbial organism testing and standard operating procedures [53]. A lab must have a method validation for each cannabis matrix, including culturing procedures [22].

There are two general culturing pathways for cannabis testing labs: culture-based methodology and DNA-based methodology. Culture-based methodology incorporates the culturing process in real-time with diagnostics. DNA-based methodology cultures the sample first and then detects microorganisms after culturing is complete by using molecular-based probes [54]. Both culturing techniques will be explored in this section.

In general, once the representative cannabis samples are brought into the lab for testing, the samples are homogenized and added to appropriate growth media that will promote growth of mold, yeast, or bacterial microbes, based on state requirements. Then the sample culture will undergo a timed- and temperature-regulated culture environment to segregate and multiply any contamination in the samples. One should note that the growth during this incubation reflects the medium used for culturing [55]. Therefore, a suitable medium is necessary to proliferate microorganisms.

Microorganisms require energy, nitrogen, and a carbon source. In addition to these sources, there may be additives that select for certain types of microbes. During the microbiological culturing process, different growth phases are occurring. The objective is to guarantee that proper nutrients are present for the appropriate microbes and to keep growth occurring to allow for adequate organism numbers for test identification [56]. Following bacterial growth phases in culture medium over time, the bacteria are acclimating to initial shock of environment and nutrients during the lag phase, with no notable increase in number of cells. During the log phase, the bacterial cells are at optimal population growth because of ideal nutrients and space. As nutrients and space in the culture starts to run out, the cell population starts to level out, with cells dying and growing at the same rate in the stationary phase. During death phase, dead cells and waste accumulate while the cell population crashes [57] (Fig. 3).

Culture-based methodology focuses on total count tests of indicator organisms, like total yeast and mold (TYM), aerobic (oxygen-loving) bacteria, and coliform (warm-blooded intestinal tract) bacteria. This methodology is limited because it cannot distinguish between harmful and benign organisms within these groups. Ultimately, it determines how many colony forming units (CFUs) are found in a sample. It can be considered a cleanliness test, and this microbial load may not necessarily indicate the safety of the cannabis product [22]. Many states allow this platform, as it typically has a faster turnaround time than the DNA-based methodology. Culture-based methods will only grow live or viable cells, which provide valuable insight for consumer safety.

The overall process of culture-based methods involves growing the samples in or on selective media with antibiotics or other additives which will only allow for growth of certain types of organisms. Incorporated on or in this selective media is

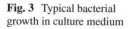

Fig. 3 Typical bacterial growth in culture medium

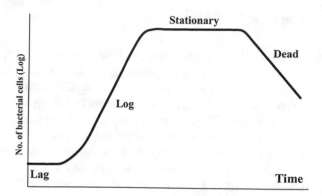

chemical technology that allows the scientist to visually count the number of micro-organisms from that selected organism type. This enumeration is done during the culturing process [58]. No further methodology is necessary, unlike the DNA-based methodology.

DNA-based methodology drawbacks differ from culture-based methodology drawbacks by being able to determine microbial species, thus elucidating human pathogens. It is a powerful diagnostic tool for consumer safety. This technology utilizes quantitative, or real-time, polymerase chain reaction (qPCR), where specific DNA pieces, or probes, find and amplify species-specific microbial contaminants using fluorescence in real-time. qPCR is extremely precise and can enumerate how many CFUs of each human pathogen were found in the sample [54]. States like California require species-specific microbial contaminant identification of *Salmonella* species, STEC and four species of *Aspergillus* fungi [31].

DNA-based methodology does have a longer turnaround time for the lab because culturing is followed by qPCR. Because time is of the essence, liquid nutrient medium, or nutrient broth, is typically used, as it yields faster growth than solid culture techniques, especially for fungi [59]. For workflow simplicity, adding the sample directly to the liquid culture also brings a level of ease and fewer steps. Because the qPCR diagnostics is precise to the DNA level, the culturing process does not require the specificity that culture-based methodology media needs. This liquid culturing uses general purpose, undefined media or non-selective media, such as buffered peptone water (BPW) for bacteria and potato dextrose agar (PDA) for fungi. A drawback for DNA-based methodology is the possibility for detection of live and dead organisms which can yield inaccurate microbial load numbers [60].

As discussed, both microbiological culturing platforms have advantages and challenges. As cannabis regulations move towards the federal level, we can expect the development of reference methods. These standardized methods will alleviate the differences among states and labs. When everyone is using the same methodology, collaboration and optimization will be possible with the ever-changing and challenging cannabis matrices. Ultimately, these advances in microbiological culturing will ensure consumer safety.

Quantitative PCR

Nucleic acids, including DNA and RNA, are biomolecules that are an integral part of all living cells. This ubiquitous nature makes them excellent indicators for the presence of life forms, including pathogenic microorganisms. The Polymerase Chain Reaction (PCR) is an amplification technique that generates millions of copies of a targeted DNA region in a few hours. This makes PCR a powerful molecular tool for the detection of specific microorganisms such as bacterial and fungal pathogens and contaminants. Target DNA sequences can be broad (for example, it can detect all yeast and mold or all *Salmonella* species) or very specific (detect only *Aspergillus niger*). PCR can also be either qualitative ("yes" it's present, or "no" it's absent) or quantitative (if present, how much is present?).

The basic components of PCR include primers, DNA polymerase enzyme, nucleotides, specific buffering ions, and target DNA template. Primers are short fragments of DNA that are complementary to specific segments of the target DNA. Primers are designed to target unique regions of the DNA, lending high specificity to PCR. These components are added to a test sample and placed in a thermocycler instrument which carries out repetitive heating and cooling cycles in three steps: denaturation, annealing, and extension.

The reaction begins with the 'denaturation' step at high temperatures, usually 95 °C, which allows for the double stranded DNA template to separate into two single strands. The reaction temperature is then dropped which allows for the 'annealing' or binding of the primers to its complementary region on the DNA. As the temperature is increased again, the enzyme DNA polymerase performs the 'extension' step by attaching near the end of the primer and begins adding nucleotides. The resulting PCR product is an exact replica of the target DNA called an 'amplicon'. Several million copies of amplicon are created as these three steps repeat over a set number of cycles. The amplicons produced during PCR can be bound to stains such as ethidium bromide or fluorescent dyes (e.g. SYBR Green), that intercalate with double stranded DNA and can be detected by the light they emit, upon excitation. Detection can also occur during or at the end of each PCR cycle using a specialized PCR technique called Real-time PCR, which boasts a quicker result than traditional PCR. Real-Time PCR uses fluorescently labeled probes that act in tandem with the primers and provide added specificity and sensitivity.

PCR based pathogen detection assays provide several advantages over the culture methods such as rapid detection, higher sensitivity and specificity and the ability to quantitate with relative ease. Results are obtained in 1–2 h after 18–24 h sample enrichment compared to 3–5 days by cultural methods. Unlike cultural methods, PCR does not depend on the physiological state of the cell. The method performs with equal efficiency even in cases where culturing cells may be challenging.

Some disadvantages of PCR include cost, user expertise, and difficulties differentiating living from dead cells without additional sample preparation. In the latter case, additives could be used to clean up dead background DNA, leaving live bacterial and fungal contaminants intact for PCR detection. Some sample types may contain materials that could cause PCR inhibition or signal interference. PCR inhibition or signal interference can often be overcome by modified enrichment conditions, illustrating the importance of internal PCR control reactions as well as validation studies to ensure the method is fit for purpose.

Although PCR methods are highly specific, the primers selected for each target microorganism are of critical importance. If primers are not specific to the target microorganism, users may see cross-reactivity due to genetically similar microorganisms, resulting in a false-positive result. On the other hand, if primers are not inclusive of genetic variations within the target microorganism, users may fail to detect the microorganism of interest, resulting in a false-negative result.

PCR detection systems have been used for industrial food and environmental applications for more than 20 years. Technological advancements in instrumentation and software for data analysis, reagents, and workflow have led to improvements in robustness, ease of use, and time to result. These attributes have made PCR a front runner in rapid pathogen detection methods.

Microarray

Several groups and companies have applied microarray technology to pathogen detection. The different approaches are distinguished based on the initial starting material (bacterial, fungal, viral, human etc.), the range of pathogens targeted, the probe design strategy and the array platform used [61]. Each group and company has also devised different sample preparation strategies, enrichment strategies, process workflows, and data analysis packages that meet the sensitivities and specificities required for their diagnostic platform.

DNA microarray technology is a high-throughput diagnostic tool for detection and identification of a wide range of pathogenic and non-pathogenic organisms simultaneously and without the need for sample enrichment [61]. DNA microarrays are flexible, sensitive, and specific with the ability to reveal a vast number of genomic content dissimilarities with determined microbial strains and is an invaluable tool for detection and identification of closely related microbial organisms [62, 63]. The construction and implementation of DNA microarrays are cheap, reliable, and accessible making it an ideal tool for high-throughput and cost-effective diagnostics.

The process of developing DNA microarray technology is divided into two sections of dry laboratory (in-silico) (bioinformatic practices) and wet laboratory (invitro) design and validation. The dry laboratory of microarray technique includes the processes of probe and primer design utilizing fully sequenced microorganisms to identify unique primer and probe regions within the genomic sequences and finalizing data analysis which are known as in-silico studies. Other stages of DNA microarray technology development such as probe spotting, PCR optimization,

DNA labeling, and hybridization are recognized as wet laboratory (*in-vitro*) practices.

To construct a DNA microarray chip, the designed single stranded probes are immobilized on the solid surface of a glass microscope slide or similar chip by an automated and robotic contact or piezo spotting/printer machine. The immobilization of the oligo probes, ranging in size from 18–70 base pairs, is done when the probes are modified at the N- or C- terminus by different types of active chemical groups such as NH_2, Succinyl, Disulfide, and Hydrazide. The majority of microarray chips are made of glass; but, the surface of a chip may be treated with different types of chemical materials including Epoxy silane, Isothiocyanate, Aminophenyl, Mercaptosilan, Aldehyde, or Epoxide to improve the stability of the covalent bonds between a single stranded DNA probe and the surface of a chip. The modified probes and the related treated chips guarantee the occurrence of a stable, powerful and optimized bond [61].

In parallel, the microbial cells isolated from samples must be lysed in order to release the genomic material and the DNA/RNA must be processed as a raw sample genotype or purified prior to PCR. If there is a low microbial cell count, as often there is for food safety, environmental and/or clinical samples, there are two different ways to prepare an appropriate amount of microbial DNA molecules for hybridization: (1) Utilization of culture medium (sample enrichment), (2) Performance of PCR (tandem PCR). Use of culture media can be difficult as the majority of microorganisms do not have selective enrichment broths often leading to unpredictable enrichment bias and the potential for false negative results, so a more practical approach is to use PCR.

During this process, the extracted DNA can be amplified in a single or tandem PCR reaction to increase the copy number and improve assay sensitivity. The dsDNA must be separated either through heat/ chemical denaturation or bias PCR that would produce ssDNA. Depending on the assay design the ssDNA can labeled with fluorescent dyes such as Cy3, Cy5, Tamra, and Texas Red post PCR or during the labeling PCR reaction. In the field of microbiological diagnostics, the use of one color (one-channel microarray) is acceptable while, regarding gene expression profiling the use of two fluorescent dyes is needed (two-channel microarray). Once the ssDNA has been labeled, microbial DNAs (target sequences) and immobilized probes go through the hybridization process. Probes are ssDNA sequences that are designed to bind to a complementary region within the amplified ssDNA sequence from PCR. During the hybridization process the labeled ssDNA amplicons are hybridized under optimal conditions and the slides are washed to remove residual DNA and fluorescent proteins. Finally, the hybridized DNA strands (probes and labeled targets) will result in fluorescent spots which are detectable by the microarray scanner.

Several promising approaches to microbial detection array design and analysis have been tested during the past few decades. The array platforms vary widely in terms of fabrication costs, range of organisms targeted, sensitivity and specificity of detection. Downstream, an essential component of any pathogen detection platform is an analysis algorithm that can make sense of the noisy data produced with

different array technologies to easily yield interpretable results for third party testing laboratories. Further advances in array technology will facilitate their broader
use for pathogen detection by incorporating automated techniques for sample preparation, PCR amplification, and faster hybridization times that will deliver results in
1–4 h rather than overnight. In addition, as metagenomic sequence analysis continues to improve and provide more detailed sequences this will lead to better array
designs for pathogenic species that do not grow under isolation.

Next Generation Sequencing

Next Generation Sequencing (NGS) is an emerging tool for use in microbial characterization. Building on the technique used in PCR, NGS technology is able to
assay for as few as a single gene target up to an entire genome in one experimental
run. NGS executes this task by running nucleic acid detection reactions in a mas-
sively parallel fashion on a chip.

Some NGS experimentation relies upon primer sets designed to specific genes of
interest in a microbial sample. When using NGS for this type of experiment, one can
amplify the 16 s (or other) region of a microbial genome and determine ALL microbial genus in a sample that contain that specific 16S conserved genomic region
being amplified. When assaying for 16 s, this is commonly called 16S metagenom-
ics. Metagenomic analyses are routinely leveraged for a deeper understanding of an
environment before and after some sort of treatment or insult, such as an antibiotic,
sanitizing treatment or probiotic treatment. Looking at populations before and after
a stimulus or insult can help the researcher tune an environment for better production of a related chemistry or the avoidance of a pathogen that harms overall returns.

Another technique that falls under the NGS umbrella is whole genome sequencing (WGS). WGS is a process by which the entirety of the genome of an organism
is extracted, fragmented and assayed to determine its genetic code. This type of
methodology is extremely useful for pathogen traceback during outbreaks and is the
core technique used for the FDA GenomeTrakr surveillance system and database
[64]. When a foodborne illness outbreak occurs, it is extremely important to be able
to rapidly traceback the outbreak organism to the source product that the organism
was carried in. Utilization of WGS for this type of work provides the granularity
needed to effectively find the source without causing unnecessary recalls that are
unrelated. Just think about it—there are 2500 serovars of Salmonella that are known,
with over 268,000 genomic sequences deposited in the NIH Pathogen Detection
Database that are used for traceback for *Salmonella enterica* alone [65, 66]! Without
WGS and the associated NIH database, traceback would be excruciatingly difficult
and ineffective.

The power of this technique does not come without drawbacks however. The
time to result for a NGS assay is generally a week, so it is not applicable for immediate screening needs. NGS is also much more expensive than a simple PCR or
immunoassay. There are equipment costs, consumable costs, labor costs and

bioinformatics costs. Datasets that are this large don't analyze themselves! When running an NGS system at full capacity, costs per test may be as low as $50/test at the time of this publication, not accounting for bioinformatics costs—a significant difference from PCR and immunoassay.

Laboratory Method Validation Requirements

When choosing a method to use in your laboratory, ensuring it has been rigorously validated and is fit for purpose should be the top priority. Similar to the lack of a national reference method, validation guidance for commercially available test methods or in-house developed test methods do not exist for the cannabis industry, which leads to conflicting information from method developers or state regulators as to what actually makes a method "fit-for-purpose". As a laboratory, there are several key questions that should be asked:

- What parameters should a method meet to be considered valid?
- How was the study performed?
- Was there an independent evaluation?
- Was the data peer reviewed?

While some states, such as California, have performance requirements that methods must meet in order to be used in the laboratory [31] other states do not provide any guidance on determining if a method is valid. These state regulations indicate that a method used by a laboratory must be validated but provide no instructions or performance requirements. How then, do laboratories ensure that the methods they are using or one that has been developed is working correctly?

Rapid methods used for detection or enumeration in food industries are validated according to one of several available guidelines, depending on the market they desire to sell their method. In the US, microbiology methods are certified according to AOAC International Appendix J guidelines which were adopted by the FDA [51, 67]. In Europe they follow requirements as set forth by ISO 16140-2 [68] and in Canada they'll follow the procedures set forth in the *Compendium of Analytical Methods* [69]. While some differences exist, the requirements set forth in the validation guidance between these organizations is consistent, have been developed through harmonized approaches, and have two main components: specific and selectivity testing, and matrix studies.

Method validations (except for methods that claim general aerobic bacteria counts) must be tested against a large panel of inclusivity (target) organisms as well as exclusivity (non-target) organisms. For most microbial targets, a minimum of 50 different isolates are required (the exception being for *Salmonella*, where 100 isolates are required). Testing a wide range of strains demonstrates the robustness of a method and indicates that the method should be able to detect a majority of strains that could be encountered by end users. For exclusivity testing, typically only 30 strains are required. For in-house developed methods, obtaining such a large panel may not be feasible but ensuring a wide range of varying strains is evaluated will greatly reduce the risk of poor performance of the method. Reduced number of strains may be used but would require justification for the change.

In addition to pure strain evaluation, each of these organizations require that a method be validated using matrix studies. Matrix studies involve the evaluation of multiple replicates (the number will vary depending on the type of test method) of inoculated or naturally contaminated matrix and comparing the results to those obtained by a reference method. While no reference method exists for marijuana matrices, cultural confirmation procedures as laid out by food or pharmaceutical regulatory bodies can be employed as a substitute where appropriate.

Not only do these validation guidelines have common protocols for conducting the matrix studies, but the specific aspects of preparing the cultures, inoculating the samples and conducting the analysis are often quite similar. Several of these approaches have been ignored by diagnostic manufacturers selling to the marijuana industry, but as state (or federal) scrutiny tightens, these generally required practices will be integrated into method validation.

- *Applicability claims.* A matrix, specific flower/plant types, extracts, or infused product, would only be considered applicable to the method if it were evaluated during the matrix studies. An entire matrix category can be claimed if enough individual matrices from that category are studied during the validation. Determining matrix applicability is made even more difficult by varying levels of compounds within different strains of cannabis flower which can impact the recovery of the target organisms. Due to this fact, method verification should be performed on strains of marijuana tested in your facility.
- *Viable culture evaluations.* During the matrix studies, a viable culture must be used to inoculate the test product. The act of inoculating a matrix with extracted DNA does not allow for the method to be evaluated from beginning to end. It excludes the impact that the matrix (and background microflora) may have on the growth of the target organism, the efficiency of the lysis process, and the ability to isolate the organism after detection.
- *Target microorganisms.* Surrogate organisms can provide method developers a quicker and safer way to evaluate a technology, but this work cannot replace the use of the true target microorganism. The growth rates of surrogates may differ from those of the target strains and surrogate strains may perform differently in the presence of matrix and its naturally occurring microflora, both of which may result in the under or over performance of a method.
- *Inoculum stressing and equilibration.* When evaluating a method, the test samples should be representative of samples an end user may encounter in the market. Inoculating a sample with an unstressed healthy culture may overestimate a method's ability to detect the target. Strains should be stressed (heat, chemical, freezing) as applicable to the product that is being evaluated. After stressing, test portions should be held to allow for equilibration of the target strain. Equilibration of the organism in a matrix may cause sublethal injury (mimicking what may occur during processing) resulting in slower growth of the organism. If no equili-

bration period is observed, a method may overestimate its ability to detect an organism in a shorter time frame.

- *Test Replicates.* Just as crucial to the inoculation process, is designing your validation to include a sufficient number of test replicates for each contamination level. The study guidelines for both qualitative and quantitative validations allow for more accurate interpretation of data, strengthening the statistical power of the data generated and reducing the variability that the results are due to chance (which can occur if fewer than the recommended number of replicates is used).
- *Confirmation of presumptive results.* The lack of a reference method for cannabis matrices makes it difficult to interpret presumptive results. It is vital to verify that a validation with no reference method includes some confirmation steps to ensure that that presumptive results are accurate. All test portions, regardless of presumptive results, should be confirmed, and this process should include an extended primary enrichment or use of a secondary enrichment to ensure all positive isolates are recovered.

While rapid microbiological methods exist for the pharmaceutical industry, they are not as prevalent, due to longer shelf life of these types of products. Instead, methods used in this market must demonstrate they are suitable, followed by evaluating long term effectiveness of antimicrobials used in the product formulations [70]. USP suitability guidelines require that for the method to be acceptable for use, antimicrobial agents used in the product are sufficiently neutralized prior to inoculation, so that a method can demonstrate acceptable recovery of the target microorganisms when compared to a control standard. The method is then evaluated over a 28-day test period to verify that the antimicrobial agents will either reduce or maintain the level of bacteria or fungal agent that is inoculated into the test product. Similar to food validation, both suitability and antimicrobial effectiveness testing require that organisms be propagated in specific ways (the use of specific media, incubation times and temperatures, the number of allowable passages from the stock culture) and most adhere to specific inoculation levels. The antimicrobial effectiveness test, by design, is similar to the equilibration requirements of food testing.

Cost Per Test

Many cultivators, consumers, and processors consider microbial testing an unnecessary expense that only adds to the final cost of the product. Every state that has approved marijuana use has adopted microbial contaminant testing guidelines, so testing will not go away even if its cultivators feel it is cost prohibitive. Factoring in labor, overhead, retesting, equipment purchase, and consumables, the average cost to test a microbiological sample with state compliant methods can approach $75 depending on the complexity of testing as described previously. If the average weighted price of high quality marijuana is approximately $1500–$2500/lb., the cost to test a 10 lb. batch is between 3–5% of wholesale flower value.

This percentage contrasts to food and pharmaceutical testing, where the average cost for microbiological tests can be as low as $15 (TAMC, TYMC) to $50 (molecular PCR analysis) while the cost per lb. of product is much less. This has allowed manufacturers to perform required regulatory testing, but also more routine quality control analysis, decreasing the odds of contaminated product reaching the market. Until the price point of analysis or per pound of product decreases, additional testing to ensure safer products will not be pursued.

Sources of Microbial Contamination and Prevention

In 2017, the Grocery Manufacturers Association (GMA) estimated that the average food recall (as the result of bacterial, fungal or allergen contamination) cost a company $10 million in direct costs alone [71]. In addition to the direct economic cost, GMA found that ~15% of people would not purchase a product again if it was associated with a recall [71]. These findings demonstrate the importance and need for accurate analytical testing of products. Imagine a $10 million dollar recall that is the result of contamination due to laboratory error (i.e. cross contamination). The impact of adhering to quality guidelines and best practices is paramount to produce accurate results and the impact of improper practices can be devastating.

Microbial contamination can occur at all stages in the cannabis life cycle. During cultivation, marijuana plants can be contaminated from fertilizer/compost, water, or humans (hand manipulation, coughing, sneezing) if sanitary practices are not followed. During laboratory analysis, product contamination can occur through cross contamination with other samples, poor laboratory conditions or poor laboratory techniques. The focus of this chapter section will discuss practices that can be employed to mitigate this contamination, from sterilization techniques that can be used post cultivation to practices that can be employed in laboratory settings to reduce the chance of cross contamination.

Good Laboratory Practices

Microorganisms can be found nearly everywhere. Marijuana flowers, plants, and most derivatives, will contain some level of microbial contamination. To perform the required microbiological contaminant analysis as described in the state regulations, laboratories follow a set of standardized procedures to minimize the potential impact of handling the products can have on the microbiological population found in these products in order to ensure results obtained are both valid and accurate [72]. These standardized procedures are often referred to as good laboratory practices (GLP) and are employed by virtually every testing facility. GLP principles are derived from two fundamental properties of biosafety, containment and risk assessment [73]. GLPs outline all components of the microbiology process, from general laboratory practices (aseptic technique), to specialized equipment requirements and appropriate facility design [73].

Risk assessment is used to determine the level of containment needed when manipulating pathogenic organisms. The concept of risk-based grouping for microorganism was introduced by the CDC in 1974 to classify organisms based on infectivity, severity of disease, transmissibility, origin and work process [73]. This was later updated to categorize organisms based on their pathogenicity for healthy human adults [74]. The GLPs discussed in this chapter are applicable to biosafety level 1 and 2 laboratories. While important to laboratories, a thorough discussion of risk assessment is not possible to include in this subsection. For further information on this topic, review the CDC's *Biosafety in Microbiological and Biomedical Labo*ratories guidance. The rest of this subsection will focus on containment practices.

Adherence to traditional aseptic microbiology techniques is the first step to ensuring containment. Often confused or misinterpreted with sterility, aseptic technique and sterile technique are not the same. Aseptic techniques are routine practices employed to prevent contamination of samples, culture media, or technicians by unwanted microorganisms [75]. These practices differ from sterile techniques, which is the prevention of any organism into a laboratory environment [75]. Marijuana plants, flowers, and most infused food derivatives will inherently have natural microflora, and the purpose of aseptic technique is to accurately enumerate or detect these organisms to determine if the products are safe for consumption. Sterility testing plays more of a role in the pharmaceutical or cosmetic/beauty care product industry (infused non-edible products, therapeutic patches, lotions or creams), and provides important information on the cleanliness of an environmental manufacturing process [76]. Understanding the type of product, you are evaluating and the intended us of that product will help in determining if aseptic technique or sterility testing is required. Basic laboratory practices such as flame sterilization, use of specialized equipment (laminar flow hoods), personal protective equipment, and use of single-use items all play a role in aseptic technique.

Adhering to GLP aseptic techniques is a key vital component to accurate laboratory testing, however, it is just a single cog in the process. Ensuring appropriate personal protective equipment (PPE) are in place and that and facility design must supplement good safe practices and techniques are just as important. PPEs are the first line of defense for a technician when handling samples. Use of appropriate gloves (both type and size), regularly laundered (i.e. clean) or disposable laboratory coats and face masks or shields (when appropriate) will all lead to lower decrease the risk of a laboratory acquired infection, as well as reducing the likelihood of cross contamination between samples. Other equipment, such as Laminar Flow Hoods (LFH) or Biological Safety Cabinets (BSC) can provide encapsulated spaces where specialized work can be performed. LFHs, when used appropriately, provide a clean air space with reduced airborne contaminants through the use of high-efficiency particulate air (HEPA) filters. These filters can greatly minimize the presence of bacteria, mold spores or other airborne particulates during sample manipulation [75]. BSCs also employ HEPA filters to reduce airborne contamination but are designed to contain all air flow within the cabinet, protecting both the operator and the surrounding environment from contamination to biohazards. Different classes of BSCs exist (I, II Type A, II Type B1, II Type B2, III) with increasing levels of safety built into the different classes dependent on the needs of the labs or their usage [77]. For cannabis marijuana laboratories, LFHs serve as an excellent barrier for sample manipulation and BSC class I or II hoods are more than sufficient will allow for the required culture manipulations.

Laboratory design, while not often discussed as primary source of GLP, is instrumental in the biosafety process. Separation of testing spaces and office or common spaces, self-closing doors, and restricted access are all commonly applied approaches used in a majority of laboratories [78]. Designated areas for hand cleaning, specifically near entry and exit doorways are critical to containing organisms within the laboratory. While these principles of biosafety may seem common sense, recent changes to laboratory design have begun to minimize their importance as the concept of open laboratory designs has seen a drastic increase [79]. Open laboratories can allow for increased and improved communication and collaboration between projects or departments, reduce the need for multiple (redundant) highly sophisticated (and expensive) equipment and can allow for the flexibility to change the laboratory to meet new testing needs [79]. However, open laboratories have drawbacks especially when dealing with containment issues. With no physical barriers between different areas or departments in a laboratory, the potential for cross contamination (either aerosolized or contact transfer) is increased. The concept of open labs is also counterintuitive to the need for individualized space for highly technical work [79].

Sterilization Techniques for Marijuana Plants and Flowers

Marijuana for medicinal purposes is an invaluable alternative to traditional methods that can be used to treat a wide range of conditions or symptoms. However, individuals with serious medical conditions that use marijuana to alleviate symptoms for pain, nausea, etc. are often in a state where their bodies are immunocompromised, increasing the risk for infection from microorganisms. For these individuals, how can the marijuana industry minimize the risk of microbial condemnation while still providing products with appropriate levels of active compounds? Sterilization is one such process that has proven to be an effective option for the marijuana industry in control microbial contaminants.

Sterilization is a term used to describe a process that will eliminate all forms of biological life (bacterial, fungal, spores). This process is different than decontamination or cleaning, which describe processes that reduce bacterial contamination but do not necessarily eliminate it [80]. There are many different sterilization techniques or processes, however not all of them are appropriate for use in the marijuana industry. Some sterilization processes degrade or eliminate chemical compounds, specifically the active compounds (THC, CBD) in marijuana derivatives.

The most common of sterilization approach is gamma irradiation which has broad appeal to the marijuana industry for its effectiveness in reducing microbial contamination, including mycotoxins, its negligible impact on valuable compounds found in the cannabis plant, and its track record as a safe, established process [81, 82]. Gamma irradiation is an ionization process that disrupts the subatomic particles, i.e. RNA or DNA, of microorganisms present in the derivatives being evaluated and results in cell death [83]. Gamma irradiation is performed in specialized facilities capable of producing and controlling the radiation process. In these facilities, products are exposed to controlled doses of radiation from Cobalt-60, an element used in the evaluation due to its shorter half-life [81]. This sterilization process is commonly used in the food and medical industries because it is performed at a low enough dosage that other chemical components in the product are not adversely affected, and the process itself does not produce enough energy for the product to become radioactive [83]. Studies have demonstrated that gamma irradiation in marijuana derivatives will significantly decrease the presence of viable microorganisms, with having a minimal impact on the water activity or concentration of valuable compounds (THC and cannabidiols), although certain compounds, such as terpenes, may be impacted [84].

While gamma irradiation is the most frequently used type of sterilization for derivatives intended to be consumed or used by humans, other sterilization techniques are continuing to grow in popularity. Electron-beam sterilization (e-beam) uses accelerated electrons to disrupt cellular activity of microorganisms [82]. This process is similar to gamma irradiation in how it impacts cellular function but differs in two distinct ways. E-beam irradiation is typically performed at a higher dose rate than gamma irradiation which results in a lower sterilization time period. Alternatively, e-beam irradiation has a much lower level of penetration than gamma irradiation [85]. E-beam sterilization has been employed by some companies in Canada because it is believed to be less detrimental to the compounds in marijuana plants and flowers. Other types of sterilization focusing on chemical processes such as those that use ethylene oxide (EO), plasma hydrogen peroxide, or reactive oxygen, provide their own advantages and disadvantages [81, 84]. EO sterilization has been around for nearly a century and at one time was the most commonly used method of sterilization for the agricultural industry and other derivatives where exposure to radiation was not possible [81, 84].

Clean Room

Sterilization of marijuana derivatives through irradiation or other techniques is a post cultivation process designed to reduce microbial growth. While this option works for some cultivators, others have sought to reduce or eliminate microbial growth during cultivation through use of cleanrooms or clean green rooms.

Cleanrooms can also be employed within laboratories to minimize the risk for environmental contamination of products. Cleanroom technology is often associated with high end manufacturing industries for medical devices and pharmaceutical products but their use in the marijuana industry is growing [86].

Cleanroom technology goes beyond just specifically fabricated rooms designed to reduce microbial growth. Cleanroom technology incorporates specialized equipment and procedures to maintain air quality (humidity/temperature) to control airborne contamination [87]. How effective a cleanroom is at reducing contamination, when compared to standard rooms, varies depending on the class rating of that room. The class ratings are assigned based on the number of airborne particle counts obtained within a cubic unit of air in that room, and many medicinal marijuana growers will use ISO Class 8 cleanrooms, the same class rating used by non-critical medical devise packaging facilities [86]. While the design of a cleanroom will vary depending on the industry (as well as the size of the manufacturing in that industry), most cleanrooms have similar protocols to maintain the air quality [86].

Cleanrooms will employ a series of barriers designed to prevent contamination from entering the room. The design of a cleanroom is the key first step in mitigating contamination. Cleanrooms are typically designed with positive air flow, so that when technicians enter the room, external air is not forced into the room through air pressure. Cleanroom designs can also include a double-door air chamber that is used for donning and doffing laboratory PPEs and disinfection of equipment that will be used in the room [87]. This adds another element to reducing environmental contaminants from entering the room.

After design, the choice of appropriate equipment is the next key component to utilizing a cleanroom. Specialized PPEs, from low-particulate coveralls, bouffants, and specialized shoe covers, as well as high level air respirators will be used to prevent contamination from a technician into the cannabis plants. Some barrier rooms will also use "sticky mats' ', floor mounted mats designed to capture organic and inorganic matter from the bottom of shoes as an extra precautionary step [87]. Cleanrooms will use specific HEPA filters containing low micron openings that are designed to capture a majority of microorganisms (bacterial, fungal and spores) and dust particles in the air. These rooms can also employ a secondary carbon activated filters to remove chemical contaminants [86].

Environmental Monitoring

Environmental monitoring is the collection and analysis of data from various sources in a facility (not just a laboratory) to determine the presence and/or quantity of microorganisms present. An effective environmental monitoring program will be embedded as part of a laboratory's larger quality program. Its purpose is to determine baseline levels of microorganisms present in a facility's air, water supply, and/ or on laboratory testing surfaces. Once baseline levels are determined, an environmental monitoring program will establish alert and action levels for that allow a facility to monitor and determine if preventative measures used in the facility are working appropriately [88].

When developing an environmental monitoring program, there is no "one-size" fits all plan, as each program will need to be tailored specifically to the facility. For example, a water monitoring program is not required if a laboratory employs the use of premade media and reagents. However, if a facility prepares some or all of its own reagents and/or media then monitoring the watering supply is a necessity.

When developing the environmental program, there are several key components that should be addressed. A strong environmental program must determine the frequency of testing and location of sampling. The facility should be mapped to identify locations to determine air, water and surface quality. Each of these sites should be assigned a number, which can make tracking the analysis simpler. When conducting sampling, rotate testing between sites on a weekly, biweekly or monthly basis, depending on the frequency requirements of your laboratory accreditation or regulatory statutes. When initiating the environmental analysis, the frequency of testing should be performed daily to generate enough data points to establish baseline levels for your facility. Once these baseline levels are established, alert and action levels can be determined and testing can be tapered off to weekly analysis, and finally monthly analysis. The frequency of testing can be maintained at monthly levels unless changes to the facility's design (new water system filter, new air filters) dictate that additional testing be conducted. An environmental monitoring program is never completely finished, as analysis will always be performed. Only the frequency of analysis will change.

An environmental monitoring program should be specific in identifying the methods of analysis to be utilized (specific organism testing by cultural or rapid method, general quantitation of bacteria by surface sponging or contact plate testing, etc.). Testing conducted should be catalogued and the data trended to determine if a) the action or alert levels are reached and b) to determine if cleaning processes for surfaces, air environment (air filters) or with the facility's water need to be modified. A final component to ensure an effective environmental monitoring program is in place is to identify the appropriate steps to be performed if deviations to the plan occur or action/alert limits are obtained. The level of detail in these steps can vary, but a requirement to notify the quality control department and reporting supervisor of that section should be included at minimum.

Laboratory Disinfection and Sterilization

When developing a facility's environmental monitoring program, the selection of the disinfectant(s) to be used is a key step to mitigating the risk of contamination with samples. A wide range of disinfectants are available in the market, however, choosing the correct product is not always an easy decision. Most disinfectants are designed for a specific purpose and understanding the claims, application and preparation of the product is essential to success. Does the product require a cleaning step prior to use, or does it claim to be both a cleaning agent and disinfectant? Cleaning, the removal of bioburden (i.e. organic soil loads), if not performed prior to the application of the disinfectant, may completely invalidate the claims of a product to reduce microorganisms [41]. Disinfectants should not be thought of as interchangeable, so end users must read the label claims carefully and prepare the disinfectant concentration correctly (if the product is not ready-to-use). Dwell times, the time required for the active to be effective, must also be strictly adhered to for the product to work effectively [41]. Disinfectants will come with either a singular active ingredient (alcohol, chlorine, hydrogen peroxide, peracetic acid, phenolic, quaternary ammonium compounds) or be a combination of active ingredients (hydrogen peroxide and peracetic acid). Regardless of the active, every unique product must be registered with the EPA or be cleared by FDA [41].

The use of steam sterilization can be an effective final step for processing biohazardous (or potentially biohazardous) materials. There are four main principles of steam sterilization: steam, pressure, temperature, and time. Ideally, a dry flow of saturated steam, with pressure above 21 psi, a temperature of 121 °C and exposure for 60 min will be effective in sterilizing all biohazardous waste [41]. These parameters are achieved through the use of autoclaves, either a gravity displacement autoclave or high-speed pre-vacuum sterilizer [41]. *Geobacillus stearothermophilus*, a heat-resistant spore-forming bacteria, is used as a control organism to verify that sterilization has occurred as it can survive extreme levels of temperature and pressure, although will be destroyed if appropriate temperature/pressure limits are achieved. While their use is common in the healthcare and food industries, autoclave sterilization is not as commonplace in marijuana facilities, as the cost to purchase, certify and maintain an autoclave can be cost prohibitive. Small test portions sizes and limited scopes of microbiological testing make it more cost effective to outsource the sterilization of biohazardous materials. As this industry grows, ensuring that biohazardous materials are properly disposed of becomes an important step in reducing a source of contamination.

Remediation

Up to 10% of marijuana products fail the microbial contamination test when analyzed post-harvest [89]. When a test product produces a result above a regulatory threshold, a facility should have an established plan outlining the next steps it must take in disposing or remediating the product. If the test sample in question is a marijuana plant or flower, the option to remediate or reduce the level of contaminants below the regulatory threshold can be performed. There are many different methods to remediate products, and some that can be performed prior to testing (see Sterilization Techniques subsection) can also be employed after a failed result.

The most common approach to remediation is the process of extraction. In some states, extraction is the only acceptable method as the processes involved high levels or heat or chemicals toxic to microorganisms. These processes are easy to perform and highly scalable but can result in a high profit loss, as marijuana flower has a higher value than an extracted derivative [90]. Other treatments such as irradiation (gamma, e-beam, etc.) or chemical remediation (gas or liquid) are also frequently utilized. Irradiation of products is a fast, effective way to eliminate pathogens that does not alter the product composition; however, some municipalities have banned this practice from being used over concerns that it does not eliminate mycotoxins [89]. Chemical treatments are also inexpensive to apply but are often time consuming and can be difficult to scale. Less used options of autoclave sterilization and dry heat sterilization are effective at remediating products but often leave finished products that are undesirable [90]. Some cultivators have turned to newer techniques, such as radio frequency which use oscillating electromagnetic fields to eliminate microbial contaminants. Facilities should review their state regulations to determine which methods of remediation are acceptable in their states and then choose the method that best suits their needs.

Future of Testing

The bulk of this chapter outlines current state regulations and best practices for analyzing marijuana. Over the last 10 years, the legal marijuana market has exploded and the next logical question to ask is what does the future hold? These questions can't be definitively answered until a federal framework for testing regulations is implemented, the roadmap for the future will remain uncertain, but the cannabis industry can look to regulations currently enacted for other industries as a guideline for what may come.

FDA Involvement

Looking to the future, the legal marijuana industry is on a pathway of continued growth, and most in the industry believe this trajectory will continue to increase with time as more states decriminalize marijuana. However, the future will remain murky until one major hurdle is overcome, the establishment of a national framework for marijuana regulations. When Congress passed the Agriculture Improvement Act of 2018 (2018 Farm Bill) in December of 2018, it legalized the cultivation and sale of hemp by removing it from the Controlled Substances Act [91]. While this decision should be viewed as a positive step for the federal legalization of marijuana, until that happens, the impact of this ruling only adds to the uncertainty that cultivators, manufacturers and laboratories face. It is the FDA's responsibility to protect the public health (per their mission statement), by ensuring the safety of the products millions of Americans consume every day. To ensure this is achieved for the marijuana industry, a federal framework of regulations needs to be established that at minimum establishes a standardized baseline for states. This framework could be built from the existing regulations of states with mature marijuana laws, while also incorporating aspects of the Food Safety Modernization Act (FSMA) which have been demonstrated to be successful in the food industry [92]. When Congress passed FSMA in 2011, it was the first major piece of food safety legislation in over 50 years. In the framework of FSMA, several rules were established, that expanded upon approaches already being adopted by industry, focusing on preventative actions instead of reactionary actions [92]. Many of the rules established can and should be applied to the marijuana industry, ensuring the same level of consumer protection that the FDA provides with its food oversite.

One of the main focuses of FSMA legislation was to expand upon current good manufacturing processes, cGMPs, by incorporating aspects already adopted by industry into law. The legislation expands upon the principles of Hazard Analysis and Critical Control Points (HACCP) (long adopted by industry) by focusing on preventative control measures designed to mitigate risks through unintentional adulteration [92]. FSMA requires manufacturers to create written food safety plans, conduct hazard analysis, establish preventive controls which need to be verified and monitored, determine when they fail and the corrective actions that should be implemented to respond to failures, all while increasing the requirements for sufficient documenting of these processes [92]. These requirements, often viewed as practical approaches within industry, are already being employed by some in the marijuana industry but mandating them on a federal level will help ensure that all products going to market will have been produced under a standardized risk assessment process.

Why choose FSMA as a starting point? The food industry has some of the most diverse set of products, and foods containing marijuana derivatives are no different. In order to ensure that infused foods do not contain pathogenic organisms or high levels of spoilage organisms, regulators could quickly turn to aspects of already established regulations by the FDA or USDA for quicker implementation. Regulatory methods, already in existence for food products, can easily be validated for marijuana plants and infused edibles. The development of regulatory methods will also allow for a more through validation of rapid methods. Development of regulations from scratch would be inefficient, especially when proven regulations already exist and could be fully or partially implemented.

Tobacco and Medical Products as a Road Map

While many aspects of FSMA can and should be adopted by marijuana producers, this should not be the only framework of regulations that the industry should look to adopt. A one size fits all approach is destined to fail as the breadth of the marijuana industry will require specialized regulations for different commodities. Guidance provided by the FDA in the Family Smoking Prevention and Tobacco Control Act (TCA), is designed to help the tobacco industry understand and comply with all regulations within the law [93]. These guidelines cover compliance policies for premarket applications, modifications to current on market products, as well as federal regulatory requirements. Baseline requirements established from the TCA regulations would be a natural fit for marijuana and marijuana derivatives that are consumed via inhalation.

Another source of federal regulations, the Code of Federal Regulations, Title 21, could serve as the basis for developing regulations for evaluating medicinal marijuana [94]. CFR 21 outlines requirements for medical products including extensive pre-market evaluation and clinical trials to determine not only the performance of the product tested but to determine which, if any, side effects are observed. While industry may have issues with the delays and cost associated with taking a product through this process, all drugs being marketed with health claims are required to follow this process. Medical marijuana should not be viewed differently as its intended consumers are often the most vulnerable, immunocompromised individuals.

Emerging Pathogens

As cultivation facilities continue to grow in size and the types of marijuana derivatives continue to increase, new pathogens or organisms of concern may emerge. *Listeria monocytogenes,* a Gram-positive bacterium known for its hardiness is one organism that may soon find itself listed as a microbial contaminant in state regulations [19]. While outbreaks associated with *L. monocytogenes* are rare when compared to *Salmonella* or pathogenic *E. coli,* the mortality rate associated with the organism are much higher, especially in immunocompromised individuals, and this organism remains one of the leading causes of death from foodborne illness [19]. Due to its ability to survive wide ranging environments, *L. monocytogenes* can be found in most food manufacturing environments, resulting in post processing contamination of products. Recent changes to food safety regulations as outlined in FSMA, require robust environmental monitoring programs which typically involve routine surveillance of *Listeria* species or *L. monocytogenes* specifically [19]. These regulations could serve as a starting point for future requirements for expanded environmental monitoring programs in manufacturing facilities.

Expansion of Test Portion Sizes

Along with new target analytes, one major change that would drastically impact the marijuana industry is an increase in the size of test portions for analysis. While state regulations may vary in identifying the size of a single batch of marijuana or the required number of samples to be analyzed per batch, there is near universal consistency in the test portion size requirements by states. For example, in Colorado the % of batch tested ranges from 0.09% to 0.03% (a batch of 10 lbs. requires 8 separate samples at 0.5 g; a batch of 100 lb. requires 29 separate samples at 0.5 g) [21]. Regardless of the size of the batch, or the number of samples required, test portion size remains constant at 1 g [21]. Due to the cost of matrix, and a strong lobbying influence from the cannabis industry, 1 g test portions have been almost universally adopted by all states, however, using a test portion size this small brings into question its value in representing the batch it comes from. As discussed in previous sections, if states move toward regulations established for tobacco, medical products or food, increases in test portion sizes to mirror the sizes in those regulations would most likely follow. Additionally, certification organizations such as AOAC International are mandating that methods which seek to be validated for cannabis and its derivatives will be required to use larger test portion sizes (10 g for flower and plant; 5 g for concentrates, 25 g for infused edibles and 10 g for infused inedible) [67]. These test portion sizes were developed in synchrony with established test portion sizes for similar industries (botanical guidelines, FDA food guidelines, USP guidance, etc.).

Proficiency Testing

One change to test facilities that is already starting to be seen, and one that is being driven by accreditation and not regulation, is the requirement for proficiency testing (PT) for laboratories. PT involves the analysis of artificially contaminated samples following the methods a laboratory uses for its routine analysis. PT test portions are prepared by a certified provider, blind coded and sent to testing facilities for analysis. Laboratories perform testing according to the methods they use routinely. Qualitative results are scored presence/absence while quantitative tests are traditionally scored based on the overall mean of the results obtained from all participants. PT testing is designed to verify that a laboratory can perform a method correctly and provides support that the results produced during routine analysis is accurate.

The driving force behind the requirement for PT testing is laboratory accreditation, most specifically the ISO/IEC 17025:2017 laboratory accreditation standard [95]. This standard, which has been broadly adopted within the FSMA regulations, requires that laboratories engage in routine PT testing for each employee and method that a lab uses [92]. Laboratories are required to obtain passing results for all methods annually in order to maintain accreditation. While some PT programs exist for the marijuana industry, they often have a limited scope of target analyte and matrix combinations. It is vital that an expansion of these programs occurs to new matrix and analyte combinations to ensure that methods are "fit-for-purpose". Although not to be interchanged with a PT sample, having established reference standards, such as those being created by The National Research Council (CRC) of Canada and the National Institute of Standards and Technologies (NIST) of the United States, will also aid in ensuring accurate results are obtained during routine analysis.

Summary

While regulations for microbial contaminants vary from state to state, there is a universal goal to keep harmful products from reaching the market. Regulations are the building blocks for microbial safety but must be supplemented with continued scientific support and advancement through scientific organizations such as CASP, ASTM, and APHL to be effective. These organizations use consensus based thinking and scientific data to ensure that rigorous standards are put in place which in turn lead to the development of improved detection methods. As the marijuana industry continues to grow, the refinement of current technologies and development of new detection tools will allow for quicker detection of pathogens while maintaining high levels of sensitivity in difficult to analyze products. These methods may be the most important step in ensuring safe products reach consumers but must be employed in combination with good laboratory techniques, including environmental monitoring programs, effective use of clean rooms and PPEs, and the use of appropriate disinfectants in order to be as effective as possible. The combination of effective testing, robust cleaning procedures, and an effective safety program helps ensure marijuana and its derivatives are safe.

References

1. Bergey's Manual Trust (2015) Bergey's Manual of Systematics of Archaea and Bacteria. https://doi.org/10.1002/9781118960608
2. Procop G et al (2017) Koneman's color atlas and textbook of diagnostic microbiology, 7th edn. Lippincott, Williams & Wilkins, Philadelphia, PA
3. Cescon DW, Page AV, Richardson S, Moore MJ, Boerner S, Gold WL (2008) Invasive pulmonary aspergillosis associated with marijuana use in a man with colorectal cancer. J Clin Oncol 26:2214–2215
4. Chusid MJ, Gelfand JA, Nutter C, Fauci AS (1975) Letter: Pulmonary aspergillosis, inhalation of contaminated marijuana smoke, chronic granulomatous disease. Ann Intern Med 82:682–683
5. Hamadeh R, Ardehali A, Locksley RM, York MK (1988) Fatal aspergillosis associated with smoking contaminated marijuana, in a marrow transplant recipient. Chest 94:432–433
6. Kagen SL (1981) Aspergillus: an inhalable contaminant of marihuana. N Engl J Med 304:483–484
7. Llamas R, Hart DR, Schneider NS (1978) Allergic bronchopulmonary aspergillosis associated with smoking moldy marihuana. Chest 73:871–872
8. Marks WH, Florence L, Lieberman J, Chapman P, Howard D, Roberts P, Perkinson D (1996) Successfully treated invasive pulmonary aspergillosis associated with smoking marijuana in a renal transplant recipient. Transplantation 61:1771–1774
9. Remington TL, Fuller J, Chiu I (2015) Chronic necrotizing pulmonary aspergillosis in a patient with diabetes and marijuana use. CMAJ 187:1305–1308
10. Salam AP, Pozniak AL (2017) Disseminated aspergillosis in an HIV-positive cannabis user taking steroid treatment. Lancet Infect Dis 17:882
11. Schwartz IS (1985) Marijuana and fungal infection. Am J Clin Pathol 84:256

12. Sutton S, Lum BL, Torti FM (1986) Possible risk of invasive pulmonary aspergillosis with marijuana use during chemotherapy for small cell lung cancer. Drug Intell Clin Pharm 20:289–291
13. Taylor DN, Wachsmuth IK, Shangkuan YH, Schmidt EV, Barrett TJ, Schrader JS, Scherach CS, McGee HB, Feldman RA, Brenner DJ (1982) Salmonellosis associated with marijuana: a multistate outbreak traced by plasmid fingerprinting. N Engl J Med 306:1249–1253
14. Shapiro B, Hedrick R, Vanle BC, Becker CA, Nguyen C, Underhill DM, Morgan MA, Kopple JD, Danovitch I, Iskak W (2018) Cryptococcal meningitis in a daily cannabis smoker without evidence of immunodeficiency. BMJ Case Rep 2018:bcr-2017
15. Centers for Disease Control and Prevention (2020) Outbreak of Lung Injury Associated with the Use of E-Cigarette, or Vaping, Products. https://www.cdc.gov/tobacco/basic_information/e-cigarettes/severe-lung-disease.html#latest-outbreak-information. Accessed Jan 8, 2020
16. Verweij PE, Kerremans JJ, Voss A, Meis JF (2000) Fungal contamination of tobacco and marijuana. JAMA 284:2875
17. Scallan E, Hoekstra RM, Angulo FJ, Tauxe RV, Widdowson MA, Roy SL, Jones JL, Griffin PM (2011) Foodborne illness acquired in the United States--major pathogens. Emerg Infect Dis 17:7–15
18. Painter JA, Hoekstra RM, Ayers T, Tauxe RV, Braden CR, Angulo FJ, Griffin PM (2013) Attribution of foodborne illnesses, hospitalizations, and deaths to food commodities by using outbreak data, United States, 1998–2008. Emerg Infect Dis 19:407–415
19. Upton R Lyle Craker, ElSohly M, Room A, Russo E, Sexton M (2014) Cannabis inflorescence. American Herbal Pharmacopeia :64
20. United States Pharmacopeia and National Formulary (2019) USP 42-NF 37. Microbial examination of nonsterle products: acceptance criteria for pharmaceutical preparations and substances for pharmaceutical use
21. Colorado State Legislature (2012) Amendment 64: Use and Regulation of Marijuana
22. Homes M, Vyas JM, Steinbach W, McPartland J (2015) Microbiological safety testing of cannabis, p 1–54
23. U.S. Food and Drug Administration (2012) Bad bug book, foodborne pathogenic microorganisms and natural toxins, 2ed edn
24. Foodsafety.gov (2019) Bacteria and Viruses. https://www.foodsafety.gov/food-poisoning/bacteria-and-viruses. Accessed Mar 30, 2020
25. McKernan K, Spangler J, Helbert Y, Lynch RC, Devitt-Lee A, Zhang L, Orphe W, Warner J, Foss T, Hudalla CJ, Silva M, Smith DR (2016) Metagenomic analysis of medicinal Cannabis samples; pathogenic bacteria, toxigenic fungi, and beneficial microbes grow in culture-based yeast and mold tests. F1000Res 5:2471
26. McKernan K, Spangler J, Zhang L, Tadigotla V, Helbert Y, Foss T, Smith D (2015) Cannabis microbiome sequencing reveals several mycotoxic fungi native to dispensary grade Cannabis flowers. F1000Res 4:1422
27. Thompson GR 3rd, Tuscano JM, Dennis M, Singapuri A, Libertini S, Gaudino R, Torres A, Delisle JM, Gillece JD, Schupp JM, Engelthaler DM (2017) A microbiome assessment of medical marijuana. Clin Microbiol Infect 23:269–270
28. Kwon-Chung KJ, Sugui JA (2013) Aspergillus fumigatus--what makes the species a ubiquitous human fungal pathogen? PLoS Pathog 9:e1003743
29. Kamei K, Watanabe A (2005) Aspergillus mycotoxins and their effect on the host. Med Mycol 43(Suppl 1):S95–S99
30. Alaska Alcohol and Marijuana Control Office (2019) Regulations for the Marijuana Control Board. 3 AAC 306
31. Bureau of Cannabis Control (2019) California Code of Regulations Title 16. Division 42
32. Nevada State Legislature (2019) Assembly Bill No. 533 – Cannabis Control Board. Chapter 453D adult use of marijuana. Requirements for Marijuana Testing Facilities
33. Aspergillus and Aspergillosis (2016) Mycotoxin and Metabolite Database. https://www.aspergillus.org.uk/content/mycotoxin-metabolites?page=13. Accessed Mar 30, 2020

34. Klich MA (2007) Aspergillus flavus: the major producer of aflatoxin. Mol Plant Pathol 8:713–722
35. McCormick A, Loeffler J, Ebel F (2010) Aspergillus fumigatus: contours of an opportunistic human pathogen. Cell Microbiol 12:1535–1543
36. Frisvad JC, Larsen TO, Thrane U, Meijer M, Varga J, Samson RA, Nielsen KF (2011) Fumonisin and ochratoxin production in industrial Aspergillus niger strains. PLoS One 6:e23496
37. Steinbach WJ, Benjamin DK Jr, Kontoyiannis DP, Perfect JR, Lutsar I, Marr KA, Lionakis MS, Torres HA, Jafri H, Walsh TJ (2004) Infections due to Aspergillus terreus: a multicenter retrospective analysis of 83 cases. Clin Infect Dis 39:192–198
38. Baxi SN, Portnoy JM, Larenas-Linnemann D, Phipatanakul W (2016) Exposure and health effects of fungi on humans. J Allergy Clin Immunol Pract 4:396–404
39. Dicostanzo, A., and Murphy, M. (2012). Strategies for Feeding Mycotoxin and Mold Contaminated Grains to Cattle. Available online at: https://wayne.osu.edu/sites/wayne/files/imce/Program_Pages/ANR/strategies_for_feeding_mycotoxin_and_mold_contaminated_grain,%20UMN.pdf (Accessed Dec 3, 2020).
40. Government of Canada (2018) Foods and Drug Act: Cannabis Regulations. SOR/2019–144
41. Kohler JR, Casadevall A, Perfect J (2014) The spectrum of fungi that infects humans. Cold Spring Harb Perspect Med 5:a019273
42. Cundell T (2019) Microbiological attributes of cannabis-derived products. Cannabis Science and Technology 2
43. United States Pharmacopeia and National Formulary (2019) Chapter. 61. Microbiological examination of nonsterile products: microbial enumeration tests
44. United States Pharmacopeia and National Formulary (2019) Chapter. 62. Microbiological examination of nonsterile products: tests for specified microorganisms
45. New York State Register (2019) (Health) of the official Compilation of Codes, Rules and Regulation of the State of New York. Part 1004 to Title 10. https://www.health.ny.gov/regulations/medical_marijuana/docs/regulations.pdf
46. AOAC International (2019) Cannabis analytical science program prospectus
47. AOAC International (2019) Standard methods performance requirements (SMPR) 2019.001 detection of aspergillus in cannabis and cannabis products
48. ASTM International (2017) Committee D37 on Cannabis: Overview. https://www.astm.org/COMMITTEE/D37.htm.
49. ASTM International (2019) Committee D37 on Cannabis: Subcommittees. https://www.astm.org/COMMIT/SUBCOMMIT/D37.htm
50. Association of Public Health Laboratories (2019) APHL Profile. https://www.aphl.org/aboutAPHL/Pages/profile.aspx
51. U.S. Food and Drug Administration (2019) Bacteriological analytical manual
52. American Herbal Pharmacopeia (2014) Cannabis Inflorescence Cannabis spp.: Standards of Identity, Analysis, and Quality Control
53. Gill A (2017) The importance of bacterial culture to food microbiology in the age of genomics. Front Microbiol 8:777
54. Silvestri E, Y. Chambers-Velarde, J. Chandler, J. Cuddeback, K. Jones, K. Hall (2018) Sampling, Laboratory and Data Considerations for Microbial Data Collected in the Field. Agency USEP. https://cfpub.epa.gov/si/si_public_file_download.cfm?p_download_id=536543&Lab=NHSRC
55. Bridson E, Brecker A (1970) Design and formulation of microbiological culture media, vol 3A. Academic Press, London
56. Cundell AM (2002) Review of the media selection and incubation conditions for the compendial sterility and microbial limit tests. Pharmacopeial Forum 28:2034
57. Wang L, Fan D, Chen W, Terentjev EM (2015) Bacterial growth, detachment and cell size control on polyethylene terephthalate surfaces. Sci Rep 5:15159

58. Taniwaki MH, Silva N, Banhe AA, Iamanaka BT (2001) Comparison of culture media, simplate, and petrifilm for enumeration of yeasts and molds in food. J Food Prot 64:1592–1596
59. McClenny N (2005) Laboratory detection and identification of Aspergillus species by microscopic observation and culture: the traditional approach. Med Mycol 43(Suppl 1):S125–S128
60. Baylis CL, MacPhee S, Betts RP (2000) Comparison of two commercial preparations of buffered peptone water for the recovery and growth of Salmonella bacteria from foods. J Appl Microbiol 89:501–510
61. Bumgarner R (2013) Overview of DNA microarrays: types, applications, and their future. Curr Protoc Mol Biol Chapter 22:Unit 22 21
62. Richter A, Schwager C, Hentze S, Ansorge W, Hentze MW, Muckenthaler M (2002) Comparison of fluorescent tag DNA labeling methods used for expression analysis by DNA microarrays. Biotechniques 33:620–628, 630
63. Lipshutz RJ, Fodor SP, Gingeras TR, Lockhart DJ (1999) High density synthetic oligonucleotide arrays. Nat Genet 21:20–24
64. National Center for Biotechnology Information (2020) Pathogen detection
65. Centers for Disease Control and Prevention (2020) Serotypes and the importance of serotyping salmonella https://www.cdc.gov/salmonella/reportspubs/salmonella-atlas/serotyping-importance.html
66. National Center for Biotechnology Information (2020) Isolate Browser. https://www.ncbi.nlm.nih.gov/pathogens/isolates/#/search/taxgroup_name:%22Salmonella%20enterica%22
67. AOAC International (2019) Appendix J: AOAC International Methods Committee Guidelines for Validation of Microbiological Methods for Food and Environmental Surfaces 21st ed.
68. International Organization for Standards (2016) ISO 16140-2:2016: Microbiology of the food chain – Method validation – Part 2: Protocol for the validation of alternative (proprietary) methods against a reference method.
69. Health Canada (2019) The compendium of analytical methods 1
70. U.S. Food and Drug Administration (2015) Pharmaceutical microbiology manual. https://www.fda.gov/media/83812/download
71. Grocery Manufacturers Association FMI, GS1 US (2010) Recall execution effectiveness
72. Sanders ER (2012) Aseptic laboratory techniques: volume transfers with serological pipettes and micropipettors. J Vis Exp 63:2754. https://doi.org/10.3791/2754
73. U.S. Department of Health and Human Services (2009) Biosafety in microbiological and biomedical laboratories, 5th edn
74. U.S. Department of Health and Human Services (2016) NIH Guidelines for Research Involving Recombinant or Synthetic Nucleic Acid Molecules. Section II-A-3 Comprehensive Risk Assessment. https://osp.od.nih.gov/wp-content/uploads/NIH_Guidelines.html#_Toc3457028. Accessed April, 2020
75. Bykowski T, Stevenson B (2008) Aseptic technique. Curr Protoc Microbiol Appendix 4:Appendix 4D
76. World Health Organization (2012) 3.2 Test for Sterility, vol QAS/11.413
77. ESCO Technologies (2009) A guide to biosafety & biological safety cabinets, vol ES1284/OW8359R_V3_5K_02/09
78. University of California (2019) Lab safety design manual. Basic laboratory design for biosafety levels 1 and 2
79. National Research Council Committee on Prudent Practices in the Laboratory (2011) Prudent Practices in the Laboratory: Handling and Management of Chemical Hazards: Updated Version. National Academies Press
80. University of Rochester Medical Center (2020) Basics on processing & sterilization. https://www.urmc.rochester.edu/sterile/basics.aspx. Accessed Jan, 2020
81. Gamma Industry Processing Alliance (2017) A comparison of gamma, E-beam, X-ray and ethylene oxide technologies for the industrial sterilization of medical devices and healthcare products

82. Ghanem I, Orfi M, Shamma M (2008) Effect of gamma radiation on the inactivation of afla-toxin B1 in food and feed crops. Braz J Microbiol 39:787–791
83. Finkiel M (2016) Sterilization by gamma irradiation. https://tuttnauer.com/blog/sterilization-by-gamma-irradiation. Accessed Jan, 2020
84. Hazekamp A (2016) Evaluating the effects of gamma-irradiation for decontamination of medicinal cannabis. Front Pharmacol 7:108
85. BOAZ Pharmaceuticals Inc. (2020) Natural Handcrafted Cannabis vs. Gamma Radiation vs. E-Beam Sterilization. https://www.boazpharm.com/natural-cannabis-vs-radiation-sterilization/. Accessed Jan, 2020
86. Walleser M (2019) Adopting cleanroom technology for safer medicinal cannabis. Cleanroom Technol 3:45–47
87. PharmOut (2014) Basic Clean Room Requirements/Designs for GMP Clean Rooms. https://www.pharmout.net/basic-cleanroom-requirements/. Accessed Jan, 2020
88. Sandle T (2019) Biocontamination Control for Pharmaceuticals and Healthcare. Academic Press
89. Kern R, Green JR (2019) It's not too late: post-harvest solutions to microbial contamination issues. Cannabis Sci Technol 2:15–19
90. DeGabrielle K (2019) Largest U.S. Cannabis Farm Shares Two Years of Mold Remediation Research. Analytical Cannabis
91. U.S. Department of Agriculture (2019) Establishment of a domestic hemp production program
92. U.S. Food and Drug Administration (2011) Food Safety Modernization Act. 21 U.S.C. 301 et seq. Public Law 111–353. https://www.govinfo.gov/content/pkg/PLAW-111publ353/pdf/PLAW-111publ353.pdf
93. U.S. Food and Drug Administration (2009) Family smoking prevention and tobacco control act
94. U.S. Food and Drug Administration (2019) Food and Drugs. Code of Federal Regulations, Title 21 parts 1 to 1499. https://www.accessdata.fda.gov/scripts/cdrh/cfdocs/cfcfr/cfrsearch.cfm
95. International Organization for Standards (2017) ISO/IEC 17025:2017: general requirements for the competence of testing and calibration laboratories

Index

© Springer Nature Switzerland AG 2021
S. R. Opie (ed.), *Cannabis Laboratory Fundamentals*,
https://doi.org/10.1007/978-3-030-62716-4

Printed in the United States
by Baker & Taylor Publisher Services